ALFRED RAWORTH'S
ELECTRIC SOUTHERN RAILWAY

For Oscar, perhaps you will be an engineer one day...

ALFRED RAWORTH'S
ELECTRIC SOUTHERN RAILWAY

PETER STEER

AN IMPRINT OF PEN & SWORD BOOKS LTD.
YORKSHIRE – PHILADELPHIA

Front Cover: 2-BIL at West Worthing in 1966. The '1' head code refers to the Brighton–West Worthing service. (Martin James)

Back Cover : The Southern Electric control room at Woking. (Author)

First published in Great Britain in 2022 by
Pen and Sword Transport
An imprint of
Pen & Sword Books Ltd
Yorkshire - Philadelphia

Copyright © Peter Steer, 2022

ISBN 978 1 52677 841 3

The right of Peter Steer to be identified as Author of this work has been asserted by him in accordance with the Copyright, Designs and Patents Act 1988.

A CIP catalogue record for this book is available from the British Library.

All rights reserved. No part of this book may be reproduced or transmitted in any form or by any means, electronic or mechanical including photocopying, recording or by any information storage and retrieval system, without permission from the Publisher in writing.

Typeset in Palatino 11/13 by SJmagic DESIGN SERVICES, India.

Printed and bound in India by Replika Press Pvt. Ltd.

Pen & Sword Books Ltd incorporates the Imprints of Pen & Sword Books Archaeology, Atlas, Aviation, Battleground, Discovery, Family History, History, Maritime, Military, Naval, Politics, Railways, Select, Transport, True Crime, Fiction, Frontline Books, Leo Cooper, Praetorian Press, Seaforth Publishing, Wharncliffe and White Owl.

For a complete list of Pen & Sword titles please contact

PEN & SWORD BOOKS LIMITED
47 Church Street, Barnsley, South Yorkshire, S70 2AS, England
E-mail: enquiries@pen-and-sword.co.uk
Website: www.pen-and-sword.co.uk

or

PEN AND SWORD BOOKS
1950 Lawrence Rd, Havertown, PA 19083, USA
E-mail: Uspen-and-sword@casematepublishers.com
Website: www.penandswordbooks.com

CONTENTS

Abbreviations 6
Introduction 7

Part 1: Innovation and Confrontation

Chapter 1	Textiles, Electric Lighting and Trams	12
Chapter 2	Raworth's Traction Patents	24
Chapter 3	Alfred Raworth Becomes a Railwayman	35
Chapter 4	A Design for an Electric Railway	46
Chapter 5	The LSWR Electrification	57
Chapter 6	Raworth RN	69
Chapter 7	A Railway Under Stress	74
Chapter 8	Alfred Raworth's Electric Railway	90
Chapter 9	Meeting the Men from the Ministry	101
Chapter 10	Making Three Fit Into One	112
Chapter 11	The Battle of Angerstein's Wharf	126
Chapter 12	The Birth Pangs of the Southern Railway	140
Chapter 13	Completing the Legacy Schemes	146

Part 2: Southern Electric

Chapter 14	'The World's Greatest Suburban Electrification'	170
Chapter 15	A New Design for an Electric Railway	183
Chapter 16	The Brighton Electrification	188
Chapter 17	Maintaining Momentum	204
Chapter 18	Two Electric Routes to Portsmouth	216
Chapter 19	The Southern Electric in the Community	234
Chapter 20	The End of an Era	241
Chapter 21	Chief Electrical Engineer	250
Chapter 22	The Southern Electric at War	265
Chapter 23	The Electric Locomotives	277
Chapter 24	Looking to the Future	289
Chapter 25	Epilogue	301
Endnotes		316
Bibliography		335
Index		337

ABBREVIATIONS

ASEA	Allmanna Svenska Elektriska Aktiebolaget	LIRR	Long Island Railroad
ASLEF	Association of Locomotive Engineers and Firemen	LMS(R)	London Midland and Scottish Railway
BBC	British Broadcasting Corporation	LNER	London and North Eastern Railway
BEF	British Expeditionary Force	LNWR	London and North Western Railway
BFB	Bulleid-Firth-Brown		
BRB	British Railways Board	LPTB	London Passenger Transport Board
BTC	British Transport Commission		
BTH	British Thomson-Houston Company	LSWR	London and South Western Railway
CEB	Central Electricity Board	LV	low voltage
CLESC	County of London Electricity Supply Company	LYR	Lancashire and Yorkshire Railway
CLR	Central London Railway	M&EE	Mechanical and Electrical Engineering subcommittee
DAB	delayed action bomb		
EMU	Electric Multiple Unit	MDR	Metropolitan District Railway
GER	Great Eastern Railway	MER	Manx Electric Railway
GNR	Great Northern Railway	MP	Member of Parliament
GPO	General Post Office	NER	North Eastern Railway
GWR	Great Western Railway	NHS	National Health Service
HV	high voltage	REC	Railway Executive Committee
Hz	hertz	RFC	Royal Flying Corps
kW	kilowatt	RNAS	Royal Naval Air Service
LBSCR	London, Brighton and South Coast Railway	RNVR	Royal Naval Volunteer Reserve
		SCADA	supervisory control and data acquisition
LCC	London County Council		
LCDR	London, Chatham and Dover Railway	SECR	South Eastern and Chatham Railway
LESC	London Electricity Supply Company	V	Volt(s)

INTRODUCTION

Charles Klapper's biography of the famed general manager of the Southern Railway, *Sir Herbert Walker's Southern Railway*, contains a photo spread illustrating the main characters in his account. Beneath a picture of a middle-aged and balding man is the caption: *Alfred Raworth – electrification genius*.

To fully assess the validity of the use of the superlative 'genius' there must also be an assessment of the describer's credentials as well as those of the described. Charles Klapper was a journalist working for the magazine *Modern Transport* during the 1930s. He reported on the progress of the Southern Railway's successful electrification programme which was known to all as the 'Southern Electric'. Klapper followed the expansion of the Southern Electric and would have received most of his information from official Southern Railway press releases. This has resulted in a reliable contemporary account but one that is likely to have been tainted with a bias towards the 'party-line' propagated by the railway's press office. On at least one occasion however – and probably more often – Klapper met Alfred Raworth. The young journalist was treated to Raworth's vision for future railway electrification. Raworth admitted that he would not be able to fully realise his vision on the Southern Railway. This underlines Raworth's value to the railway company: he appreciated that he could only provide what his 'customers' – Sir Herbert Walker and the Southern Railway directors – would accept, and used his expertise to provide exactly what they desired at a minimum price and delivered it all on time. The Southern Electric used a ground-mounted live conductor rail at a relatively low voltage, a method of electrification which was not of Raworth's choosing, but he employed new and innovative technology to electrify routes deemed to be main lines using this 'third rail' system at a time when most engineers saw it as only being suitable for the shorter distance suburban routes. Sir Herbert Walker, the most respected railway manager of his generation, believed that without Alfred Raworth's expertise the development of the Southern Electric would not have been possible.

To modern eyes the cost of the incremental extensions to the Southern Electric seems to be incredibly inexpensive – even when the usual inflation rates are applied to the final figures. Cost comparisons with contemporary schemes are not helpful since much has changed since the pre-war years. For most of the 1920s and 1930s when Raworth took charge of the electrification infrastructure provision on the Southern Railway, the British domestic economy was weak with unacceptably high unemployment across the land. This – sadly for the workforce – helped keep wage costs low. Also, electrical plant manufacturers had very few orders and were likely to be persuaded to supply equipment at low rates – often at a level that maintained the business without making any significant profit. For example, at the end of the 1930s the Southern Railway was purchasing high voltage cable from Pirelli General at a lower cost per metre than it had paid at the beginning of the decade. Safety legislation was not as constrictive as it is today when projects must comply with the Construction Design and Maintenance (CDM) Requirements which are subsidiary regulations below the overarching Health and Safety at Work (HSE) Regulations. In the 1930s it was legal for the Southern Railway to pay 'danger money' to operatives working in the vicinity of the deadly 'live' conductor rail; and it was possible to carry out more work on tracks open to traffic and not closed off as part of a formal track 'possession'.

Notwithstanding all the above, perhaps what today's railway operators can learn from the Southern Railway's projects is that all its work was carried out 'in-house'. This is not the place to examine the present-day arrangements for electrification and other major projects on the railways, but when I worked with Network Rail providing additional electricity supplies for what many still call the Southern Electric, my impression was that our effectively 'state-owned' railway's role in respect to large projects was little more than that of an agency for dispensing lucrative contracts to large private companies. I did feel slightly ashamed of taking such a cynical view until I read the following by the *Private Eye* railway correspondent 'Dr B. Ching' whose comments, while specifically aimed at the High Speed 2 (HS2) proposals and the opinion of those managing HS2, might just as well apply to other large railway projects:

> Other countries' state-owned rail firms design, build and operate lines as their networks evolve, but Britain earned the nickname 'treasure island' after outsourcing replaced British Rail. Even the renationalised Network Rail (NR) relies on construction contractors, but NR's inadequate supervision of them causes cost rises and delays … HS2 limited … warned as long ago as 2009 … [that] European rolling programmes provided 'continuity of work and hence the opportunity to develop stable skilled teams and managements'. Britain's skill shortage [has] inflated costs through the 'extent of additional supervision' [and] 'multiple subcontracting' required 'multiple layers of technical and commercial supervision', each layer adding 'overheads and profits'.
> (*Private Eye* No.1515 7/20 February 2020, p. 14)

Sir Herbert Walker would have concurred with Dr B. Ching. Unlike Network Rail today, the Southern Railway owned both the track and operated the train services. Almost all the construction work was carried out using the railway's own labour force and contractors were only used when necessary and then were intensively managed by railway staff. The Southern Railway enjoyed a continuity of labour availability as staff moved on to the next phase as each incremental extension was completed. Long-term contracts were made with manufacturers to ensure continued equipment availability made to exact specifications. For these reasons it is easy to see how the Southern Railway was able to complete electrification schemes 'cheaper' and to schedule – something that the present-day railway often seems incapable of achieving.

The centenary of the opening of the very first section of what was to become the Southern Electric fell in 2015. Over the years, the London and South Western Railway (LSWR) and the Southern Railway came under attack from many directions including from the Board of Trade, the Ministry of Works, the Ministry of Transport, Members of Parliament, county councils, the local and national press and even the Women's Institute – but the third-rail system weathered the storms and has survived. While others procrastinated over the virtues of electrification, the Southern Railway got on with the job, brushing criticism aside to provide much of the south of England with what it needed – reliable public transport. The Southern Electric is still with us today, in spirit if not in name. Generations of commuters may have cursed the overcrowded trains but the whole economy and social life of Central London depends on the commuter railways and the Southern lines still carry the lion's share of this traffic. In 1950, when private car ownership was relatively small, almost 17 per cent of all passenger mileage on British Railways was by Southern Electric trains.

In order that the reader may understand some of the complex issues surrounding the early arguments as to how to electrify a railway, I have found it necessary to digress into the realms of electrical theory. Most aspects of electrical power engineering are easy to understand at a basic level – it is only when you dig deeper that you find a complex mathematical core. Therefore I have applied what one of my college lecturers would have termed a number of 'boy's explanations' to impart the basic concepts to allow the lay reader to appreciate the arguments. I trust that this will not confuse the non-technical reader or appear to patronise those with some technical knowledge.

Alfred Raworth's personal story is part of the history of the Southern Electric and the Southern Electric is an important element in the history of the Southern Railway which existed between 1923 and 1947. Several books have been written on the history of the Southern Electric. For many years successive editions of G.T. Mooney's *Southern Electric* was the standard reference book and John Glover later brought this history up to date. More recently, David Brown has provided us with his excellent *New History of the Southern Electric* and there have been many publications describing the electric trains such as Kevin Robertson's *First-Generation Southern EMUs*. An electric railway is more than a fleet of electric trains running to tight schedules; underpinning it all is an electricity distribution system which has received minimal literary interest. One publication which has given equal weight to the subject of the electrical power distribution and to the trains is Colin Chivers' *The Riverside Electric*, recounting the earliest days of what was to become the Southern Electric. He describes the pioneering work by the LSWR to electrify its suburban services emanating from Waterloo and gives detailed descriptions of the Durnsford Road generating station and the trackside substations. Alfred Raworth played an important part in this project as assistant to the railway's electrical engineer, Herbert Jones. Since the publication of *The Riverside Electric* by the South West Circle, Alfred Raworth's personal notebook has been deposited at the National Railway Museum allowing a deeper insight into the design process for this early electrification project. Another recent excellent publication, *Southern Nouveau and the Lineside* by George Reeve and Lesley Tibble, describes all manner of Southern Railway modernist buildings and station fittings but, sadly, does not venture into Raworth's closed world of the electrification control rooms, substations and track paralleling huts, all of which are from the same architectural genre. This book will attempt to fill this gap with details and illustrations. Many of the original structures do survive, but with extensive modification and some change of use. There is a lack of contemporary photographs (substations often appear as only a backdrop to photographs of the trains), so I have included new drawings to explain the formation and function of the substation equipment. For those who wish to make models of Southern Electric substations, please note that due to the lack of available information most of the dimensions have been interpolated (in the mathematical sense) from photographs and Google Maps.

Another gap in railway literature is an examination into the impact that national politics have had on the history of railway electrification in Great Britain. In the early twentieth century the government expected the railways to use a system of electrification that their 'experts' deemed the most suitable. This brought the Southern Railway into conflict with the Ministry of Transport when the Southern wished to go its own way and use the third-rail system. This book will recount the story of the various government committees that were set up to examine the design and extent of any proposed railway electrification. National politics also intervened with railway nationalisation, politically driven modernisation plans and the infamous Beeching Report.

Electrification is expensive, and the Southern, as with all the other railways, would constantly seek financial assistance from the Treasury with only a limited degree of success.

While I take full responsibility for any mistakes or omissions in this book, I must give my wholehearted thanks to many individuals who have given assistance in providing information and photographs. These include (and I apologise to anybody that I miss), Alan Booth, James Boudreau, David Brown, Robert Caroll, Graham Feakins, John Harvey, Martin James, Mike King, Hugh McAulay, Chris Raworth (unfortunately no relation to Alfred), John Scott-Morgan, Peter Waller, Peter Winchester, and the Dulwich College Archivist Soraya Cerio. Wherever possible permission has been obtained for the use of photographs, but this has not always been possible despite my efforts to contact potential copyright holders. Apologies therefore to anybody who can claim ownership of any of the images. I must also acknowledge the assistance of the many unnamed staff at the National Archives, the National Railway Museum (search engine), the British Library, the London Metropolitan Archives, Trinity College Library, Cambridge, and the volunteers who man the preserved Woking Control Room during the annual open days. Last, but not least, the invaluable assistance of my wife, Christine, who has assisted with the research and proofreading.

PART 1
INNOVATION AND CONFRONTATION

CHAPTER 1

TEXTILES, ELECTRIC LIGHTING AND TRAMS

On 30 December 1932, the Worthing Corporation hosted a gala dinner at the Burlington Hotel in their town.[1] This auspicious occasion was to celebrate the completion of the Southern Railway's electrification scheme from London to Brighton and Worthing. The Southern Railway's electrified lines – marketed as the Southern Electric – had now progressed from being the 'world's greatest suburban electrification' serving the suburbs mainly to the south of the River Thames, to achieving the status of a 'main line' electrified railway connecting the capital with the south coast. The route may not have been far in main line terms – London to Brighton was only 50 miles – but the Southern Railway was to operate non-stop and limited-stop electric express services along the route. The new, purpose built rolling stock was to offer the travelling public a level of comfort and amenity that would astonish today's travellers on the Brighton line, and this high standard was to be equalled elsewhere in the country over the subsequent years, but it was never to be surpassed.

When the meal was over, the mayor introduced several speakers. The first of those to stand up was the Southern Railway chairman, Mr Gerald Loder. The Southern Railway directors were not professional railwaymen; Loder – soon to be ennobled as Lord Wakehurst – was foremost a wealthy lawyer and a former Conservative Member of Parliament (MP) for Brighton, whose private passion was the growing of specimen trees at his country estate, Wakehurst Place. Loder was not an inspiring speaker, but he generously congratulated many individual Southern Railway staff members who had key roles in the project, including the chief engineer and one of his assistants, the signaling engineer who was responsible for the new signalling installation. He also gave thanks to the electrical engineer of new works, Alfred Raworth, who he said had been responsible for 'all the technical and more difficult details'.[2]

Loder seemed to be slightly unsure as to what had been the exact role that Alfred Raworth had played in the electrification scheme and what precisely constituted those very 'technical and difficult' details. But the Southern Railway's general manager, Sir Herbert Walker, fully appreciated Alfred Raworth's expertise in designing and project managing not only the electrical power arrangements for the Brighton electrification scheme, but also his part in creating the Southern Railway's suburban electric network. Sir Herbert, a man who was originally destined to be a doctor but had become the most respected railway manager of his generation, was later to confide that without Alfred Raworth's engineering expertise the development of the Southern Electric would not have been possible. Surprisingly, Sir Herbert was absent from the Worthing celebrations, so it fell to his assistant general manager, the popular Gilbert Szlumper, to dutifully read out the speech that the general manager would have made had he been present. In case any of the diners at the Burlington Hotel were unsure as to what the 'technical and difficult' aspects of the electrification scheme were exactly, Sir Herbert's text outlined the new electricity supply arrangements and stated that Alfred Raworth should take all of the credit for designing the ingenious power supply control system installed at Three Bridges. From the basic design concept Raworth had, in association with the manufacturers – the Swedish company ASEA – worked out the detailed arrangements to install a system that, in Sir Herbert's opinion, had significantly advanced the progress of the electrification of the railways. Sir Herbert was more interested in commercial viability than engineering perfection and his vision of railway electrification was out of step with contemporary thinking, and thus electrification progress in his mind related to the system now in use on 'his' railway. The Southern Railway with his leadership had pursued a policy of using track-mounted conductor rails operating at a relatively low voltage, in opposition to the views of the Ministry of Transport and other main-line railways, who believed that the way forward was to use overhead wires operating at a higher voltage. Raworth's innovative electricity distribution and control system had overcome many of the disadvantages of using the 'third rail' system, which many engineers had believed to be totally unsuitable for longer routes such as from London to the south coast.

As the Southern Railway's electrification expert, Raworth had not only to be an innovative engineer, but also, in order to gain the wholehearted confidence of the astute Sir Herbert, needed a high degree of commercial expertise and 'political' acumen to press forward the cause of railway electrification. These were qualities which he had inherited from his late father, the distinguished consultant engineer and entrepreneur John Smith Raworth, who had – as would any dutiful Victorian father – arranged for young Alfred's education, apprenticeship and then employment as assistant to himself in the business with the family name: Raworth's Traction Patents. It had never been intended that Alfred should become a railwayman but, quite literally due to an accident, John Smith Raworth had been forced to seek out alternative employment for his oldest son.

Alfred Raworth's family had their origins far away from Brighton in the North of England. Many surnames relate to the location where the family originally lived. The likely origin of the Raworth family is thought to be the tiny hamlet of Rowarth, a village which is perched on the edge of the Peak District near New Mills, within Derbyshire but close to the Yorkshire border. Most of the Raworth family spread across North Derbyshire and into Sheffield, which was later to become heavily industrialised and be known as the 'Steel City', while others ventured to the smaller nearby town of Dronfield.

The generally accepted modern pronunciation of Alfred's surname is *Ray-worth*. Several Raworths had settled in faraway London in the 15th century and, of these, those who had not yet achieved full literacy would often insert the usually unwritten 'y' when they were obliged to sign their name.[3] In his book *Locomotive Panorama Volume Two*, E.S. Cox – who was the executive officer (design) in early British Railways days – twice misspells Raworth as *Rayworth* when writing about Alfred, so it might be assumed that that is how the name was pronounced by many of those who knew him at that time. But I have regularly heard the name pronounced *Row* (as in to have a quarrel) – *earth*. Given the spelling of the tiny hamlet which is likely to have provided them all with their name, there must be the suspicion that in the 19th century, where this account begins, the Sheffield Raworths used this now generally defunct pronunciation. Many Southern Electric enthusiasts do however use the alternative pronunciation, suggesting that it was possibly used by Alfred's colleagues at the Southern Railway.

Alfred's father was born in February 1846 in Sheffield,[4] the son of Benjamin Joseph Raworth who had married Epenetes Walker in 1845 at Frome in Somerset. Epenetes' father, Thomas

Walker, was a Wesleyan minister and the origin of her unusual Christian name is certainly biblical. The name means 'praiseworthy' and refers to the story about the exceptionally pious convert who was the first Christian from Asia. After the marriage Benjamin brought Epenetes home to Sheffield where the Raworth family were in the iron and steel trade, for which the city had by then become world famous. Benjamin Raworth was a merchant dealing in iron and steel products, while other members of the Raworth family had manufacturing businesses. Sadly, in 1850, Alfred's grandfather died and the widowed Epenetes together with John and his two younger brothers had to fall back on the support of both the Walker and Raworth families in and around Sheffield.

Nonetheless, due to the relative wealth of the extended Raworth family John was, unlike most Victorian children, fortunate to be given a good education. He first attended Sheffield Grammar School and later Chesterfield Grammar School. The reason for the move south to Chesterfield was because John's mother took up the position of housekeeper at a school for girls which her sister Grace Walker ran in partnership with another unmarried lady, Mary Pocock. The school was at Tapton House in Chesterfield. John Raworth had first attended the school as a visitor in 1851[5] when he was aged 5 and even at such a tender age he would have been delighted to discover that Tapton House had been lived in from 1838 until as recently as 1848 by none other than the iconic George Stephenson. The owner of Tapton House, George Wilkinson, had leased the house and grounds to Stephenson who was by then a very wealthy man. George and his second wife had moved to Tapton House from Alton Grange in Leicestershire, but it was not a happy time for him as his second wife died in the year that they arrived in Chesterfield. George married for the third time in 1848, but he himself died later that year and is buried at nearby Holy Trinity church.[6] Upon George's death the lease passed to his son, the equally famous Robert Stephenson, who sub-leased the property to the Misses Walker and Pocock.[7]

By 1861, Epenetes, John and his youngest brother Harrison were living at Tapton House while his 11-year-old brother Benjamin lived with their retired grandfather about a mile away from the school's main gate at Saltergate in the town.[8] Did the splendour of the house indicate to John Raworth that there was a fortune to be earned from a successful engineering career? Who can say, but a feature of John's future career was his incessant applications for patents relating to his own designs for textile machinery, steam engines and electrical machinery, and he used these, together with his private consultancy work, to accrue wealth beyond any remuneration that he was to receive from his employers. But, at the time he resided at Tapton House, probably the most important thing to young John was the thrill of being in the same rooms and walking through the same grounds as George Stephenson had.

Possibly inspired by his time at Tapton house, John became fascinated by all things mechanical. He craved an engineering career but at the grammar schools he had received only a classical education. What he needed was an apprenticeship to learn about the basic engineering fundamentals, such as the cutting, drilling, bending and forging of iron, to prepare himself for a career as a professional engineer. Epenetes, with her strong non-conformist background inherited from her father, would have wished young John to be in a good Christian environment and would have been pleased when in 1862 it was arranged for John, at the age of 16, to be apprenticed for three years to Mr Edward Hayes of Stony Stratford in Buckinghamshire, a man who was not only an engineer with the highest reputation but also a former part-time teacher at the local British and Foreign Bible Society School.[9] The spiritual influence of his mother, grandfather and Edward Hayes resulted in John Raworth being well versed in the scriptures such that in later life when speaking at any technical or board meeting he would invariably enliven his words with an appropriate biblical quotation. Hayes had set up his own business in Stony Stratford by opening a small forge named Watling works, originally intending to make agricultural machinery. Subsequently, benefiting from his previous experience at the nearby London and Birmingham Railway's Wolverton locomotive works, he began to develop and build small steam engines to drive the equipment that he manufactured. He soon earned a reputation for innovative and high-quality work.

At the end of his apprenticeship in 1866 John Raworth, with all his new knowledge, entered the engineering profession. He was appointed as second draughtsman in the locomotive drawing office of R. and W. Hawthorne at Newcastle upon Tyne. Two years later in 1868 he moved south and slightly west to Manchester, the Lancashire city that was for many years to be the base for his engineering and business activities, and it was from there that his career blossomed. The commercial environment of Victorian Manchester nurtured John Raworth's career development by honing his technical and entrepreneurial skills. Manchester was not just a bleak industrial conurbation consisting of grim 'dark satanic mills', since following the success of the industrialists the ever more prosperous middle class were having built for themselves fine houses in elegant residential areas, and the city became a centre for classical music and the arts.

John Smith Raworth certainly wanted to climb the social ladder. While he may not have craved the culture, he did not wish to be enslaved to dreary industry and intended to own and run his own companies and to live in one of the new upper middle class residential areas. With this ambition, vibrant Manchester was the place to be. He could not have chosen a better first employer and sponsor than John Hopkinson, who was the owner of the Wren and Hopkinson London Road Ironworks and who was later to become an alderman of the City of Manchester. Lancashire not only spun and wove cotton, but also made the spinning machinery and weaving looms, and foremost among these manufacturers was Wren and Hopkinson, which supplied equipment not only for the local mills but also enjoyed success in the export market. It is an irony therefore that following the removal of all restrictions on the export of textile machinery in 1843, it was the likes of Wren and Hopkinson that contributed to the later decline of the local cotton industry, as foreign spinners and weavers stole the trade from the Lancashire mills, often using British-built machinery. But any serious decline was many years hence when John Smith Raworth arrived in Manchester in 1868.

John Raworth was initially appointed as one of the assistants to John Hopkinson who, with his sons, took a keen interest in anything electrical, and it was while working at Wren and Hopkinson that John Raworth too became fascinated by the potential offered by the new scientific discoveries in this field. This was the era when the fertile minds of men such as Faraday, Maxwell, Kirchhoff and Fleming were formulating all the main laws and theories appertaining to electrical engineering.

After two years John was promoted to a senior position, designing and developing the cotton spinning and weaving machinery. Wren and Hopkinson were looking forward to the days when all such machinery would be powered by electric motors and John Raworth's practical and innovative skills mixed well with John Hopkins' theoretical knowledge of electricity.

In 1878, John Raworth was appointed as the Lancashire and Yorkshire agent for Siemens Brothers. This company was the British offshoot of a German company which in 1858 had become a separate and independent entity known as Siemens, Halske, and in 1865 the name was changed to Siemens Brothers. In 1863 a cable factory – which was later to supply cables for the electrified railways in South

London – had been established at Charlton, near Woolwich in south-east London, and in subsequent years Siemens Brothers manufactured a variety of other electrical goods including motors, dynamos and telephone equipment. The move to Siemens was, at first sight, a strange one since the entrepreneur John Raworth had already set up two business enterprises himself, but by accepting this employment he was able to increase his network of industrial contacts, vital for his forthcoming consultancy business.

Raworth's first independent venture following three years' employment with Wren and Hopkinson had been the establishment of a machinery manufacturing business at St Mary's Gate, Manchester. John, with his brother Benjamin, had taken out several patents for improvements to machinery for use in the textile industry. John ran the business until 1877 when two of his staff, Messrs C.M. Dorman and R.A. Smith, took over and developed it into the world-famous Dorman Smith Switchgear Limited which, with now over 130 years' experience, continues to provide electrical contractors with switchgear for low-voltage electrical distribution and circuit protection. John Smith Raworth is considered the founder of the company.

His second venture was in 1877 when he and his brother Benjamin had acquired a mill in Carruthers Street, Ancoats, in the City of Manchester. They set up a new joint stock company to be known as John Smith Raworth and Company Limited.[10] This was an era when joint stock companies had become prolific following the Consolidation Act of 1862. The Act allowed companies to be established with limited liability, giving them some protection from the cyclic fluctuations in the national economy and furthermore, theoretically at least, resulted in the separation of the owners of the company – those who had bought shares – and the managers. This was a significantly different arrangement from the family-owned companies that had been the norm. The new company's capital was intended to be £6,000 made up of 600 shares at £10 each. John and Benjamin were to be paid £2,500 for their patent rights and in addition each received 50 shares worth £500. Benjamin was to

J.S. Raworth with his staff which included the founders of the Dorman Smith Company. Raworth sits in the middle of the centre row with the founders of Dorman Smith. C.M. Dorman seated to his extreme right with R.A. Smith sat between them. This photograph was taken in about 1884. (Chris Raworth Collection)

become the managing director at an annual salary of £200 per year. Part of Benjamin's conditions of employment was that he was to work exclusively for John Smith Raworth and Company. The Raworth brothers each purchased another 75 shares, making them the major shareholders. Therefore, this new venture, despite being a joint stock company, was nonetheless very close to being a 'family' business. Other shareholders were to be found mainly from within the local industrial community, including a glass manufacturer, a rubber manufacturer and a chemist, as well as a sprinkling of members of the moneyed class and a pawn broker. As well as the patent rights, the new company had more tangible assets – Wellington Mill with its machinery. Many of the machines listed in the inventory were given descriptions prefixed with the words 'Raworth patent'. John and Benjamin probably intended that Wellington Mill would be a showroom to demonstrate their patented machinery in a genuine manufacturing environment. But John Raworth was not particularly interested in operating the mill himself; his interest was intended to be purely financial and, leaving Benjamin in charge, he departed to take up his post with Siemens.

An early and important project that John Raworth was employed on while working for Siemens was the introduction of electric lighting in ships.[11] Electric lighting in general was in its infancy, but there was enormous potential for ship-board installations to avoid the use of dangerous gas and candles in the restricted and confined environs of a ship. The early designs of dynamos needed to be rotated at considerable speed by the small steam engines to produce enough electricity. To achieve the required speed the early steam engines had to be geared up using belts and pulleys so that as much as 10 per cent of the steam engine's power output would be lost due to the inefficiency of this gearing. John Raworth used his steam engine knowledge acquired with Edward Hayes and at R. and W. Hawthorne to design and patent his 'Universal' high-speed steam engine which could be directly connected to the dynamo without gearing. Raworth originally used his Universal engine for his ship lighting installations but its applicability to public electric lighting supply was soon realised. At the same time as he was involved in the ship lighting he was also in at the very beginning of the provision of public electricity supplies when he became responsible for the operation of a small generating station in Manchester which supplied electric lighting to the Piccadilly locality using his Universal engine. This is where John Raworth's talent began to manifest itself. He was now able to describe himself as an 'electrical engineer', since he was a competent mechanical engineer with a good working knowledge of electricity and the ability to take an electrical concept and transform it into a practical application.

Meanwhile there was trouble at Wellington Mill. Exactly why John Smith Raworth and Company was failing is not known. The decline of the Lancashire cotton industry due to foreign competition was not yet a serious problem, however mills generally were getting larger and it is possible that Wellington Mill was not able to compete with the giant establishments being built all around it at Ancoats. The only real clue is that the company appears to have been undercapitalised as only 400 shares out of a possible 600 had been sold. In September 1879, the company went into voluntary liquidation, leaving Benjamin in charge as administrator.[12] Once this had been completed Benjamin joined the staff of the *Engineer* magazine and rose to become joint editor in 1906.

While working for Siemens Brothers, John Smith Raworth, now 35 years old, married. In the spring of 1881 Margaret Cannington Kershaw, nine years younger than her new husband, became Mrs Raworth. Margaret was the daughter of James and Mary Kershaw and was the oldest of four children. James Kershaw was a chemist who lived and worked in Birkdale, then a fashionable village on the coast near Southport in Lancashire. Several of John's new in-laws had been minor shareholders in John Smith Raworth and Company and had possibly suffered some financial loss due to the failure of his company. Perhaps Benjamin had recovered all or most of their investments – the patent rights alone would have had considerable value – or perhaps John Smith Raworth just had a lot of charm! On the night before his wedding John Raworth stayed at 31 Part Street in St Meols, which is another suburb of Southport. He was the guest of three ladies who were all former residents of Tapton House: his mother Epenetes, his aunt Grace Walker and Grace's former business partner, Mary Pocock.[13] In 1865, George Wilkinson discovered that Robert Stephenson had sub-leased his property without his consent, and this resulted in a dispute that could not be resolved and sadly Tapton House School had been forced to close.[14]

The remuneration from his employment with Siemens and royalties from his by now numerous patents for textile machinery and improvements to his Universal engine enabled John to set up home with Margaret in Heaton Mersey, part of the then new and fashionable suburban area located to the south of Manchester. Heaton Mersey, then a location popular with those who had earned their money through ownership of expanding local industry, lies on the north bank of the River Mersey only a few miles downstream from its source. The Raworths may have been the very first occupants of a newly-built house, since between 1852 and 1892 this previously rural area was being developed with spacious new houses built along tree-lined streets for appreciative affluent buyers. On 20 January 1882, their first son, Alfred, was born into this comfortable and privileged middle-class world.

Just over a year later, on 7 May 1883, Alfred had a new baby brother, Arthur Basil, who was to choose to be known as Basil Raworth during his adult life. On 1 July 1883, the Raworths travelled to Birkdale for a family reunion, when the joint christening of Alfred and Arthur Basil took place at St James' church.[15]

In 1886 John Raworth left Siemens to become superintendent engineer to the Anglo-American Brush Electric Light Corporation[16] which had premises at 112 Belvedere Road, Lambeth. His new employer had a disreputable past, for in 1882 it had been involved in the creation of a stock market 'bubble' when £7,000,000 had been raised, exploiting gullible investors who had little or no understanding of electrical plant manufacture or application.[17] At the time Raworth arrived the company was in a legal battle concerning the patent rights for the dynamos it was manufacturing. The final battleground was to be at the Court of Session, Edinburgh, where Brush's English lawyers – who had been able to keep their client safe in the English courts – were unable to be heard in the Scottish court and Samuel Varley, the claimant, emerged as the victor. This resulted in the Anglo-American Brush Electric Light Corporation going immediately into voluntary liquidation and reforming as the Brush Electrical Engineering Company to avoid liabilities.[18]

The site of the Brush factory, and John Raworth's new place of work, was on the south bank of the River Thames where today the Shell building stands, and the London Eye rotates. At Lambeth, lighting equipment – both arc lamps and incandescent lights – were made and by 1886 when John Raworth arrived production had expanded to include the manufacture of motors, switchgear, small transformers and the disputed dynamos. Business was brisk, and it was evident that the Lambeth site was too small and larger premises were required, so the decision was made to purchase an existing factory in Loughborough, Leicestershire. The site was the Falcon works owned by the Falcon Engine and Car Works Ltd, which had earlier traded as Hughes Engine and Tramway Works Ltd. Such a facility as the Falcon works would also enable the Brush Company to later expand into the production of electric trams. John Raworth, as the Brush Company's superintending engineer, organised the removal of the Lambeth works to Loughborough and the alterations to the old Falcon works to meet the company's new requirements. The now listed Falcon works at Loughborough, still displaying a 'Brush' sign despite changes to company ownership, is probably the only monument to John Smith Raworth to have survived into the twenty-first century.

After a short residence in Kirkstall Road, Streatham – when Alfred's first sister Gladys was born – the Raworth family moved to their long-term permanent home in an area of social standing very similar to that of Heaton Mersey, at 46 Christchurch Road, Streatham in South London. Meanwhile in 1893 Margaret Raworth gave birth to their second daughter, Phyllis. This was the same year John's mother and Alfred's grandmother died and the administration of Epenetes Raworth's estate was entrusted to Benjamin Raworth. Her effects were valued at almost £110, not a large legacy even in 1893.

In 1891 John Raworth had been promoted to general manager of the Brush Company. He did not remain in this post for long as he relinquished his position to become a full member of the Brush board of directors and took up independent work as a consulting engineer. He set up his practice in Westminster, basing himself at what was later to become the home of Raworth's Traction Patents at Queen Anne's Gate. In addition to his membership of the new Institution of Electrical Engineers he also became a member of both the Institution of Civil Engineers and the Institution of Mechanical Engineers and was an active member of all three of these revered organisations.

During the late nineteenth century, horse-drawn trams appeared in many British towns and cities. Horses needed rest and refreshment, so thoughts naturally turned to some form of mechanisation. Due to the difficulties caused by smoke and noise when operating steam-powered trams through the crowded streets, the tramways were jumping a technology, avoiding steam, and moving straight from horses to electricity. So, in his role as a private consultant and as a member of the Brush

A twenty-first century view of the Brush Falcon Works at Loughborough – still retaining the Brush signage. (Rob Newman – geograph.org.uk/p/2309721)

board, John Raworth increasingly became involved with electric tram undertakings. In October 1896 a new company, the British Electric Traction Company, was registered naming John Smith Raworth as a director and one of its engineers.[19] In 1898, the British Electric Traction Company entered partnership with Brush Electrical Engineering to form the Electrical Power Distribution Company. This new company was established with the aim of providing electricity supplies to municipal areas with a degree of synergy with British Electric Traction, which was to obtain the electricity supply for its trams from the Electrical Power Distribution Company wherever possible. In 1901, in a bid to expand activities into the business of both building tram cars as well as operating them, British Electric Traction took over Brush Electrical Engineering, the deal being achieved by an exchange of shares, one British Electric Traction share for 12 Brush shares.[20]

Brush was taking commercial advantage of the considerable amount of common technology used in both tramway and electricity distribution systems. Alone among the three main utility providers at the time, gas, electricity and water, the electricity supplier had, and still has, not only to send its product out to the customer but also to collect it back again. This is because electricity only functions when a complete electrical circuit is made. Tramway systems at the time typically consisted of a 500-Volt (V) direct current supply from an uninsulated overhead line suspended above the street (but some systems did have a clever arrangement using a slot in the road with a shuttering device for safety). The 'out' route for the electric current was the overhead wire and the 'return' was the steel rails set in the road. Imagine the overhead wire as an extended positive (+) battery terminal and the rails as an extended negative (-) battery terminal. Over time, the adequate provision of a return path for the electric current was to prove just as controversial and problematic as the means of providing the 'live' connection for electric transport systems.

By 1906 British Electric Traction was operating 15 per cent of the nation's tramways. By 1908 it had a controlling interest in 50 tramways that between them operated 387 miles of tram line and 54 miles of motor bus routes.[21] As well as being on the board of British Electric Traction, John Raworth also became the chairman of both the Devonport and Southport Tramway Companies, and a board member of the North Staffs Tramway Company. With the same almost missionary zeal that his son Alfred was later to apply when promoting railway electrification in southern England, John Smith Raworth from his influential position set himself the task of facilitating efficient tramway operations in areas where the economics were marginal. Tramway operations in small towns and on the outskirts of the larger towns and cities faced several difficulties due to the terms of the 1870 Tramways Act, which gave local authorities the rights to compulsory purchase the private tramway companies after 21 years of operation, and often they were very keen to do this as the trams provided useful additional revenue to fill up the council's coffers. In many towns and cities, the electrification of tramways was being hindered either by the local corporations' refusal to either purchase the tramways before the lease expired and proceed with conversion to electricity, or conversely refusing to extend the leases to allow the private companies to invest in electrification. This was an obvious disincentive for the tramway companies to invest in new equipment or larger trams. John Raworth became a champion for the cause of providing electric tramways to many small communities long before what he termed the 'municipal traders' in the large cities provided their populace with electric trams.

The smaller concerns did not own their own generating stations and had to rely on others for the supply of electricity, and this was often from the same local authority that was coveting their business. The unit cost for electricity was high, hence the

1898 commercial arrangement between British Electric Traction and the Electrical Power Distribution Company to cooperate wherever possible to avoid having to use the municipal electricity supplies. The existing electricity suppliers did however have an operational problem in that the electricity generating and distribution networks had been designed to provide a basic 'electric light' service during the hours of darkness and there was the expectation of a virtual closedown during the day, even to the extent of shutting down the generators and relying on batteries. They did not wish to incur the extra expense of running generators to feed the spasmodic high starting currents that were demanded by the trams. Whatever the rights and wrongs of the electricity companies' charging arrangements the need for the most economical use of electric power was obvious.

Meanwhile, in his private life, John with his wife Margaret had to make decisions about their sons' education. It might be supposed that they would have wished the boys to follow a similar career path to their father's – that is, a good basic education followed by an apprenticeship. But John Smith Raworth was now more than just an engineer; he was an entrepreneur and a multiple company director with his hands placed firmly on the managerial levers of several organisations. He was a man of status who had to confidently mix with the titled and moneyed classes, and he wished the same for his two sons. A good technical education was deemed necessary for Alfred and Arthur Basil, but only after a public school education that was then the norm for the type of men that they would have to rub shoulders with during their future business life. So, in 1893, at the age of 11, Alfred was sent away to Lowestoft on England's east coast to attend St Aubynes' Preparatory School.[22]

At so late an age he must have had some previous education, and strangely Alfred's younger brother Arthur at the earlier age of 9 first attended this school the year *before* Alfred in 1892. The delay in sending young Alfred to his preparatory school invites speculation as to why this was so. Financial considerations can be ruled out of course; John Raworth was a rich man, and even if he had short-term financial difficulties Alfred as the elder son would have received preference over his younger brother Arthur.

One possibility is that his parents did not consider that Alfred was emotionally ready to be whisked so far away from his home to suffer the tortures of a Victorian preparatory school. John may seem to have been the stereotypical Victorian father, likely to believe in the 'stiff upper lip' and 'making a man out of the boy', but he was an intelligent man who had earned a reputation for clear thinking and common sense. He may well, encouraged by Margaret, have agreed to a postponement of Alfred's education awaiting a degree of maturity to develop before the young boy's spirit was totally crushed at a boarding school. Was Alfred quiet and timid? His later reputation as a ruthlessly professional manager does not sit well with the image of a nervous and introverted small boy. But while those who were later to have dealings with Alfred Raworth found him to be initially terse and difficult to communicate with, when colleagues did become close to him, they were often to remark on his kindness and willingness to provide them assistance in respect of technical matters. Often an inwardly shy person will be guilty of putting up a defensive front to hide an inner natural shyness.

Another possibility was that Alfred was a weak and sickly child who needed to be kept at home for the benefit of his health. Childhood mortality in the Victorian era was high, and while Alfred had survived the most dangerous years, his parents may have thought that keeping him at home for a year or two longer was necessary. There are hints – but only hints – that Alfred Raworth, a relatively small man, did not enjoy the best of health throughout his life. These indications include the facts that when he volunteered to join the forces during the First World War he was not to be posted to a fighting unit; that when he arrived in New York in 1919 the immigration official noted his sickly appearance; that due to his illness a very important meeting in 1921 had to be postponed; that at the close of 1943 he was admitted to hospital for an undisclosed illness; and that subsequently it was 'health issues' that were given as the reason for his relatively early retirement from the Southern Railway. Each of these instances, taken alone, is easily explainable without having to rush headlong towards the conclusion that Alfred suffered a lifetime of poor health, but they suggest that possibly poor general health may have been a factor in his late leaving of home to start his formal education.

There is yet another possible explanation. At some time in his childhood Alfred Raworth received a severe facial injury which left him with an asymmetric jawline and a scar in the point of his left ear. Later in life Alfred would brush aside any enquiries as to the reason for his disfigurement by claiming the acceptable cause of a bad bounce of a cricket ball[23] – thus avoiding having to recount any personal and distressing detail. Perhaps a traumatic childhood injury was the reason why young Alfred was kept away from school for so long.

Leaving St Aubynes' School in 1896 Alfred then attended Dulwich College from January 1896 until July 1900. Dulwich College was, and remains today, a premier public school situated in London's south-eastern suburbs. It may have been chosen due to Alfred's continuing shyness or poor health, since he did not have to be a boarder as the school's location was within easy daily commuting distance from Christchurch Road, Streatham. Given its location most pupils were 'day boys'. Putting any daily travel considerations to one side, Dulwich College was as good a public school as John Smith Raworth could have found for Alfred – and Arthur Basil who joined his brother at the school a year later – for unlike many public schools at the time, students were able to study mathematics and science and not just the 'classics' which had been John Smith Raworth's educational diet at his grammar schools. At the time that the Raworths attended Dulwich College the school was organised into 'Sides'. These were the Classics Side, the Modern (Language) Side and the Science Side. As would be

Dulwich College in 2021 showing the oldest buildings, largely unchanged since the time that Alfred Raworth and his brother attended. (Author)

expected Alfred was in the Science Side. In 1900 Alfred represented the college as a member of the shooting VIII. Once again, his brother Arthur went ahead of him, being in the gymnastics VI during 1899, of which in 1900 Alfred also became a member.[24]

With a respectable public school background behind him, it was now time for Alfred to start his engineering education. An apprenticeship was easily arranged with Browett, Lindley and Company back in Manchester, as John Smith Raworth was a friend of the owner Thomas Browett. Alfred lodged with the Kirkpatrick family at their home in Northern Etchells in Cheshire. He was taken through all stages of the production of the large steam engines for electricity generation which the company manufactured. He witnessed the process from design through manufacture and erection to final testing. At the same time Alfred's formal education was extended by attendance at the Salford Technical School. In April 1901 while at Browett, Lindley and Company Alfred joined the Institution of Electrical Engineers as a student member and at the Glasgow Exhibition he was awarded a £5 premium by the Institution for a report he wrote on steam engines.

In 1903, after his two years at Browett and Lindley, John Smith Raworth arranged for Alfred to have a second short apprenticeship, this time with Brush at Loughborough. It was here that Alfred was introduced to steam turbines as well as other Brush products such as pumps. John Raworth must have felt that Alfred was still short on academic knowledge and arranged for him to have additional private tuition while at Loughborough. At this time, his address in the Institution of Electrical Engineer's list of members was given as 'Falcon Works, Loughborough, Leicester'.

Meanwhile, Alfred's father continued with his crusade for low-cost tramways. He designed his small four-wheel, single-decker 'demi-cars' for use on routes where the traffic was light, and costs needed to be kept low if any profit was to be made.[25] Typically, trams at the time were short four-wheeled vehicles but were double-deckers. The demi-cars were single-deckers with seating for only 26 passengers. The first of these was built by Brush at the Falcon works in 1903. It weighed 4.5 tons compared to the 11 tons of a standard double-deck tram. As John Smith Raworth was the chairman of the Southport Tramway Company it is no surprise that the first demi-tram was to be built for the Southport fleet and given the number 21. It was experimentally used on the off-peak services between Southport and Birkdale. Passengers were charged one old penny to travel between London Square Southport and Birkdale railway station.[26] The tram was designed to be operated by the driver only, thereby eschewing the need for a conductor. The driver would collect the fares from the passengers as happens today on a modern bus. Another of Raworth's patented innovations which was incorporated into all the demi-cars was described as the 'One-Man Bar'. This was simply a waist-high gate around the driver's position, and this acted as an emergency brake as it could be raised by passengers to apply the brake should the driver, the only tramway employee on board, became incapacitated. (See plate 1 (bottom))

On completion of his second apprenticeship at the Falcon works, Alfred's father launched his son into the world of commerce and engineering by appointing him as his assistant to be ready to take his place in the soon to be established company: Raworth's Traction Patents.

CHAPTER 2
RAWORTH'S TRACTION PATENTS

In 1904 John Smith Raworth made a second attempt to set up a successful joint stock company to exploit his many patents. The most important of these was his design for a regenerative control system whereby the tram's motor could be switched to a generating mode to retard the vehicle's speed when required. Although many cite Raworth as the inventor of 'regenerative braking', he never used the term, his design was for a control system not simply a brake system. For many years, engineers had seen a potential practical application for a machine that could be used as both a motor and a dynamo. In tramway terms it was hoped to find a practical way to use the hard-earned kinetic energy gained by dragging a heavy tram up a steep hill which was wastefully dissipated as heat at the brake blocks during the descent. Innovators suggested that as the vehicle rolled downhill, due to the forces of gravity all the necessary retardation could be achieved using the drag caused by the traction motors arranged so that they generated electricity. Furthermore, any electricity so generated could be sent back up the trolley pole onto the overhead wire, offsetting the electrical energy provided by the generating station.

Outwardly this concept seemed to be straightforward. The contemporary direct current motors and dynamos both exploited Faraday's discoveries which linked magnetic fields and the flow of electricity to physical motion. He had shown that if a wire made of a material such as copper and part of a continuous loop was passed through a magnetic field, an electric current would flow through the wire. This was the basis for the dynamo. Using similar apparatus Faraday also showed that if an electric current from a different source was passed through the same wire within the magnetic field, if this was unconstrained then movement of the wire took place. This was the basis for the direct current electric motor.

Contemporary motors and dynamos based on Faraday's discoveries consisted of a rotating shaft – the armature – upon which were mounted coils of copper wire – the armature coils. The armature was shrouded in a set of electromagnets – that is, more coils around iron cores. These field coils produced the desired magnetic field. Connection between the armature and the electrical source or load was by segmented moving contacts on the armature shaft bearing on fixed contacts, this part of the machine being known as the commutator. The fixed contacts were termed brushes, as originally actual steel wire brushes had been employed, but very soon spring-loaded renewable graphite blocks were to be used. The commutator also performed a switching function to correct the polarity of the input or output, since as the armature rotated through the fixed magnetic field, the direction of flow of the electric current was opposite for each half turn of travel.

The two sets of coils, field and armature, were connected either in line – series connection – or in parallel – shunt connection. In some instances, such as the dynamos being built by Brush, a compound arrangement, part series, part shunt was used.

The genius of John Smith Raworth – and I do not apologise for using the word 'genius' given the man's lifetime of inventive achievement – was that he was able to take what is only a broad and very simplistic concept, and to then devise and build a practical working regenerating motor system using his own design of controller and, significantly, by economically modifying existing machines. As early as 1885 Frank Sprague (1857–1934) the 'Father of Electric Traction', had patented his version of regenerative braking in America. His system involved a switching arrangement to increase or decrease the number of turns in the field winding coil to achieve an output voltage when the vehicle was coasting.[1] But as far as Raworth was concerned his most significant rival was the Johnson-Lundell Company. Its system of regenerative braking used either two or four motors which had two commutators on the armatures and had the feature of an automatic mechanical brake being applied as the tram slowed down to a speed below which the regenerative braking was effective.[2] The British subsidiary of this American company was based at Southall in West London and had experimentally fitted and extensively tested their design on a Newcastle upon Tyne tram in 1902.[3] The Johnson-Lundell Company was later to challenge John Raworth's patent rights.

John Raworth had the laws of physics on his side, for a direct current motor will always generate within itself an opposing electrical force working against the applied current. In other words, the machine as it runs is already trying to turn itself into a dynamo. This opposing force is termed the back-electro motive force: the 'back emf'. This back emf can be increased if the motor's magnetic field is increased. So, John Raworth had to boost the magnetic field to increase the back emf and in so doing flip the motor into dynamo mode. The usual arrangement for a traction motor was to have the field coils connected in series with the armature coils. The applied current and hence the speed was regulated by a controller consisting of a variable resistor. Raworth designed and patented a new design of controller which could switch into use additional field coils connected in parallel with the existing coils when regeneration was required. If the tram continued to move forward due to rolling down an incline, generation would be achieved. Consequently, when the output voltage – the electrical 'pressure' driving the current – from the motor running as a dynamo exceeded the voltage on the trolley wire above, an electric current would flow from the tram onto the wire. (See plate 1 (top))

Raworth initially used his regenerative control design on his new demi-trams. Thus these 'minis' of the tramway world would not only use less electricity due to their reduced size and weight but would also make a further 25 per cent saving in electricity by using Raworth's innovative control system. It was this saving of the precious electricity not the saving on brake block material that was the main aim of the Raworth patented system. After the first Southport car, three more demi-cars were built, two for Halifax Corporation and one for the Glossop Tramway Company.[4]

Following the early success with the demi-trams, John Raworth decided to commercially exploit his patents and repeat what he had attempted in Manchester by setting up another joint stock company. Perhaps he should have learned from his experience with John Smith Raworth Limited and from the misfortunes of the company of which he was still a director, Brush, which had previously come to grief after relying too much for its prosperity on what we now term 'intellectual property'. Once again, this new venture, Raworth's Traction Patents, was destined to have a relatively short and difficult history. Despite the name Raworth in the company title, this was not a 'family firm'. The board of directors consisted of men from tram manufacturing companies and tramway operators. By this means Raworth had influence on the production and use of his patents – something that the rival Johnson-Lundell Company did not have. John Raworth's old friend Thomas Browett, formerly managing director of Browett, Lindley and Company Ltd, by then in semi-retirement and living in Kensington, was to be the new company chairman. The registered address of Raworth's Traction Patents was given as 2 Queen Anne's Gate, Westminster – John Raworth's London office for his consulting practice.[5]

The choice of Queen Anne's Gate was a good one as this address would impress potential clients – the building survives today in an elegant street of town houses. It is only a five-minute walk from Whitehall in one direction and Buckingham Palace in the other. The rear garden and the magnificent bay windows look out across Birdcage Walk to St James' Park. The fact that number two is tucked around the corner at the west end of the street in no way diminishes its air of importance. This large and apparently private house was to be used by Raworth as a prestige office only, and he probably only leased one or two rooms. Therefore, this was not an address at which any electrical manufacturing process was ever to take place. Today what is outwardly a magnificent private town residence discreetly advertises the fact that it is still the home to a commercial enterprise.

The actual equipment manufacture was to be contracted out to several manufacturers such as the Brush Electrical Engineering Company, the British Electric Car Company, the United Electric Car Company and Westinghouse. As well as the making of the controllers and associated wiring and auxiliary switches, the major task was the revision of the motor field windings to comply with Raworth's tight specification for regenerative control. In the early days of traction motor production, a certain amount of mechanical slack was incorporated in the armature mounting as different and changing forces were exerted on the armature at different speeds and loading during the duty cycle. The alignment of the brushes against the commutator was often a compromise to accommodate the resulting armature movement. A badly designed motor would consequently generate sparks at the brushes. When a motor was to be used with the Raworth system – being both motor and dynamo – this difficulty was more acute and the sparking more likely. It followed that not all machines were suitable for conversion, but the Raworth adage was that if it was a good motor that did not have a propensity to make excessive sparks at the brushes the design was suitable for modification. Several motors were to be used during the time of the new company's existence; the original manufacturers of these machines were Brush, General Electric and Westinghouse.[6]

The new company was authorised to raise capital to the value of £30,000 to be made up of 15,000 preferred ordinary shares at £1 a share and 15,000 deferred ordinary shares also at £1 a share. Raworth's Traction Patents intended to pay out a 6 per cent dividend on the preferred ordinary shares with any net profit left over to be paid out to the holders of the deferred ordinary shares. The price that Raworth's Traction Patents was to pay John Smith Raworth for his patents was £18,600. This was to be made up of £3,600 in cash and the remainder in deferred ordinary shares. Raworth allocated some of his deferred ordinary shares to members of his immediate family. His wife, Margaret, received 50 shares, his sons Alfred and Arthur received 500 shares each, but his two daughters, Phyllis and Gladys, only received 20 shares apiece. While the allocation of more shares to his sons than his wife and daughters might seem to modern sensibilities misogynistic, Alfred and Arthur (now insisting that he was to be called Basil) were expected to work for the new company and were thus stakeholders.[7]

There was an early incident involving a demi-car which was, with hindsight, a harbinger of the problems to be encountered in later years. On 14 October 1904, exactly one week before Raworth's Traction Patents was incorporated, one of the demi-cars sold to Halifax Corporation, number 95, was descending a 1 in 14 gradient under regenerative control. Due to some completely unconnected occurrence, the circuit breaker at the power station that fed the tram line opened automatically. This resulted in the loss of regenerative braking as the electric circuit that the tram was feeding into was no longer complete and, with all braking power removed, the tram ran out of control. At the bottom of the incline, it struck a glancing blow with sister tram number 96 and careered on until it struck a horse-drawn dray. The driver of the runaway was badly hurt as was the dray driver who was flung under his cart. Three passengers in the stationary tram were shaken but not seriously injured. The Board of Trade duly carried out an investigation appointing Lieutenant Colonel Druitt to officiate. His finding was that since the Raworth control system and the emergency brake were both in perfect working order, the blame should be squarely put upon the driver who had not been adequately trained in the use of the emergency brake.[8] The incident did however highlight a flaw in the Raworth system, and further development incorporated a feature that any such disconnection would automatically divert the generated current internally through a resistor to bring the speed of the tram down to about 2mph. The tripping of the circuit breaker back at the generating station might be a rare event, but the trolley pole accidentally removing itself from the overhead wire was a stronger possibility and the revised design also catered for that eventuality.

If there were to be no other Raworths on the board of directors as might be expected given the company name, at least employment was to be found for the younger family members. Alfred was already working as his father's assistant, helping with the development of the equipment. Once he had finished his studies Alfred's brother Basil was also to join the new company. Another Raworth joined in October 1905; he was John Ernest Raworth, aged 32. John was the son of Harrison Raworth and therefore a first cousin to Alfred. His work was part-time as in 1905 he was also working as a patent agent, being the chief assistant and salaried partner to Mr G.J. Lorraine. On joining Raworth's Traction Patents as an assistant engineer, he was given permission by the board to concurrently continue working with Lorraine and later he was to become

It is thought – and it is very likely – that the dapper young man with the fashionable beard (as worn by royalty and a convenient cover for any facial scars) is Alfred Raworth. He is demonstrating the regenerative control system to one of Raworth's Traction Patents' customers, Glossop Tramways, here represented by the manager, Mr Knowles. (Walter Gratwick/Estate of C.C. Hall/Scottish Tramway and Transport Society)

a partner at the patent agents Raworth and Moss.[9]

Alfred became an associate member of the Institution of Electrical Engineers and an associate member of the Institution of Mechanical Engineers. One of his sponsors for both applications was, naturally, his father. Alfred was to represent his father on many occasions, for example on 3 November 1904 he read a paper entitled 'Raworth's System of Regenerative Control' at the annual conference of the Association of Tramways and Light Railways Officials held in Wakefield Town Hall, and later he was to confidently present a paper to a meeting of the Leeds branch of the Institution of Electrical Engineers.

At the formation of Raworth's Traction Patents, the initial outlook was very promising as there were orders for two new demi-cars for the Barrow Tramway Company and a third for the Yorkshire Woollen District Tramway Company, a British Electric Traction owned company operating in Dewsbury. The firm also had orders to install regenerative control in 'large cars': one for the Southport Tramway Company; one for Bournemouth Corporation; eight for the Yorkshire Woollen District Tramway Company; seven for John Smith Raworth's Devonport Tramway Company; five for the Gravesend Tramway Company; and a massive order for 40 from the Birmingham and Midland Tramway Company. Later, more demi-cars were provided to various tramway operators: the Chester Corporation bought a single car, Erith Urban District Council purchased two, the Darwen Corporation bought three, and Plymouth Corporation ordered six.[10]

On 29 March 1905, the company secretary, John Hodgson, gave a progress report to his directors. He reported that the staff had been concentrating their efforts on the equipment for the Birmingham and Midland Tramway Company order and that this contract was attracting a lot of interest from other tramway operators, indicating that more business would be forthcoming. Hodgson also reported that the costs of providing the regenerative control equipment had not exceeded the original estimates.[11]

Following the establishment of Raworth's Traction Patents, development of the regenerative control for trams continued. As early as the autumn of 1903 John and Alfred Raworth carried out further experiments on trams operated by the Devonport Tramway Company. It was soon realised that while the regenerative system was practical using the two tram motors connected in series as described in the original patent application, the range of speed control that was available was very limited. The drivers, particularly when the cars had to ascend hills, were experiencing sluggish performance and the commutator brushes on the motors were sparking vigorously. Other operators claimed that the Raworth trams were so slow on the upward inclines that they were holding up other trams and disrupting the timetable.[12]

The original patent design used motors permanently connected in series, but to obtain maximum controllability to match the performance of contemporary conventional trams it was necessary to develop the control system further, to follow what was at the time the usual arrangement of switching the motors between series and parallel operation. The parallel motor running was used for starting the trams when the full voltage was applied to both motors – not just half the voltage when they were connected in series when 'cruising' at the top service speed.

Eventually, in late 1905 Alfred was to carry out most of the development of the 'R3' controller which switched the motors into parallel operation when necessary. In the 12 months following this development 14 trams operating on the Woollen District Tramways system had been successfully fitted with this new controller. The R3 controller was a complex piece of equipment which achieved the desired aim of using the control lever positions, or 'notches' as they are universally known, to select the desired speed and either produce acceleration or regeneration. Alfred was later to honestly admit that the development of the R3 controller had been very difficult, and that it was fortunate that the earliest work to prove the concept of regenerative control had been carried out with only the series connected motors, otherwise he opined that the difficulties of designing a series-parallel controller compatible with the Raworth regenerative control concept from the start might have stifled the invention altogether.[13]

Using the improved Raworth system, it was anticipated that a saving in the cost of electricity of as much as 30 per cent was achievable on very hilly routes. In March 1906, a trial to compare a regenerative tram with a conventional tram was carried out on the Bristol Tramways and Carriage Company's network in Bristol. The trial runs, carried out on a rainy day when the rails would have been slippery, showed a 24 per cent saving in electricity consumed by the regenerative control equipped vehicle. Similar trials were done on the Devonport and District Tramways Company's system in March 1906 indicating a saving of almost 29 per cent; and on South Metropolitan Tramways Company's routes in the Crystal Palace area during July 1906 where an average saving of nearly 27 per cent was achieved. But these were the 'on board' savings, for Raworth's Traction Patents claimed that there were further savings to the operator due to the reduced output from the generating stations and a reduction in the system losses in the electricity distribution system. Furthermore, there would also be significant savings on the maintenance of the tram's conventional mechanical braking systems.[14]

Alfred, by continuing to assist his father with further development of the tram control equipment, was able to share the patent rights for the 'improvements' and he later acquired a patent dated 1906 solely in his name. John Raworth would later claim that the control of a car with a regenerative motor was much easier, and ironically given what was to follow, a safety feature to reduce tramway accidents. The rationale for this belief was that should a driver carelessly accelerate over the brow of a hill, rather than panic, if using regenerative control all he had to do was to select a position on the controller and this would set the tram's velocity down the incline. Furthermore, the regenerative braking enabled the tram to be slowed down without any of the wheels locking and creating a skid, which was likely to occur using a mechanical brake. The Raworths claimed that as the regenerative braking was supplied by the same forces that drove the tram forward, if for any reason this was faulty then the tram would not be able to move off anyway.[15]

It was not just tramways that were attracting the entrepreneur that was John Smith Raworth, a man who was always on the lookout for new business opportunities. In 1905 Alfred was to make his first venture into the railway world when Raworth's Traction Patents purchased two unwanted electric locomotives from the Central London Railway (CLR) for £1,000. The CLR (popularly known as the 'two-penny tube') was one of the earliest London Underground 'tube' lines and is now part of the Central line.[16] The first section to be opened in 1900 ran from the Bank of England station in the City of London to Shepherd's Bush station in West London. The two locomotives were from a fleet that had been purchased from the American General Electric Company in 1899 to operate this underground railway and had unfortunately been the source of massive vibrations that shook the buildings situated above the tunnels. The reason for these vibrations was that the motors had no gearing as the armature shaft was effectively part of the driving axles. The reason why these 'gearless' motors were adopted was because it was believed that the loud whining from the gearing would be very disagreeable to the passengers when either in the confines of one of the deep-level stations or when travelling in the trains. This whining however would have been a small price to pay to avoid the even worse vibration nuisance on the surface above. These locomotives were of the 'steeple cab' variety which was common at the time. They were an elegant, symmetrical design having a central raised cab from which the driver had a clear view of the road ahead when running in either direction by looking out across the two sloping bonnets (or should they be described as 'hoods' as they came from the USA?). All the CLR steeple cab locomotives were taken out of normal service in 1903; the CLR retained two for shunting purposes and disposed of the rest to Raworth's Traction Patents and to a scrap dealer.[17]

John Raworth sold on the locomotives to the Metropolitan Railway with the intention that Raworth's Traction Patents would convert them to regenerative control locomotives at the railway's workshops at Neasden in North London. Alfred Raworth was entrusted with this project which involved the complete rebuilding of the locomotives.[18] To accommodate the new equipment, the appearance of the locomotives was drastically changed during the rebuilding process – only the main frames, the bogies and motors (which would have had to be modified) remained from the original locomotives. Buffers and couplings to comply with Metropolitan Railway standards were fitted. The body of the locomotive was made from several stark plain steel panels which were riveted into place, far less elegant than the original panels which had ornate wooden moulding. The locomotives never received the Metropolitan maroon livery but acquired only a 'works grey' coat of paint. Given that this did not result in a subsequent introduction of regenerative control on the Metropolitan Railway it must be construed that the experiments did not produce the results that John Smith Raworth had promised. The Metropolitan Railway did not adopt regenerative control, as when it obtained what were effectively new electric locomotives from Metropolitan Vickers in the early 1920s, it reused electrical equipment that had been recovered from its old locomotives.[19] Perhaps the lesson that the Metropolitan Railway learnt from the Raworth experiments at Neasden was: 'keep it simple!' The two locomotives languished at Neasden until they were scrapped in 1913.

Regenerative control had its detractors among the tramway community who were still sceptical about regenerative braking. Advertising material being distributed by Raworth's Traction Patents at the time implied a lack of appreciation by the operators as to the benefits of the regenerative system as many still had to be converted to the idea. Concerns ranged from continuing doubts about acceleration performance to concerns about the fact that the motors being used both for traction and retardation would overheat. John Smith Raworth was adamant that while the motors did indeed have to operate at a higher temperature, and that a temperature rise from around 110 degrees Fahrenheit to 140 degrees Fahrenheit on the windings was typical, this was well below the level that the insulation would start to be damaged at around 212 degrees Fahrenheit and below the 160 degrees Fahrenheit at which the traction motors on the London Electric Railway were operating at the time.[20]

But the most worrying concern that was voiced by the engineer for the City of Birmingham and by others was that the electrical output from the trams when operating in the regenerative mode was causing the voltage on the overhead wire to exceed the Board of Trade's allowable maximum of 650V.[21] To the modern electricity distribution engineer this concern has a familiar ring. Electricity distribution systems are designed to feed outwards from a source, either a generating station or substation to an electrical load, not the other way around. The system cannot necessarily cope with what is now being termed 'distributed generation'. When there is generation on the system away from the source, it must produce an output at a higher voltage than at the connecting point, otherwise current simply will not flow out into the electricity distribution network. The Raworth trams had been blamed for many minor problems such as indicator lamp bulbs blowing at the generating stations but were also blamed for a catastrophic failure at Yardley power station in Birmingham. Fortunately, no injuries were received despite large pieces of machinery being flung across the powerhouse. Raworth's Traction Patents claimed that this issue had been resolved when further modifications were completed that ensured that the tram was disconnected from the supply when the voltage on the trolley pole reached 600V.[22]

Birmingham City Council's engineer continued to be unimpressed and he revealed that the 40 trams with regenerative control operating on his network in Birmingham had not been a success. All the trams had the regenerative control equipment removed by 1906 and there was a formal ban on regenerative braking imposed on 1 January 1907.[23] The Birmingham engineer had solid practical experience but nonetheless seems to have been out of step with other engineers, for despite concerns being expressed in matters of detail by the attendees at Alfred's Institution of Electrical Engineers meeting at Leeds, no one present suggested that the system should not be used, and there was general approval that many significant difficulties had been resolved.

In 1906 John and Alfred Raworth had applied for the patent relating to the use of regenerative control incorporating the switching from series to parallel operation which was given the title 'Improvements in controllers for electrically propelled vehicles' and in May 1907 the father and son again formally applied for another patent based on their work on the R3 controller which was given the title 'Improvements relating to the control of electric motors'.[24] Probably this second patent application spurred the Johnson-Lundell Company to begin legal proceedings in August 1907 against Raworth's Traction Patents for alleged infringement of its patents relating to its system of regenerative braking. The Johnson-Lundell Company's action at the time is understandable, for having developed a workable system of its own it was then unable to find any buyers, as John Smith Raworth and his cronies had taken almost full control of not only the market but also the tram manufacturing base. There followed a long and acrimonious legal battle, literally to the commercial death, a fight from which only the lawyers could ever be the true winners. During the proceedings, John and Alfred's own patent was published in May 1908. The Johnson-Lundell Company had to mortgage its Southall premises to meet ever-mounting legal costs and eventually, in October 1908, facing further insurmountable costs, had to withdraw the action and concede defeat. This theoretically left John Smith Raworth as the undisputed inventor of regenerative braking and this accolade is credited to him by many modern sources. Raworth's Traction Patents may have won the battle but was impoverished by the legal fees. To cut costs in April 1908 the company moved from its prestigious Westminster office to the former centre of John Smith Raworth's business activities: Manchester. The announcement to the world hinted at expansion not contraction when Raworth's Traction Patents opened its new office at 22 Cooper Street, which is situated in the very centre of the city.[25]

During his northern travels for Raworth's Traction Patents the independent Alfred Raworth met Ruby Robinson, the daughter of a retired cloth exporter, Frederick Robinson. The Robinsons lived in a house named 'The Rosiery' in the spa town of Ilkley, situated on the edge of the moors north-west of the City of Leeds. They had previously lived in Shadwell, a north-eastern suburb of Leeds, but had settled in Ilkley following Frederick's retirement. Ruby was the eldest of seven children.[26] The marriage between Alfred Raworth and Ruby Robinson was solemnised by the canon of Rippon at St Margaret's church in Ilkley on 6 May 1908. The two fathers, John Smith Raworth and Frederick Redfearne Robinson, were the witnesses.[27] Six years later the groom's brother – Basil – was to marry the bride's sister – Winifred.[28]

Alfred and Ruby set up home together in a large house called 'Helostia' in the village of Timperley, which is near Altrincham, seven miles south-west of Manchester in the county of Cheshire. They soon had a baby daughter, Marjorie, who was born in 1909. Living with them in middle-class comfort at Helostia was Alfred's sister, Gladys, and two live-in servants, one of whom was a nurse to attend to little Marjorie.[29]

Despite Alfred's new marital bliss, the outlook was bleak. Business was very sluggish, particularly after the Birmingham ban. There was a further knock to the company's esteem when Raworth's Traction Patents was summoned to the High Court of Justice, Chancery Division, to be forced under the requirements of the Companies Consolidation Act of 1908 to reduce its authorised capital from £30,000 to £26,625 with the subsequent reduction in the share prices from £1 to 15 shillings (75p) for the preferred shares and 12 shillings and sixpence (62.5p) for the deferred shares. Prior to the judgment signed by the Registrar (Companies Winding Up), both John and Alfred Raworth had signed affidavits appertaining to the company's business health. Raworth's Traction Patents' solicitors duly placed a notice in the *London Gazette* to this effect and noted that while all the deferred ordinary shares were taken up, there were 2,141 unsold preferred ordinary shares.[30] Once again, John Smith Raworth's venture was under-capitalised, and worse, had very few orders for new equipment.

However, in what must have seemed at the time a boost to the Raworths, the tramway operator at Rawtenstall, a small town in the Rossendale Valley in Lancashire, purchased equipment to be fitted to all 16 of its double-deck trams when their system was electrified, but unfortunately this was to be a false dawn since it was to be the last order of any significant size.[31]

At 11 o'clock on the evening of 13 November 1911, a serious accident occurred involving a tram equipped with the Raworth regenerative control belonging to the newly electrified Rawtenstall Corporation system. It was a frosty night, and the tram rails were very slippery. One of the four-wheeled double-decked trams was travelling between Haslingden and Accrington and was at the time of the incident on the section of the route owned by the Accrington Corporation Tramways, explaining the presence on board of Inspector Slattery – an employee of Accrington Corporation. The tram was descending Manchester Road, Rosendale, when it began to speed up, despite the driver attempting to use the regenerative control to steady the descent. Inspector Slattery made attempts to alert the driver to the problem and getting no response used his own initiative and rushed to the rear of the tram, which was by then rocking and swaying about, to wind on the handbrake. Driver Cuff meanwhile, who obviously knew that there was a problem, was operating the sanding gear to deposit sand onto the track in a vain attempt to combat what he thought was greasy rails, but to no effect. Inspector Slattery noticed that the usual humming sound heard when a Raworth tram was in the generating mode was absent. The tram, which should have maintained a steady 10mph down the incline on the notch selected had sped up alarmingly and upon reaching the passing place, a distance of 356 feet (110 metres) from the start of the descent, it jumped the points and in a flurry of smashed glass and tearing woodwork collided with another tram waiting to proceed in the opposite direction. Sixteen persons suffered injury and the local newspaper, with the usual journalistic hyperbole, graphically described the damage to both vehicles.[32]

The Board of Trade, of course, conducted the usual inquiry into the accident with Lieutenant Colonel Druitt presiding – the same inspector who had investigated the runaway of the Halifax demi-car number 95 in 1904. After receiving evidence from the driver, the inspector, and the joint manager and engineer of the tramways department of Rawtenstall Corporation, Lieutenant Colonel Druitt concluded that the cause of the failure was the connection to a shunt field coil on one of the motors which had 'fused' and that the tram continued to draw current from the overhead line rather than generate electricity. Therefore, rather than retard the tram, the result was the acceleration of the runaway vehicle. The shunt field windings on each motor were made up of four individual coils connected one after the other with connecting wires between each coil. The failure occurred on the connection between two of these coils. Despite the term 'fused' being used by Lieutenant Colonel Druitt, it can be assumed that the connecting wire had broken, not necessarily due to a high current implied by the term, but more likely due to excessive vibration or to a poorly made connection. With the shunt circuit broken no current would be available to increase the magnetic field but the series field coils of lesser strength would have still been connected to the supply line, and in this configuration would be akin to a conventional series traction so the bewildered driver Cuff had a runaway on his hands!

In his appendix to the Board of Trade report, the electrical advisor, Mr A.P. Trotter, confirms that the wire which broke was a thin one and liable to break or become disconnected, and if it had shown any indication of possible 'fusing' this would have been due to sparking at the time of the break. In Mr Trotter's opinion, the way that the wires had been terminated was a defect in construction. Furthermore, it would be difficult to detect this defect by inspection. Consequently, Lieutenant Colonel Druitt recommended that regenerative braking should only be used on level routes. This was not a ban on the use of regenerative braking, but the recommendation effectively meant that only a tram operating on, say, the level promenade along a seafront could still use the Raworth system. As so many of the northern towns which operated tram systems were very hilly and one of the advantages of the Raworth system was that electricity could be saved using regeneration on the descents from these hills, the Lieutenant Colonel's recommendation was every bit as good as an outright ban. Questions remain however: the tram should have been equipped with a switch to cut off all power in the event of a failure of the controller and neither the driver nor the travelling inspector saw fit to use this switch. Also, the excessively high value of electrical current driving the tram onwards should have operated the tram's automatic circuit breaker.[33]

Not followed up by Lieutenant Colonel Druitt was the high number of failures in Birmingham which led to the ban there, and possibly such a failure had occurred before but with less serious consequences. Furthermore, the accident might have been avoided, and many of the issues that arose during normal operations resolved, had there been better communication between the Birmingham engineer and the Raworth organisation. One of the Birmingham engineer's concerns was that he was unhappy with the response given to some of the searching technical enquiries he had made to Raworth's Traction Patents. He concluded that the company was evasive when concerns were raised about the system. This friction highlights the cultural difference between a commercial undertaking not wishing to disclose too much technical information for fear of competition from others and the more public service attitude of the Birmingham engineer seeking the best technical solution with little concern about such matters as business confidentiality. His concerns were that the electricity supply could not accommodate the regenerated current from the trams and that the new Raworth equipment was not robust enough for the tram environment, being unable to cope with the vibration and violent jerks as the vehicles clattered about the streets. Many operators had found that the Raworth system, after initial success working well for a short period, inexplicably suffered a deterioration in performance after about a year's operation.

So, with the company financially weakened after the legal battle with the Johnson-Lundell Company; the allegation from the Board of Trade's electrical advisor that the Raworth's Traction Patent's regenerative equipment had a defect in construction which might lead on to further unwelcome litigation; a history of failures resulting in a major customer refusing to use the regenerative control equipment; a recommendation from the Board of Trade that effectively prohibited the use of the Raworth system; and no interest in the regenerative control system being used by any of the railway companies following the Metropolitan Railway experiments, John Smith Raworth 'with a heavy heart' clearly realised that it was time to close down Raworth's Traction Patents.

An extraordinary general meeting of the board was held at 1 Kingsway 'London' on 15 December 1911, only one month after the Rosendale crash and three weeks after Lieutenant Colonel Druitt's damning report. At the meeting, Thomas Browett, in his capacity as the company chairman, issued the resolution that the company be voluntarily wound up, and a liquidator appointed.[34] The company was formally wound up on 14 January 1913 – a sad end to such an exciting and far-sighted project.[35] The Raworth regenerative control system had failed not because of any inherent deficiency in the concept, but on the simple matters of poor detailed design and a suspicion of poor workmanship.

This was not the end of regenerative breaking. After a decent interval, regenerative systems more akin to the Johnson-Lundell system began to be introduced into not only trams but also the new trolley buses. Today, in the twenty-first century with the advent of power electronics, regenerative braking is commonly used for trains, modern trams and even Formula One racing cars where the 'harvested' energy acquired during braking is stored in a battery for use when a burst of extra power is called for.

Alfred was now a householder with a young family to support, and did some 'patent' work in his own right. In September 1911 he took out a patent for a design of a 'cycle tractor attachment'. Cousin John Raworth no doubt assisted him with the patent procedure. The patent related to apparatus by which an ordinary pedal-powered bicycle could be adapted to internal combustion propulsion by mounting a motor in a frame which was fitted at the rear hub of the cycle. A small trailing wheel, described as 'castor-like', was used to ease steering the bike and to help support the weight of the small single cylinder internal combustion motor. Transmission from the motor to the wheel was the friction between a single belt-driven pulley bearing directly onto the bicycle's rear tyre.[36]

For the Raworths this was also a time of family sadness. John's brother, Benjamin, died in 1911. John Smith Raworth went into semi-retirement but remained a director of several companies including British Electric Traction. No longer a member of the Institution of Mechanical Engineers he also left the Institution of Electrical Engineers; but he remained a member of the Institution of Civil Engineers; mainly because he liked the social contact with the civil engineers and therefore, in later life, he was to describe himself as a civil engineer.

Retaining his duty to ensure the success of his sons, John Smith Raworth began to put out feelers in the direction of his many business contacts. Basil Raworth, having achieved associate membership status of the Institution of Electrical Engineers, joined General Electric in Hong Kong, later becoming their branch manager in the colony. Alfred however was destined to remain in England for in March 1912, only 10 weeks after the folding of Raworth's Traction Patents, the 30-year-old Alfred Raworth became a railwayman.

CHAPTER 3

ALFRED RAWORTH BECOMES A RAILWAYMAN

The railway company in which Alfred Raworth was to take up employment was the London and South Western Railway (LSWR). This railway, while never acquiring the romance or the glamour of the main lines to Scotland or the railway enthusiasts' devoted following of the Great Western Railway (GWR), was nonetheless one of the most important and oldest lines in England. At its peak, the LSWR boasted 980 route miles of railway line. Traffic on the LSWR was predominantly passenger rather than freight due to the extensive suburban network of lines used by London commuters.[1] Passenger revenue was over twice that of freight revenue, and this was atypical for railways of the period, for generally it was the railway goods traffic receipts that were the greater in that pre-motorway age. Originally promoted as the London and Southampton Railway, the first section of the LSWR was opened in stages from 1838. The LSWR's London terminus at Waterloo was not opened until 1848, so initially from a temporary station at Nine Elms the railway proceeded purposefully westwards as far as Basingstoke before turning due south to pass through historic Winchester, and then onwards to the developing port of Southampton, which was at the time little more than a fishing village. At Bishopstoke, between Winchester and Southampton, from the soon to be renamed Eastleigh station, a branch line was built through Fareham and then out along the peninsula bounded by Portsmouth Harbour and the Solent to reach another town which was also at the time little more than a fishing village: Gosport. Travellers from Portsmouth to London were expected to first take a ferry across the harbour to Gosport. The ill will this created in the ancient city of Portsmouth has been given as the reason why the railway pragmatically changed its name to the London and *South Western* Railway but it has also been inferred that the name was changed as expansion was planned *westwards* from Basingstoke towards Exeter, and even possibly as far as Bristol.[2]

Portsmouth had to wait until June 1847 before the London, Brighton and South Coast Railway (LBSCR) meandered its way across Sussex and after arriving from the Chichester direction constructed a fine terminus conveniently sited in the centre of the city. Meanwhile, further north, the LSWR built a line from a junction at Woking south as far as Guildford, another historic town. From Guildford, the LSWR extended its line south to Godalming and built a branch from Guildford to Alton via Farnham which, rather awkwardly, required the reversal of trains from London at Guildford station to gain access to this new line. At Guildford, an interloper arrived – the Reading, Guildford and Reigate Railway. This line was acquired by the South Eastern Railway (SER) and ran from Reigate to a junction with the LSWR just south of Guildford near Shalford and then used the LSWR tracks to pass through Guildford and out along the Alton branch as far as Ash, where it made its own way north, running under the LSWR main line near Farnborough, and then through Wokingham to join the GWR at Reading.[3] This line was opened in 1849 and for the LSWR it was to become an uncomfortable presence in its territory. Enter onto the scene one Thomas Brassey, a highly successful railway contractor who, after receiving the support of financial backers from Portsmouth, constructed a steeply graded and sinuous line through the South Downs from where the LSWR line ended at Godalming to a junction with the LBSCR at Havant – just outside Portsmouth on the LBSCR coastal line. Brassey then offered to lease his line to any taker, and the SER took just enough interest in this offer for Brassey to commence an embankment at Shalford Junction on which he proposed a connecting spur to allow the SER trains from the Reigate direction to turn south onto the LSWR line to reach Portsmouth. The LSWR could not allow any further incursion by the SER into its territory and so in 1859 was virtually forced to accept the leasing offer for the new line from Brassey. This resulted in fierce altercations with the LBSCR over the use of the line from Havant into Portsmouth (often referred to as the 'Battle of Havant'), but the two companies did eventually pull together and in 1876 joined forces to extend the railway line beyond the city centre through the old, fortified walls to a new station, Portsmouth Harbour.[4] This new station was, and still is, by the jetties used by both the Isle of Wight and Gosport ferries. Thus, the Portsmouth Direct line from Waterloo to Portsmouth Harbour was formed, and this line was to later play an important part in Alfred Raworth's career when he embarked on the electrification of the Southern Railway's main lines.

The LSWR expanded further to the west. From Basingstoke a new main line was built to Exeter giving access to Devon and Cornwall where eventually the LSWR processed an extensive network comprising of a main line to Plymouth and several meandering rural lines across North Devon and Cornwall. These lines were extended by degree and eventually Padstow was reached in 1901, just short of 260 miles from Waterloo. There was further extension beyond Southampton to Dorchester, at first using a circuitous route, but later a more direct line to serve the growing Bournemouth conurbation. Beyond Dorchester the LSWR used the GWR line to reach Weymouth. Over the years the LSWR built many infill lines including a new more convenient line branching from the main line just beyond Brookwood to serve the burgeoning military town of Aldershot, and this line then continued to connect with the original Farnham and Alton line which was also extended to Winchester. Between Ash and Farnham, the original line has now been abandoned.

The railway so far described is a typical main line railway of the Victorian or Edwardian era. The LSWR had its own locomotive and carriage works at Nine Elms which was later moved to Eastleigh. The coastal destinations of Southampton, Bournemouth and Portsmouth generated huge traffic revenues and the railway could take advantage of the increase in leisure travel by serving the smaller seaside resorts such as Seaton, Sidmouth, Ilfracombe and Bude. But in terms of the passenger count, it was the extensive and complex network of suburban lines radiating from Waterloo that contributed significantly towards the railway's prosperity. Before the coming of the railway the land to the south and west of London had been very rural, dotted with small market towns and idyllic villages, but the improved transportation that the LSWR provided resulted in the area becoming part of the suburban sprawl, as commuters took up residence away from the smell, smoke and noise of the metropolis. Due to the convenient connection to London, expansion continued into the twentieth century, with housing built on what had been heath or agricultural land.

The main line from Waterloo to Woking of course carried, and still carries, substantial suburban commuter traffic but there is also a secondary main line spine which, with its loops and branches off was known as the 'Windsor Lines'. This secondary main line runs as far as Wokingham where it joins the former SER line to reach Reading. Significant loop lines were built: the 'Hounslow loop' through Brentford and Hounslow and the 'Kingston loop' from Twickenham through Teddington and Kingston to the main line at New Malden. Another connecting line runs between Virginia Water and Weybridge on the main line which passes through Chertsey and Addlestone. Chertsey, perhaps relatively insignificant today, was an important town in Surrey in the age before the railways, being a registration district, and a parliamentary constituency. Nevertheless, the coming of the railway resulted in the doubling of Chertsey's population to 12,000 in the first 50 years following the arrival of the LSWR – a typical example of the effect of the railways in the London outer suburbs. From Ascot a line runs through Camberley to Frimley, where once there were spurs in both directions onto the main line, but now a single line continues to join the line to Aldershot.[5]

Not all stations were on loop lines, for there are termini on the former LSWR network, the most important of these being Windsor and Eton Riverside. The LSWR Staines to Windsor line, which opened in 1849, beat the GWR to Windsor by only a few weeks. Both railways' stations remain in use today – the LSWR Windsor and Eton Riverside and the GWR Windsor and Eton Central are now served by South Western Railway trains and the modern Great Western Railway trains respectively. The former LSWR station has a splendid overall roof that would do justice to any London terminus and the station building itself is in the magnificent Tudor Gothic style designed by the LSWR's favourite architect, Sir William Tite. Every detail of the station, including of course the royal waiting room – which, with the proximity of Windsor Castle, was probably never used – was subject to the scrutiny and approval of Prince Albert. Nestling as the station does below Windsor Castle the prince insisted that the railway planted a row of trees to hide the railway from their majesties' view from the castle, but the same trees should not be so tall and dense as to obscure the more pleasant sight of Eton College on the far side of the River Thames.[6] There is also a branch off the Kingston loop to Shepperton. This station, which was opened in 1864, was never intended to be the terminus and even today one feels that it is not yet quite complete, patiently awaiting the installation of some more tracks in the westerly direction.

There are other suburban lines – one of these branches off the main line near Surbiton (or, more precisely, Hampton Court Junction) and runs to Hampton Court with only one intermediate station at Thames Ditton. At Hampton Court the LSWR built another fine station in 1849 which faces off the magnificent Tudor Hampton Court Palace which proudly stands on the opposite bank of the River Thames.

The situation south and east of the main line is less complex. There are two routes which despite being built as long ago as 1882 have forever been referred to as the 'New Line'.[7] The main route runs from Hampton Court Junction – a location that is now endowed with two 'flying' junctions – one heading north to Hampton Court, the other heading south along the New Line to Guildford through Cobham. The fact that this line arrived at Guildford 37 years after the direct line through Woking justified, originally at least, the 'New' name tag. Another LSWR line leaves the main line at Raynes Park to connect with a line owned by the LBSCR at Epsom from whence a line that was once jointly owned by the LSWR with the LBSCR continues to Leatherhead. From Leatherhead, the LSWR built a line to connect with the New Line at Effingham Junction. This Leatherhead to Effingham Junction section was part of the overall New Line project and provides the other New Line route from Waterloo through Leatherhead to Guildford. The line from Guildford to Effingham Junction, despite the removal of goods yards and signal boxes, still retains much Victorian atmosphere with its mainly original station buildings.

During the nineteenth century, the LSWR had a virtual monopoly within its suburban territory, and like other monopolies rather took advantage of its position and provided a somewhat sloppy service to its captive customers who, living out beyond the metropolitan boundary, had little choice at the time other than to use the railway for their daily commute and other journey needs. Much of the blame for this has been attached to the railway's general manager between 1870 and 1884 who seemed to have possessed a blind spot in terms of what we would now call 'customer service'. He finally resigned after a spate of complaints about the unpunctuality of trains. His replacement was Charles Scooter, who did much to improve matters but was somewhat inhibited by a board of directors who were reluctant to invest in track or station improvements to alleviate the operational difficulties due to the increase in traffic.[8]

At the turn of the century however, the LSWR did invest in new carriages for its suburban services. From 1901, to supplement the older stock, the railway's distinctive salmon and umber livery was applied to new 51-foot (15.5 metre) long bogie vehicles arranged in what the LSWR described as 'blocks' of four carriages specifically for the shorter suburban journeys.[9] To suit this duty the luggage space was minimal, and no lavatories were provided. This contrasted with the main line stock which had ample luggage accommodation, and access to lavatories for at least some of the passengers. While inter-coach corridor connections were rare at the time, stock with lavatories often had side corridors within the coach, or some full-width compartments had a direct connection to a lavatory. Without any need for side corridors or half a compartment length for a pair of lavatories facing into compartments, the block carriage set trains could carry many more passengers. At the time, the LSWR had to cater for three classes of travel: first, second and third. By 1910, 145 of these sets were built and provided modern and comfortable transport for the commuters from the suburban area.

The LSWR was notoriously satirised in 1889 by Jerome K. Jerome in his book *Three Men in a Boat* when the hapless confusion at what could only have been Waterloo station was lampooned and by 1911 the fortunes of the railway were in decline, despite the improvements that were being made to the infrastructure (notably the commencement of the complete rebuilding of Waterloo station into the form it is today). The railway had been paying out a 6 per cent dividend to the shareholders, but a reduction in the suburban traffic revenue was beginning to eat away at the investors' return on their capital, as the suburban customers began to desert the railway in favour of the trams and buses.

The LSWR was not alone. All of London's suburban railways were losing traffic in the early part of the twentieth century due to a variety of reasons, not only from competition from the new alternative modes of transport but also due to the general decline in the economy, with retail prices rising and wages remaining stagnant. The public seemed to be less inclined to travel. As the LSWR was not known for the speed or frequency of its services it was no surprise that traffic was being lost as the new road services began to operate in what the railway considered to be its territory. Electric trams were now competing with the LSWR right across the area served by the inner suburban lines in Chiswick, Putney, Richmond, Brentford and Hounslow, and threats were being made to instigate tram services to Hampton Court, Kingston and even as far out as Staines.[10] In addition to the competitive threat from trams and buses, the LSWR was beginning to experience line occupation difficulties between Clapham Junction and Waterloo. This was in part due to the reluctance of the LSWR directors in the previous century to sanction expenditure on extra tracks. It was the track occupancy problem that prevented the LSWR from simply running more steam trains to improve the service in response to the tramway competition.

Many, including the LSWR chief engineer, J.W. Jacomb Hood, speculated that the solution might be railway electrification. In May 1902 Hood presented a paper to the LSWR Debating Society entitled 'Electric Traction in its Relation to Existing Railways'.[11] Noting the growing opinion that the days of the steam locomotive were limited due to its inefficient steam generation and need to carry its own fuel, he nonetheless did not foresee the imminent elimination of steam traction. He did however voice some interesting ideas as to the circumstances where electrification might soon be advantageous. First, he speculated that it might be economical to electrify minor branch lines using a similar argument to that of Alfred's father when he had been promoting electric tramways for routes with less traffic potential. The LSWR chief engineer proposed versatile 'self-propelled' electric 'cars' – surely a type of tram. Alfred Raworth was to pick up the baton for this argument later in the 1930s and 1940s. Hood also proposed electrifying the LSWR suburban lines, correctly predicting a service level which was to be a reality just over a decade later. To do this he speculated that it would be necessary to increase the number of tracks to four as far out as Twickenham on the Windsor Lines and, bizarrely, he said that there would need to be reversing loops at Waterloo to keep the traffic flowing.

It is perfectly possible to operate a monopoly and still provide a good service and, with such a good service, the LSWR could fend off any of the new and unwanted competition in its suburban area – all it needed was a good firm manager to take charge. The LSWR found such a man when they appointed Herbert Walker as its new general manager on 1 January 1912.

Herbert Ashcombe Walker was born in 1868 in Paddington. He was the son of a doctor and the original intention was that the young Walker was to follow in his father's footsteps. However, for what official sources have quoted as 'family reasons' (and here we must surely suspect a lack of finance), Herbert gave up his full-time medical studies to earn his living, taking up employment with the London and North Western Railway (LNWR). This position was obtained by the usual 'good connections' route, in this case via high ranking LNWR officers who were Dr Walker's patients. He started his railway career at the age of 17 as a clerk in the London Division offices of the LNWR but his rise through the ranks was to be meteoric for he was soon to be transferred to the staff of the superintendent of the line to whom he became outdoor assistant in 1899. Four years later he was further promoted to the post of assistant district superintendent in North Wales and then, after only 10 months there, he became assistant district superintendent of the Southern Division of the LNWR. In 1902 he visited the USA to find out what could be learned from American traffic methods and to report on how their procedures might be applied to good effect by the LNWR.

Herbert Ashcombe Walker. (Railway Magazine Archive)

In 1911 he was appointed as the outdoor goods manager for the Southern Area of the LNWR. His abilities may well have been recognised by the LNWR but not fully rewarded. After years of experience across the LNWR system he was to represent the railway on several occasions including at many public inquiries when he was officially presented as an 'assistant to the general manager'.[12]

It was at such public inquiries that the latest LSWR chairman, Sir Hugh Drummond, took notice of Walker and when the LSWR general manager post became vacant he wrote directly to the LNWR's 'assistant to the general manager' with the offer of the post of general manager of the LSWR. Herbert Walker duly accepted this belated recognition – even if it was from another railway.

He arrived at the LSWR with a huge experience of railway operation and his views about the desirability of electrification were well known throughout the wider railway community. In the early part of the twentieth century opinion was divided as to the virtues of railway electrification. Many held the view, correctly, that electric trains were more expensive to operate when the repayment of capital with interest was factored into the running costs. But many railway managers were beginning to take a more progressive view, including the general manager of the Lancashire and Yorkshire Railway (LYR), John Aspinall, who maintained that when a railway line was electrified on a route paralleling a steam railway or a tram route, passengers always transferred to the electric trains. Herbert Walker fully agreed with this view and took the stance that the extra revenue earned would cover the full costs of electrification and provide a healthy return on the capital invested.[13]

It is probable that Sir Hugh Drummond agreed with this view and selected Herbert Walker – at least in part – for this reason. The LNWR was proposing the electrification of its suburban services from Euston, and Herbert Walker would have been enthusiastic about these plans even if not directly involved. He was young to be a railway general manager, being only 44 years old. Tall and broad he looked the typical Edwardian gentleman, with hair neatly centre-parted, perching his 'pince-nez' spectacles on his nose above a neat and tidy moustache. He was to prove to be an autocratic manager, but he retained the common touch – always remembering the names of quite lowly staff when he met them during his tours of inspection of the railway. He took up residence in the wood-panelled general manager's office at Waterloo which was to be his lair for the next quarter century. Always a man to feel the cold, the fireplace in his office was kept well supplied with best locomotive coal.

Walker and Drummond needed to convince the LSWR board of directors of the need for electrification. The directors were not railwaymen and did not necessarily fully appreciate the day-to-day practicalities of running a railway. They represented the 'establishment' – often politicians, owners of large family businesses, members of the aristocracy or landowners. Railways were rarely their main interest. Their chairman, Sir Hugh Henry John Drummond Bt., was born in Clovelly, North Devon, in 1859, the third son of the fourth holder of the Baronetcy of Hawthornden in the City of Edinburgh which had been created in 1828. This was a relatively low place in the UK aristocracy, but it allowed him to use the courtesy title 'sir'.[14] Sir Hugh first became a railway director in 1900, joining the board of the LSWR, but in the 1911 census, the year that he was elected chairman of the board, he gave his occupation as 'bank director' so he saw his primary business interest as being banking. Earlier, in 1902, when the Exeter Bank printed its own banknotes, the signature of 'Hugh H.J. Drummond' promised to pay out the cash.[15] Sir Hugh was a very wealthy man – at his London home his wife and daughter were served by nine servants including a butler and two footmen.[16]

The need to convince the LSWR directors of the desirability of electrification is, at this distance in time, hard to understand because on its own tracks the Metropolitan District Railway (MDR) was operating trains using a 600V direct current system of electrification. These highly successful electric services were operating over LSWR rails from Richmond to Turner's Green and from Wimbledon to Putney and then into Central London using the MDR's own lines. The LSWR had made an agreement with the MDR dated 4 December 1903 to install and maintain the necessary cables and conductor rails along these lines. The LSWR also constructed (and owned) substations at Wimbledon Park and Kew Gardens, while the MDR provided the substations to feed the LSWR lines at Studland Road Junction and at Putney Bridge. The agreement was that the LSWR would finance all the new plant in its territory and in return would receive from the MDR an annual sum equivalent to 4 per cent of the capital that had been outlaid by the LSWR.[17] A 4 per cent return seems a bit low for a '6 per cent railway' but, times were changing, and a 3 per cent share dividend was becoming the norm. As part of the package the LSWR had to employ staff to man the substations, to install an improved signalling system suitable for the more intensive service that was proposed and to provide telephone communication between the LSWR substations and the MDR's Lots Road power station. The agreement did permit the LSWR to operate its own electric trains over this newly electrified line subject only to the payment of the cost of the electricity consumed.

Furthermore, the LSWR already owned an electric railway, a 'tube' line, the Waterloo and City Railway. This underground railway had been promoted because the LSWR's terminus at Waterloo was rather too distant from the City of London for the convenience of commuters. After consideration of an extension to the railway into the City, the LSWR board had decided to construct an underground railway instead. The Waterloo and City Railway was opened in 1898. A new company, the Waterloo and City Railway Company, had been formed, but was hardly independent, as of the eight board members, five were also LSWR directors. In 1906 the Waterloo and City was sensibly absorbed into the LSWR.[18]

To build this underground railway the LSWR employed as its consultant an engineer who was to make a considerable contribution to this electrification story: Dr Alexander Kennedy. He took over after the death of the original consultant and was later to be an advisor to the London County Council for the electrification of its street tramways as well as continuing to work for the LSWR on its later electrification schemes. Alexander Blackie William Kennedy was born in Stepney, the son of the Reverend John Kennedy, MA. He was trained as an engineer in the shipbuilding industry but in 1889, at the early age of 27, he was appointed chair of engineering at University College London. During his academic career, he became interested in the then new field of electrical engineering which he studied avidly. Kennedy resigned his professorship to become an electrical engineer and established a thriving private practice as a consultant. Later, he made his assistant engineer, Sidney Donkin, a partner in the firm creating the world-renowned firm of Kennedy and Donkin.[19]

John Mowlem and Company, a family concern from Dorset which made its name building many of London's landmarks, was awarded the contract to construct the two tunnels from Waterloo to the City Station – renamed 'Bank' in 1940 – just under a mile and a half. Each tunnel was 12 feet 9 inches (3.87 metres) in diameter. The line was electrified using a 500V direct current 'live' rail in the centre of the track. As with all electric systems a 'return' path for the electricity was required to complete the circuit and the Waterloo and City line, like the street tramways, used the running rails for this purpose. To achieve this, all the running rails were bonded together at 100-foot (30.5 metre) intervals, and to reduce the electrical resistance of the return path, two additional reinforcing cables were installed for just over half a mile from Waterloo towards the City and this was connected to the rails. The power station for the railway was built at Waterloo station and had five Davey Paxman boilers feeding six Bellis and Morcom high speed engines, each of which drove a Siemens 200kW dynamo.

The trains, each comprising four 'cars' (American terminology, they were distinctly un-British in appearance), were manufactured in what appears to have been 'kit form', by Jackson and Sharp in America and assembled at the LSWR works at Eastleigh.[20]

In 1898, the year that the line opened, a man who was later to become one of the most important figures in the electrification project joined the Waterloo and City Railway. His name was Herbert William Jones, a tall thin man sporting a dapper pointed beard. After Jones attended Monmouth School and Faraday House Electrical Engineering College, he became a pupil with the original London 'tube' line, the South London Railway. Subsequently he was to be the technical assistant to Ludwig Epstein, a citizen of Austria then living in Old Broad Street 'London' whose work included the development of batteries. Jones returned to railway service to work as engineer in charge at the Stockwell generating station of the South London Railway. Later, with a reference from the City of London Electrical Training Institution he was appointed as the chief assistant to the electrical engineer on the Waterloo and City Railway and in December 1900 promoted to become the Waterloo and City Railway electrical engineer, taking full charge of the electrical department.[21]

Herbert Walker was to prove to be a vigorous wind of change through the cold corridors of the offices at Waterloo station, fighting the cause for the electrification of the suburban lines. The hyperactive new general manager had no doubts about the viability of electrification and, on the 25 February 1912, only two full months after his arrival, he put a recommendation to the directors attending the regular Locomotive, Carriage and Estates subcommittee meeting that due to his proposed electrification an electrical department must be established and that the present Waterloo and City Railway electrical engineer should be placed in charge and have responsibility for all aspects of electrical engineering apart from that within the Eastleigh works and at Southampton Docks. The proposal was accepted, and Herbert Jones received a pay rise of £100 a year raising his salary to £600. Jones was to report directly to the chief engineer.[22]

At the next subcommittee meeting Herbert Walker returned to report that due to the anticipated workload of preparing and then carrying out the electrification project, Jones required a suitably qualified temporary assistant. Surprisingly, Walker requested that an outsider to the railway industry be appointed and he recommended that Jones' assistant should

Herbert William Jones. (Railway Magazine Archive)

be Alfred Raworth. Furthermore, Walker requested that Alfred Raworth should be paid £275 per annum – more than Jones' present assistant.[23] It has been inferred that Alfred's appointment was a favour from Walker to John Smith Raworth. It was darkly suggested Herbert Walker *knew* John Smith Raworth. He probably did – anybody who was involved in either electricity supply or electric traction would, at the very least, have known of him by name and reputation. Walker would have been more impressed by John Raworth's reputation of being able to get things done in an economical manner than for his technical virtuosity. It is in such delicate matters as these when, say, John Smith Raworth and Herbert Walker might have discussed Alfred's future in a friendly way as they relaxed in the plush leather seats of a gentleman's club somewhere, and by the very nature of such arrangements no record was ever kept as to what was asked for or what was offered. John Smith Raworth was naturally concerned about his elder son's future following the demise of Raworth's Traction Patents and it was completely in character for him to vigorously lobby influential people in his efforts to secure a future for Alfred. However, the appointment *was* made on merit: raw nepotism can be discounted as despite John Raworth's mother being a 'Walker', Herbert Walker's grandfather was Stephen Walker, an Oxfordshire doctor born in 1785; while Alfred Raworth's grandfather was Thomas Walker, the Methodist minister, born in 1790.[24]

The committee accepted Walker's proposal. Even though only three of the directors attended this subcommittee meeting, significantly these included the chairman for the main LSWR board, Sir Hugh Drummond.

On accepting the appointment Alfred and his young family moved to 'Meadow Cottage' in Stoke D'Abernon, which is a small village in leafy Surrey with a long history dating back to pre-Norman times. When Alfred Raworth took up residence the village consisted of a scattering of cottages, the post office, The Plough public house and the ancient St Mary's church. The River Mole lazily winds its way past the south of the village, and in the centre of Stoke D'Abernon is the LSWR's New Line railway station which was then in easy walking distance of any cottage in the village. The station is named 'Stoke D'Abernon and Cobham' in deference to the larger village to the north of the railway.

Herbert Jones and Alfred Raworth immediately set to work preparing an overall plan for the LSWR's suburban electrification. All Raworth's early work for the LSWR was clearly based on the expectation that the system of electrification would be to use a 'live' ground mounted conductor rail charged at around 600V direct current. This made engineering and commercial sense since it would make the LSWR lines compatible with the other electrified lines to which the LSWR was physically connected, such as the MDR and the LNWR and additionally – should it ever be needed – with the lines running underground in London.

Raworth prepared several highly detailed tables in his personal notebook, all neatly handwritten in his small but readable script.[25] Using his intricate tables, he was able to indicate the electrical power requirements for trains in each section. This allowed determination of the potential line occupancy and train service frequency. Raworth split the overall scheme into eight sections for his traffic calculation purposes. Some of these sections overlapped, but he listed as if separate lines the Kingston 'roundabout'; the Hounslow loop line (Chiswick to Twickenham); the Waterloo to Hampton Court route (main line and the Hampton Court branch); the Shepperton branch; the main line from Waterloo to Guildford via Woking; the 'New Line' Waterloo to Guildford via Claygate; the 'Epsom line' from Worcester Park to Guildford and what he described as the 'Wimbledon Park and Staines lines', these being East Putney to Wimbledon and Feltham to Staines.

Herbert Walker was later to justifiably gain the reputation for providing the travelling public with frequent and regular trains with service patterns that eliminated long waiting times for potential passengers desiring to travel. It is surprising therefore that Walker's initial suggested service pattern as shown in Alfred Raworth's notebook was far less ambitious than what Jones and Raworth offered to their general manager as early as May 1912. Walker's suggested plan would have used 23 new electric trains during the day resulting in a total daily mileage of 9,383 miles. His more ambitious electrical engineers were able to advise him that they could provide enough electrical power to exceed his expectations for they proposed a service provision which would require 40 trains in service across the 18-hour day with a total mileage of 20,150. A final detailed plan, closely resembling that offered by Jones and Raworth, was agreed with the general manager on 9 August 1912.

The impressed Herbert Walker went back to the Locomotive, Carriage and Estates subcommittee, and once again citing the suburban electrification scheme, recommended that Alfred Raworth be immediately appointed as a permanent salaried staff member with his salary increased to £325 a year. The request was granted, but the promised pay rise did not materialise until January 1913. Raworth had now moved away from his father's world with all its concerns about patent rights and emphasis on being 'professional' through membership of the engineering institutions. He effectively resigned from his associate memberships of the Mechanical and Electrical Engineering Institutions by falling into subscription arrears.

In June 1912 Alfred Raworth was tasked with investigating the possible use of 600V General Electric 'gearless' locomotives as used on the New York Central system. These 1,500kW locomotives had four 550 horsepower motors and Raworth calculated that they would be able to haul up to 12 coaches weighing 35 tons each at an average speed of 51mph.[26] These locomotives were a larger version of those that Alfred had modified and adapted for regenerative control at Neasden. No mention appears in his notebook as to the possibility of using his father's control system on these locomotives. This was not part of the 'suburban' electrification – the proposal, not to be adopted, was to replace steam for electric power on the long-distance trains for the first 20-mile dash to Woking where locomotives would be changed.

The expectation was that the LSWR scheme would be a straightforward electrification of the existing LSWR suburban lines, but Herbert Walker, being aware of developments on the railway from which he had so recently been employed, wanted an expanded scheme investigated. In later years, the Bakerloo line tube trains would continue to run beyond their narrow tunnels and venture out onto the newly electrified LNWR main line. Herbert Walker had a similar vision and proposed incorporating the existing Waterloo and City line into a new 'tube' railway from Waterloo to Finsbury Park and to integrate this with the existing LSWR suburban lines. This scheme would require the construction of another half-mile long tunnel at Waterloo and over three miles long from the Great Northern and City underground line to the LSWR Waterloo and City line. At Waterloo, a connection would be made from the 'tube' line to the main line. The new 'Waterloo' tube would have eight stations along its length. The civil engineering costs were outside Raworth's sphere of competence of course, but in August 1912 he was requested to compare the costs for operating conventional three-coach suburban trains compared with four-coach tube trains. Raworth's calculations showed there would be an additional rolling stock cost of almost £90,000. His further investigation as to train timings revealed that this obviously expensive project would not provide commuters with a significantly improved service to justify the expenditure. For example, where the LSWR was in direct competition with the MDR such as at Wimbledon, the journey to the City using the MDR took 30 minutes, by LSWR then Waterloo and City line (change at Waterloo), 26 minutes and a direct without change using the new tube route would take 23.5 minutes – only a minimal journey time saving after a huge capital outlay.

The general manager's electrical engineers were designing a 600V third-rail electrification, but other railways across the nation were using, or were proposing to use, other forms of electrification. Walker was confident that what Jones and Raworth were proposing was the best solution for the LSWR, but he had to have the approval of the directors who were not professional railway operators or engineers. The board members, through their contacts with the directors of other railway boards, would have been aware that there were alternative electrification systems in use throughout Great Britain. For example, their near neighbour, the LBSCR, was operating a completely different system and that railway's directors were impressed by the results. The LSWR needed to be assured that its engineers and operators were proposing the best possible scheme for the railway before any commitment was made to capital expenditure for such an ambitious project.

The directors needed confirmation that the 600V third rail was the best option and agreed that Herbert Jones, assisted by Alfred Raworth, should be given the task of producing a report to recommend the system of electrification best suited to the railway's requirements. This must have exasperated Walker, and Jones with Raworth would not have wanted to waste time producing an unnecessary report. The directors agreed that Jones should, like several others before him, visit the USA which was at the time leading the way in railway electrification. In New York, Jones would be able to meet practical railwaymen and engineers who had first-hand experience of both high voltage overhead systems as well as the lower voltage conductor rail operation. Herbert Jones left Alfred Raworth back in the office at Waterloo and duly set off from his home in Surbiton to travel to New York on his fact-finding mission. He sailed from Southampton aboard the White Star Line's *Majestic*, arriving in New York on 13 September 1912. He stayed at the Hotel Martinique in the city and declared to the immigration officials that New York was his ultimate destination. One might suspect that it may have been a jolly crossing on the *Majestic* for Jones' fellow passengers included a troupe of British actresses and actors who were to perform in a production at the Comedy Theatre on Broadway.[27]

CHAPTER 4
A DESIGN FOR AN ELECTRIC RAILWAY

It is not surprising that the LSWR directors were nervous about the choice of electrification for their lines as there was much controversy and debate as to how a railway should be electrified. Compared to simply building more steam locomotives, electrifying a railway was a major undertaking for railways in the early years of the twentieth century. Not only had the railway to provide new trains, conductor rails or overhead wires and substations but also, due to unavailability of enough electricity from the local electricity supply companies, consideration had to be given to building railway-owned generating stations and the installation of extensive networks of cabling to distribute the electricity. There were proponents within the electrical engineering and railway fraternities who supported the use of conductor rails as opposed to overhead contact wires, and dissent as to whether direct current or alternating current should be used to power the trains.

If a railway were to have a central generating station it would be necessary to distribute the electrical energy for a considerable distance. There was a consensus that for distributing such large amounts of electrical power over long distances, alternating current should be employed at as high a voltage as possible. The 'system voltage' is the pressure driving the electrical current through the wires and is the equivalent to water pressure driving water through a hose. Electrical power is measured in watts, the value of which is obtained by multiplying the voltage – measured in volts – by the current – measured in amps. Therefore, the highest practical voltage for a given power transfer will result in the lowest current flow. Less current means that smaller wires can be used, resulting in a saving in expensive copper over long distances. Normally the power is measured in thousands of watts – that is kilowatts (kW).

Alternating current is generated in an alternator, which like its direct current cousin the dynamo, exploits Faraday's discoveries relating motion, magnetism and electrical current. In a dynamo, a conductor moving through a magnetic field will produce an electric current; similarly in an alternator an armature coil is rotated, and as it becomes more under the influence of the magnetic field the current rises to a peak and then falls back to zero on completion of half a turn of movement. Unlike in a dynamo when the commutator will perform a switching operation to correct the polarity, the simple non-segmented moving connection (referred to as a slip-ring) on the armature permits continued current production as it moves through the second half of its full rotation and, due to the effectively reversed magnetic field, the electricity then flows in the opposite direction.

This may not seem very practical at first sight. If the alternator output is displayed on an oscilloscope it will appear as a pure sine wave. This beautiful shape is identical to a pure sound wave and is thus one of nature's fundamental forms. Unlike a dynamo where several coils – known as poles – are arranged around the armature to produce outputs that can be combined to smooth out the waveform's peaks and troughs, the output from a single alternator coil must be kept separate from any other coil. It is usual to construct such alternators

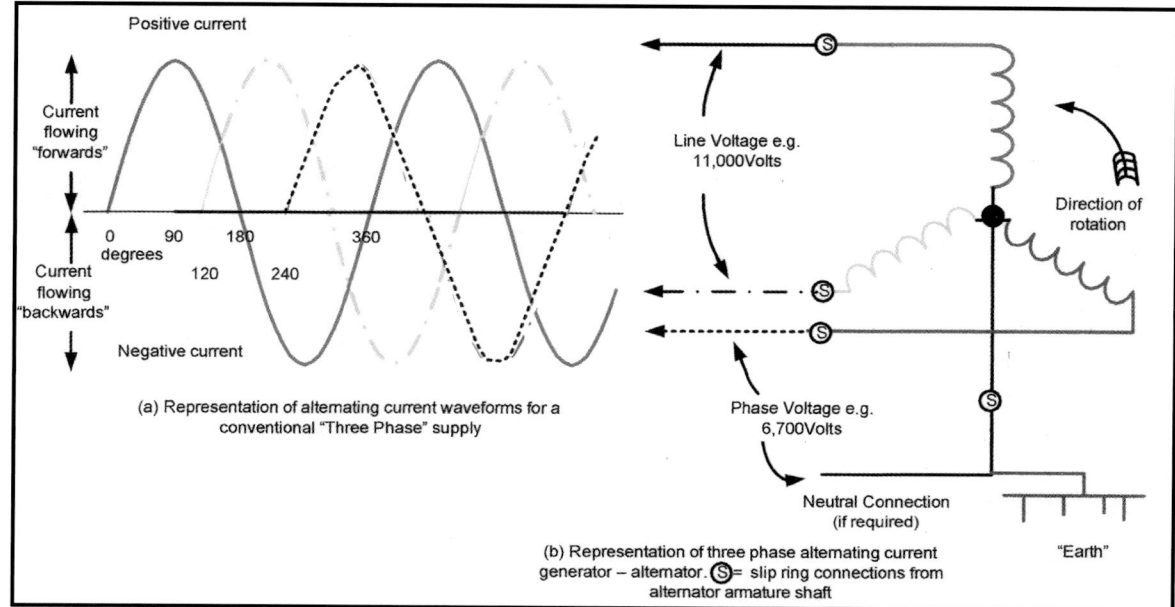

All that you need to know about three-phase electricity. (Author)

with three outputs from coils mounted 120 degrees apart around the armature shaft.

With reference to trigonometrical tables, calculator or internet connection, a full sine wave representing the first phase can be plotted out on graph paper if an oscilloscope is not to hand. If below the first waveform a second is drawn on the graph paper but this starting below the 120 degree point of the first waveform, this will represent the second phase, and a representation of the third waveform can be drawn commencing below the 240 degree point of the first. It is then possible to verify that at any point along the time axis – allowing for the curve below the centre line being a negative value – the amplitudes of each waveform when measured and added together will always equate to zero.[a]

This is more than just a neat mathematical conjuring trick, because if the load current drawn on each phase is the same, that is, it is a balanced load, it is not necessary to provide a separate return conductor. This is because the reverse or negative half-cycle will provide the return path for the current in the positive half-cycle. Should the electrical load not be balanced a separate return conductor – the neutral wire – must be provided. In a three-phase alternator the 'live' ends of each winding are connected to the output terminals via the slip rings and the other ends of the three windings are joined (at the 'star point') and bolted to an earth connection.

The voltages driving these three currents are also 'sinusoidal', and, depending on the electrical load being supplied, will more or less follow precisely the current waveform. Here is another intriguing feature. Returning to the three waveforms on the graph paper, if any two voltage waveforms are added together the resulting plotted line will be another pure sine wave about 60 per cent larger than the original single waveforms. Without a neutral wire on a balanced system, electrical equipment must be connected between the live phase lines, thus this larger voltage will be driving the current through any electrical plant. The lower voltage between any neutral wire or earth is described as the 'phase' voltage. The larger 'line' voltage is

---

a   If you do not wish to draw a graph this can be verified from the calculation of sine. Random example: SIN 20 degrees = 0.34, 20 degrees + 120 degrees = 140 degrees, SIN 140 degrees = 0.64, 140 degrees + 120 degrees = 260 degrees, SIN 260 = – 0.98, 0.34 + 0.64 – 0.98 = 0.

always the value quoted when describing a three-phase system. For example, an 11,000V system will have a phase voltage of about 6,700V. The number of sinusoidal waveforms generated per second represents the frequency of the electricity supplied and was in the early part of the twentieth century measured in 'cycles-per-second' (or more colloquially as 'cycles'), but the modern unit of frequency is hertz (Hz).

One important advantage of using alternating current for the longer distance 'transmission' of electricity at the time was that the high voltage of 11,000V could easily be obtained in the alternators using the then available technology. But more importantly, when the electricity reached the substations it could be reduced to a more useable voltage required to operate the trains by using one of electrical engineering's simplest and most efficient items of plant: the transformer. The basic concept of the transformer is easy to understand even if there is a deep mathematical heart should you wish to dig deeper. Imagine a coil of insulated wire wound onto an iron or steel 'core' which is a ring (or frame – it must have an open centre and be continuous) and an alternating current is applied to the coil. A magnetic current (known as magnetic 'flux') will flow around the iron core. If another coil is wound over the same core an electric current will be induced to flow through this second coil. This hypothetical transformer will have an input and an output voltage. The voltage output compared to the input voltage will be in proportion to the number of turns wound onto each coil. For example, should there be twice the number of turns on the input coil compared to the number of turns on the output coil the output voltage will be halved. The overall power is only minimally decreased due to a small amount of heat being generated, so there must be a proportionate increase in the value of the secondary current to pass through the same amount of electrical energy. Transformers, which are almost 100 per cent efficient, will only work with alternating current because it is only the constant *change* in the applied voltage that can induce the magnetic flow as the applied alternating current voltage gracefully sweeps from positive to negative and back again.

A practical transformer for a large power application such as railway traction will be housed in a large steel tank and the core and windings immersed in a thin mineral oil which has cooling and insulating properties. On large transformers this oil will be circulated through radiators to dissipate the waste heat generated. The coils are usually referred to as the high voltage (HV) and low voltage (LV) windings. Three-phase transformers will have the HV and LV individual phase windings connected such that all three 'neutral' ends of the coils are bolted together to an earth point ('star', connected as for an alternator); or alternatively they are joined end to end to form a closed loop, with the three phase terminals at the points where the windings make their connection ('delta' connected).

As noted, there was some dissent among electrical engineers as to whether to use alternating or direct current motors to power the trains. While it is possible, and often convenient, to produce an alternating current motor which is an 'alternator in reverse', the practical restriction when using these 'synchronous' motors is that the machine can only operate at the same speed as the alternator providing the current. Putting it very simply, the then contemporary alternating current motors generally worked by setting up electromagnetic fields and the motion was derived from the attraction or repulsion of these fields. These motors are termed 'asynchronous' motors and their rotational speed can be adjusted – exactly as required for railway traction purposes. At the beginning of the twentieth century most railway traction engineers preferred direct current motors – like those used in trams – because they were deemed to be more efficient, were cheaper to purchase and thought to require less maintenance. Furthermore, compared to the contemporary alternating current motors the direct current motors provided better acceleration and weighed less.[1]

So, if direct current was desired for traction purposes the alternating current being supplied to the substations had to be 'rectified'. A rectifier is a device that will only pass an electric current in one direction. Thus, a simple, single-phase, 'half wave' rectifier will convert an alternating current into a series of positive-only direct current 'humps'. By clever connections it can be arranged so that other rectifying devices arranged in a 'bridge' arrangement can fill in the gaps to produce a reasonably smooth direct current output. Today, semi-conductor diodes are everywhere, producing low power direct current from our alternating current mains supply to power all manner of small domestic equipment. But, at the turn of the twentieth century, the railway engineers needed to convert massive amounts of power from alternating current to direct current to run their trains, and at the time even thermionic 'valve' technology had hardly emerged from the scientific papers or laboratories and solid-state semiconductors would have been the stuff of science fiction.

The early twentieth-century railways had to use mechanical means to rectify alternating current and eschewed the obvious choice of using motor-generator sets – that is to use an alternating current synchronous motor driving a large direct current dynamo. The alternative chosen was the use of rotary convertors, pieces of electrical engineering now sadly obsolete for large power applications. These impressive machines were narrow when viewed from the side but were so tall – over 10 feet (3 metres) high – that pits had to be provided in the substation floors to accommodate the lowest portion of their circular form. Overhead cranes had to be built above them (and the associated transformers) for installation, maintenance and removal when required. A rotary convertor is basically a synchronous motor having an alternating current input fed conventionally into the rotating phase coils via slip rings on the armature shaft, the difference being that these machines, with their low electrical resistance, also had an output at the other end of the armature shaft where there was a segmented commutator arrangement for each phase like that on the direct current machines. This segmentation on the commutator was arranged such that the output current would be allowed to flow in one direction only.

These machines had an Achilles heel in that they were prone to 'flashovers' at the commutator if they were operated at too high a speed, particularly when sudden overloads occurred. For this reason, it was expedient to limit the speed of the convertors, and consequently that of the generating station's alternators, to 1,500 revolutions per minute (rpm); 1,500rpm equates to a frequency of 25 cycles per second – the railway engineer's preferred frequency. This frequency would now be deemed too low as the standard UK frequency is now 50 cycles per second. Even at the lower speed the rotary convertors required frequent maintenance as blobs of melted copper had to be regularly filed down on the commutator segments following excessive sparking. The rotary convertors must have been a wonderful sight to see in action: a noisy, vibrating machine, smelling of hot oil, rotating purposely with neither any visible means of propulsion as would be seen on a generator, nor any apparent connected load as would be seen coupled to a motor.

The three-phase transformers used to reduce the 11,000V alternating current to the required voltage for rectification had the phase windings connected in an ingenious way to output six phases. This was achieved by making an extra connection halfway along the low voltage winding to produce an additional output which was the mirror image of the original[b]. With six phases available, the 'filling in' of the half-waves of the output was used to obtain a nearly 'smooth' direct current ready to be connected to either the conductor rails or the overhead wires above the track.

In 1912 it was usual to use a live conductor rail to supply the trains. The most common arrangement was – and still is – for the live rail to be positioned just outside, and slightly above, the running rails and supported every five sleepers by squat porcelain insulators. Pickup 'shoes' are provided on the trains to collect the electric current usually off the top of the conductor rail. As for the trams, some 'return' arrangement is required, so to complete the electrical circuit the negative terminal on the motors was usually connected to one of the train's axles via brushes transferring the current onto the steel wheels and thus to the running rails, thereby completing the electrical circuit back to its source – the substation – at which there is a cabled connection to the track. This arrangement in the early part of the twentieth century was usually described as an 'earth return' system. The use of the track for the return path was a practical and cheap engineering solution but, as with most simple ideas, there are snags.

It continues to be a source of wonderment to myself that our planet Earth is such an efficient conductor of electricity and this simple fact is often completely taken for granted by most electrical engineers. When an earth connection is required, it is usual to drive a copper rod as deep into the ground as is possible to obtain an intimate connection with the surrounding soil. However, a fortuitous connection spread over a greater surface distance can also provide just as good an earth connection. For example, several miles of a railway line where the steel rails are laid on damp wooden sleepers which nestle in waterlogged stone ballast above a good loam soil. The return electric current will always, as if with some internal intelligence, seek out the shortest and least obstructed route back to the substation, in other words it will quite literally 'take the line of least resistance', hence the earth return description.

You might expect that the steel rails would provide the best return path, but the rails were not continuous as they were then laid in short lengths bolted together with fishplates. This type of rail connection has some electrical resistance which for an electrification scheme would be remedied by the application of 'bonds', that is, metallic electrical connections across the rail breaks. If the bonding was properly applied this should have been enough to guide the current safely back to the substation. But if the connection of the rails to earth was a good one, 'stray' currents would pass into the ground to find buried steel gas pipes or lead water pipes and hence an 'easier' return path, possibly even taking a cross-country short cut. These stray currents had possible harmful effects. In the early twentieth century, many applications such as telegraphy or signalling equipment used the mass of Earth as their return path; low current applications could operate very effectively by connecting the return wire onto a spike which was driven into the ground and used in conjunction with a similar earth rod at the source. Stray traction currents could potentially cause maloperation of these, perhaps essential, pieces of equipment. To try to alleviate this difficulty and provide a lower resistance return path with less risk of stray currents, many railways installed an additional rail laid in the centre of the tracks.

An electrification system as described so far – direct current using a live conductor rail with earth return – was not the most common system to be found in Britain in 1912. In London, the companies which were later to form the

---

b   For those who really need to know: the low voltage windings were zig-zag delta connected with each winding centre tapped to give another phase 180 degrees from the original.

London Underground system such as the Metropolitan, Metropolitan District and the new 'tube' lines operated by the London Underground Electric Railway Group used the fourth-rail system[c]. Herbert Walker's former employer, the LNWR, was at the time developing its 630V direct current electrification using an outside live conductor rail with the return being via a fourth rail.

One area where direct current conductor rail electrification proliferated was Merseyside. Liverpool already had an electric railway, the Liverpool Overhead Railway built in 1883 at high level above the docklands (the 'dockers' umbrella') when, in 1900 the general manager of the LYR, John Aspinall, took the apparently obligatory sea crossing to America to look at electrification systems in the USA. On his return, he recommended that the LYR electrify its Liverpool Exchange to Southport line using the fourth rail system and this was duly completed in 1904. The LYR expanded its electrified network by electrifying the line from Liverpool Central to Ormskirk.[2] A near neighbour to the LYR in Liverpool was the Mersey Railway. This line ran from Liverpool and tunnelled under the River Mersey to serve the commuter belt on the Birkenhead side of the river. Travel on this steam-operated railway was not a particularly pleasant experience through the smoky depths of the tunnel – so unpleasant in fact that it is alleged that many commuters preferred to risk the elements and pace the decks of the Mersey ferries instead. The railway was rescued from financial disaster by the American electrical manufacturer, Westinghouse, which in pursuit of business in Great Britain had established the British Westinghouse Electric Manufacturing Company with its works at Trafford Park, Manchester, in 1899. Westinghouse funded the electrification of the Mersey railway which was again a four-rail system operating at 600V direct current, and the electric trains began operating in 1903.[3]

Contemporary with the Liverpool schemes was the Tyneside Electric scheme. The North Eastern Railway (NER) lines to the north of the River Tyne were having the same problems as the London railways in the form of competition from new electric trams. The NER believed that it was losing 2 million passenger journeys a year and decided to electrify its lines north of the Tyne from Newcastle Central out to Whitely Bay and Tynemouth via three routes. Work began in 1903 and was promptly completed in 1904. The main differences that set the 600V direct current Tyneside electrics apart from the Liverpool and London schemes was that there was no fourth rail; and unlike most other railways, the NER did not build its own generating station but took its supply from the local Newcastle upon Tyne Electric Supply Company.[4] As for the Liverpool and London systems, multiple unit trains were used, all with a decidedly American appearance with their clerestory roofs and open saloon accommodation rather than separate compartments.

In 1912, the LSWR directors and engineers were aware of the electrification system being used by one of their close neighbours, the LBSCR. This railway's electrified lines were using alternating current from overhead line equipment. This installation is worthy of special attention as it was to become in later years a rival to the LSWR third rail when the Southern Railway was formed in 1923. The original reason for the LBSCR's decision to venture into the world of electric railways was the competitive threat not so much from the

---

c In part to overcome the earth return problems there is a major variation to the fourth-rail system on the London Underground which uses an insulated fourth rail to provide a completely independent 'return' circuit. The nominal operating voltage is 600V; however, London Underground's usual practice is to run the outer rail at a positive voltage and the inner rail is energised at a negative voltage. The ratio of the positive to negative voltage is about two to one, and the 'potential difference'– that is the voltage between the two rails – is 600V.

trams – which were nonetheless eating into the railway's revenue following the taking over and electrifying of the tramways by London County Council[5] – but from two other completely separate threats. The first was a proposed new 'high speed' railway which was to make use of nearly 20 miles of tunnels and just over seven miles of viaduct to run express trains from London to Brighton in 50 minutes; and an even more ambitious plan to construct a 150mph monorail promising a mere 20-minute journey from the capital to the coast. It is due to these threats that from the very start the LBSCR was looking to develop a 'main line' system – even though in 1912, long after the over-ambitious competition threats had evaporated, it was only operating its electric trains over suburban routes.[6] Given the controversy that was later to follow, it must constantly be borne in mind that the LBSCR design was only ever intended to be for main line use. Main line electrification was however close to the top of the LBSCR's list of priority projects in 1912, and after the First World War it did publish plans to electrify to Brighton and Eastbourne.

The LBSCR system was designed by its consulting engineer Philip (later Sir Philip) Dawson and was very much 'Phillip Dawson's Electric Railway'. Sir Phillip ranked with Sir Alexander Kennedy as one of the most important and respected electrical engineers of the era. He was born in 1866, was a partner in the firm of Kincaid, Waller, Manville and Dawson Consulting Engineers and a member of the three main professional engineering bodies: the civil, mechanical and electrical engineering institutions. Dawson worked on dock, electricity supply and electric traction projects both in Great Britain and overseas. For his railway work he was awarded the George Stephenson medal by the Institution of Civil Engineers and the Gold Medal by the Institute of Transport. He was a competent linguist being able to converse in French, German, Dutch, Portuguese and Russian.[7] He always advocated the use of single-phase alternating current systems for main line applications believing that it was the best means of railway electrification, but his was a controversial view and for this well-publicised opinion he attracted a lot of criticism from his peers which he accepted in the spirt of genuine professional debate but was slightly upset when the attacks were anonymous.[8]

The LBSCR directors on the advice of Dawson made a decision to use high voltage alternating current influenced by contemporary experience around the world – for there was a growing consensus that this was the only satisfactory method of electrifying a main line railway.[9] Dawson recognised that the most economical way to generate alternating current electricity was in three-phase generators,[10] but despite this he also insisted on a single-phase traction supply. The LBSCR obtained its '6,700V' single phase alternating current supply at 25 cycles per second from the London Electrical Supply Corporation's Deptford generating station at Stowage Wharf. The electricity supplier used their standard three-phase 6,600 Volt alternators which they had specially modified for the railway supply. The LBSCR supply used the line voltage - nominally 6,600 Volts from these alternators - but a single-phase connection was provided by connecting one of the phases directly to earth. Connection to the railway could then be achieved by using single-phase cables comprising of a phase conductor connected to one of the phase terminals of the alternator and a neutral conductor connected to the earthed phase output. Any power station engineer today will shudder at the effects - particularly the mechanical stresses - of such an 'unbalance' by using only two phases of a three-phase generator. But Phillip Dawson seems to have been sanguine about this uneven loading of a three-phase generator. His recommendation for any tram or railway system taking a single-phase from a three-phase source was to partially balance the loads by using the third phase for auxiliary supply purposes such as electric lighting.[11] The relationship between the power company and the railway was particularly cosy with one of the LESC directors being Viscount Ponsonby whose father, Lord Bessborough, was chairman of the LBSCR. The trains with their single-phase motors used the running rails as the return – an earth return system again – but the alternating current, unlike a direct current arrangement, would not adversely affect any third party's equipment.

With the guidance of Philip Dawson, the LBSCR awarded a contract for the supply of the electrical equipment for its electrification project to the German manufacturer Allgermeine Elektricitats Gesellschaft – a Berlin-based organisation much better known as AEG. Given the untried nature of the system (at least as far as British railways was concerned), initially in 1909 only the short four-mile section of the South London line between Battersea Park and Peckham Rye was equipped and used for trials. The trains had to carry their own 'substation' on board – some isolating and fuse switchgear and a transformer to reduce the voltage from 6,700V to the 750V required to operate the motors.[12] This made the alternating current trains heavier than their direct current equivalents.

In December 1909, the LBSCR began operating the new public electric service along the whole of its South London line from Victoria to London Bridge, a route that was a typical inner suburban line beleaguered by tramway and omnibus competition. The effect on the local tram lines was devastating, for in 1907 the trams along the routes Victoria to London Bridge had 8 million passenger journeys which fell to only 3.5 million in 1909.[13]

In 1911 the LBSCR main line from Victoria through Clapham Junction to Balham was electrified and from there through Streatham to Crystal Palace, as well as the link from the South London line at Peckham down to Tulse Hill where connections were made in both directions onto the main line from Victoria. In mid-1910 the LBSCR directors had approved these extensions and Phillip Dawson was tasked to complete this ambitious electrification project in time for the Festival of Empire due to take place at Crystal Palace in 1911. It was a tremendous achievement to complete the work in nine months ready for the huge crowds that attended the event.[14]

The other significant alternating current scheme which had received a lot of publicity in railway circles was the Midland Railway's scheme to electrify its branch lines from Lancaster to Morecambe and Heysham. In 1908 the Midland had used 6,600V equipment obtained from Siemens and Westinghouse.[15] This was the brainchild of the Midland Railway's electrical engineer James Dalziel and was a trial for the potential Midland Railway electrification schemes. One of Dalziel's dreams was for a Derby to Manchester electrification through the Peak District as well as a St Pancras to Bedford scheme.[16]

On his return from New York Herbert Jones prepared his report for the LSWR directors.[17] Given that Jones believed that only the 600V system was suitable for the LSWR's suburban network and perhaps under some pressure from Walker to propose the 'right' scheme, the objectivity of this report must be questioned. First, Jones – contrary to what Alfred Raworth was to advocate in later years – made it clear that he was unconvinced that wholesale electrification of the railway was ever to be viable but conceded that for heavily loaded trains operating at increased scheduled speeds with short distances between stops, electric traction had the advantage over steam power.[18]

His report assessed the available electrification options. The first option to be considered was the hoped-for choice, that is, a direct current third-rail system using an operating voltage of between 500V and 1,200V. Jones, keeping to his wish for a 600V system, maintained that the extra cost of the on-train equipment and the additional rail protection outweighed the savings from being able to use less substations spaced further apart if the relatively high voltage of 1,200V was to be used. Jones proposed that the return path should be through the running rails. Given the problems that were to follow, he deftly made no mention of the 'earth return' complication of such a system. In his report he did however stipulate that the live conductor rail should be located beside the track so as to be compatible with the MDR to permit both companies' trains to run over each other's tracks.

A higher direct current voltage using overhead contact wires was then considered. The voltage range for such systems was between 1,200V and 2,400V. Jones was not impressed, stating that the current carrying capacity of the thin overhead wires would be considerably less than that of any substantial third rail, and that the system voltage proposed was not high enough to compensate for this difference.[19] It was on this issue that Jones, and later Raworth, were later to come into conflict with the views of Charles Merz who was probably the most forward-thinking electrical engineer of the era. Merz was a quiet, unassuming man, a Quaker, and unlike his learned contemporaries he refused any honours despite a lifetime of work in Britain and across the then British Empire in the fields of public electricity supply and railway electrification, working both as a private consultant and in government service. He is now best remembered as the 'Father of the Grid' for his work in promoting an integrated national electricity supply for Great Britain. In 1899 Merz had decided to set himself up in business as a consulting engineer and he invited his friend William McLellan to be his assistant and later offered him a partnership in what would become the world-famous firm of Merz and McLellan. At the time of Jones' report Merz and McLellan were designing for the NER a scheme to electrify one of that company's freight lines mainly used for coal traffic between Shildon and Newport. This was intended as a prelude to electrification of the East Coast Main Line between York and Newcastle using 1,500V direct current overhead wires. The NER directors approved the electrification of the Shildon to Newport line in 1913 and the work was finally completed in 1916. Ultimately Merz and McLellan would build 1,500V overhead electrification schemes overseas to such an extent that their system became virtually the world standard not only for main line but also suburban electrification schemes. At the time of Jones' report, it was not possible to inspect or evaluate a working 1,500V overhead electrification railway line.

Jones' report was equally dismissive of any alternating current system which many contemporary engineers championed. It was conceded that a single-phase system such as Phillip Dawson's for the LBSCR might be suitable for hauling heavy passenger and goods trains over long distances but this, Jones maintained, was the type of traffic most suited to steam traction. He also conceded that the additional costs of installing overhead lines compared to laying down a third rail had been overestimated, but the savings from the reduced cost of electricity for any alternating current schemes would be eliminated due to the heavier trains. Furthermore, the cost of building the trains required for the LSWR suburban electrification would be doubled – £482,000 for coaches for a direct current system, compared to £1,052,100 for an alternating current system. Apart from the cost, Jones further damned any system like that of the LBSCR on the basis that the system was technically inferior for suburban train working due to the motors' operating characteristics.

Jones concluded his report by recommending that a direct current, third-rail system was the most appropriate for the LSWR suburban lines. He proclaimed that the third-rail system was well-tried and had been used for 20 years on tube and elevated railways and for over six years on a line that he had visited in New York, the Hudson and Harlem Division of the New York Central Railroad[20] – a line whose traffic pattern neatly matched the busy LSWR suburban network within 30 miles of Waterloo.

The report, despite being tainted with the suggestion that he was merely doing Walker's bidding, made sound engineering and commercial sense, and even with over 100 years of hindsight it cannot be criticised. There is unfortunately a 'however'. Once the die had been cast and the engineering decision made to use third-rail electrification, Herbert Walker was for the next 25 years to refuse to countenance any deviation from this method for running *his* electric trains. This was despite better solutions being offered by engineers, including Alfred Raworth, as Herbert Walker was to later base his decisions not so much on engineering niceties but on pragmatic operational and commercial considerations. This narrow policy ultimately resulted in the extensive use of the third rail across southern England, not only on the suburban lines in common with other mass transit systems around the world, but also on the main lines where its application was to become unique.

Jones was unsure about the railway going to the expense of building a generating station, claiming that it would be more economical to take power from a local supplier than for the railway to generate its own electricity from a powerhouse that would have to be built solely for railway use. But upon surveying the local suppliers he discovered that the London electricity companies were all small and those who generated the required alternating current used frequencies of 50 or 100 cycles per second – not the preferred 25 cycles per second.

As an alternative to the railway building a single large power station, Alfred Raworth had investigated the possibility of building four, smaller, diesel powered generating stations. It is probable – but not mentioned in Raworth's notebook – that these generating stations would have provided a 600V supply direct to the conductor rails, avoiding the need for a high voltage alternating current distribution network. The sites he proposed were at Clapham Junction (eight 1,400kW sets), Strawberry Hill (five 1,150kW sets), Hampton Court Junction (six 700kW sets) and at Guildford (three 500kW sets).[21] Objections to this proposal were the cost of procuring and operating so many diesel sets; the perceived unreliability of diesels at the time; and doubts about obtaining a reliable supply of heavy fuel oil. Jones investigated the possibility of taking an electricity supply from the MDR's Lots Road power station at Chelsea. This he reported would have been economical for the first stage of the LSWR electrification – presumably because the MDR had some spare capacity – but more expensive for later stages of electrification. In his report Jones claims that at stage two of his plan, the annual cost of electricity from Lots Road would be £101,000 and only £91,820 from an LSWR-owned powerhouse. The ultimate decision was that the LSWR should build its own generating station at Durnsford Road, Wimbledon. This site was on railway owned land wedged between the branch to Putney and the main line.

Sir Alexander Kennedy and Partners favourably reviewed Herbert Jones' report, and consequently on 6 December 1912 the LSWR confirmed that the full board had approved the proposals and announced its intentions to the world.[22] Herbert Walker recommended to the LSWR board that Sir Alexander Kennedy and Partners be retained as consultants for a fee of £16,000 per annum.[23]

The first stage of the ambitious project was to electrify four routes amounting to 73 track miles. The initial section to be electrified was the Kingston roundabout, that is Waterloo to Waterloo via Twickenham and the Kingston loop using both routes from Clapham Junction (Windsor lines and the main line). Included with this first proposal was the intention to electrify the spur at Putney from Mount Pleasant Junction up onto the line used by the MDR and thereby allowing LSWR electric trains to reach Wimbledon via Southfield (the station for the famous Wimbledon tennis tournament). The other lines in this first stage were to be from Barnes through Hounslow (the Hounslow loop) and back to Twickenham via Whitton; the Hampton Court branch – which incorporated the short section of LSWR main line from Malden to Hampton Court Junction; and the Shepperton branch from Strawberry Hill on the Kingston loop. The board's decision was based on sound commercial considerations but there was also a public relations issue that needed addressing, for while the LBSCR electrified lines and the MDR electrified lines were not in direct competition with the LSWR suburban lines, the public were beginning to make unflattering comparisons between the LSWR steam services and the faster and more frequent electric services on those other railways in the locality.[24]

(See plate 2) The second stage was to be the main line from Hampton Court Junction to Guildford via Woking and the routes loosely referred to as the 'New Line', that is from Hampton Court Junction to Guildford via Cobham and Raynes Park to Effingham Junction, increasing the total length of track equipped with the conductor rail to a grand total of 173 miles. The press reaction was generally favourable: the LSWR's proposals made sense in terms of what was later, after the impending First World War, to become the burning issue of 'interchangeability'.[25]

The LSWR confirmed that it would not be using electric locomotives for its trains but would use 'motor trains' instead. Obviously, the company meant multiple unit electric trains, but the term 'motor train' was also used at the time to describe LSWR 'pull-push' steam trains.

Anticipating the additional traffic, the LSWR also announced that it intended to widen the lines from Clapham Junction into Waterloo by providing an additional track, making a total of 10 running lines between Vauxhall and Clapham Junction – this would be achieved by encroaching on the Nine Elms goods yard. It also announced plans for a 'flying junction' at Hampton Court Junction in order that trains from Waterloo could bridge over the other main lines to reach the Hampton Court branch line.

CHAPTER 5

THE LSWR ELECTRIFICATION

The LSWR's sudden press announcement of its electrification scheme was the first time that either the general public or anybody in government circles in Whitehall were to hear officially or unofficially of the railway's firm proposals. This was despite much speculation that the LSWR intended to electrify its suburban railway lines. It was only through a news item in *The Times* on 13 December 1912 that an incensed official working for His Majesties Office of Works became aware of the scheme.[1] This was not the way that this civil servant wanted to find out about a development on which, in his view, he or Parliament should have been consulted. The railways in Britain were tightly controlled from Westminster having all been established through Parliament using the private member's Bill procedure which involved the scrutiny of all such schemes by both the House of Commons and the House of Lords. Both Houses would set up special committees for this purpose. The railway promotors often anticipated gaining business from existing stagecoach and canal companies and they needed to be able to build their railways over private land which was often owned by powerful individuals who were not predisposed to accept the intrusion of any railway line. The private member's Bill system allowed for all opposing vested interests to air their objections and when the bills became Acts of Parliament the railway company obtained compulsory purchase powers for land (a significant concession in a society which respected private property) and permission to raise the necessary capital through the issue of shares. The system allowed scrutiny of the schemes to protect prospective investors' capital (not always successful as some schemes were over-ambitious and verged on the fraudulent), and for the parliamentarians to impose certain conditions upon the railways to protect the interests of any aggrieved parties.

The London and Southampton Railway had chosen a relatively uncontentious route to the south coast, so its Act of 1834 had passed through Parliament with little difficulty. The Act allowed the railway to use locomotives or 'other power', if whatever was used 'consumed their own smoke'. Therefore, it seemed quite reasonable to the LSWR directors that the loose wording of the Act would allow electric multiple unit trains on the original line.[2] But it was a leap in reasoning to assume that the wording was applicable to all lines operated by the LSWR. The LSWR also appears to have turned a complete blind eye to another important piece of legislation that was to catch out the South Eastern and Chatham Railway (SECR) a decade later, namely the requirement for the railway to obtain an Order under the Railways (Electric Power) Act of 1903. This piece of legislation was intended to remove the need for a private member's Bill to approve railway electrification proposals by delegating the process to the Board of Trade, which was empowered to issue an Order after a widespread notification process and the holding of a formal public inquiry. The Board of Trade's powers included the responsibility to approve the use of electricity for motive power and the construction of generating stations and substations on the railway's own land.

The Board of Trade was expected to review any public safety issues or other concerns. Reading the Railways (Electric Power) Act today it can be reasonably argued that the LSWR should have made such an application to the Board of Trade.

The day after *The Times* news item appeared, the Board of Trade received a letter from the civil servant at the Office of Works complaining that the LSWR scheme could have detrimental effects on the National Physics Laboratory at Teddington and was also likely to be detrimental to the amenities of the local royal palaces and parks. The concern was well founded: local tramways with their rails embedded in the roadways were an earth return system and were generating stray currents in all directions and the National Physics Laboratory with its delicate scientific instruments was so close to the LSWR Kingston loop that the laboratory used a small map of how to get to its premises from Teddington station as part of its letterhead. On receipt of the letter the officials at the Board of Trade were not overly concerned since they were aware that electric trams already ran in Teddington and they reasoned that the damage, if any, had already been done. They considered writing back to the Office of Works stating that whether the LSWR had enough powers to carry out its scheme was a legal issue and as such they did not have the necessary authority to make any determination.

After some deliberation however, a Board of Trade official wrote back to the Office of Works confirming that the LSWR had not made any application for an Order as per the Railways (Electric Power) Act 1903 to carry out such a scheme.[3] The reply confessed that since the LYR in Liverpool and the NER on Tyneside had proceeded with electrification without any such consent Order, precedents had been set and the Board of Trade lamely believed it was powerless. It did however write to Herbert Walker taking the tack that it had been noticed that there would appear to be no mention of electrification in any of the clauses in the current LSWR Bill passing through Parliament. (Most railways had an annual 'housekeeping' parliamentary Bill to tie up all sorts of legal issues.) But in his written reply to the Board of Trade, Herbert Walker agreed that there had been no application to electrify LSWR lines because the directors had been advised that they already possessed the necessary powers to undertake the proposed scheme. Furthermore, the characteristically bombastic Walker utilised his usual tactic when confronted with controversial issues, hoping that his assertiveness would provide a veneer of authority to camouflage his dubious position. He boldly assured the Board of Trade that as far as any detrimental effects on third parties' equipment was concerned there are 'no grounds whatsoever' for any apprehension. Walker more reasonably dismissed the suggestion that the generating station would have any detrimental effects on the royal parks since only one generating station was to be built and this was on a site adjoining the railway and far removed from any royal palace or park.[4]

The Board of Trade meekly accepted this reply, but the National Physics Laboratory was much more concerned and sceptical, and on receiving a copy of Walker's letter to the Board of Trade was unable to accept that the electrification would not affect the scientific work as Walker had insisted, and demanded more information including details of the 'return' arrangement for the direct current traction supply. To resolve the problem and placate the laboratory's concerns, Sir Alexander Kennedy representing the LSWR met with Doctor Glazebrook from the laboratory. At the meeting Glazebrook accepted that the electrification of the railway was necessary[5] and Sir Alexander made a rather vague promise that the LSWR's running rails might be insulated in the vicinity of the laboratory if found necessary. But the railway electrification was a *fait accompli*, and the LSWR was later to prove unhelpful and this issue was never to be properly resolved. A decade later the National Physics Laboratory claimed that the electrification had caused serious disturbance to its electrical equipment and that the railway refused to assist in rectifying the issue.[6] Later, in 1919, an official from the then new Ministry of Transport revealed that because the LSWR had not sought either an Act of Parliament or an Order as per the Railways (Electric Power) Act 1903, the Board of Trade had never carried out any inspections of the LSWR electrification work as would have been required by such Act or Order. This had resulted in a complete lack of any independent scrutiny to check on the safety of the electrical apparatus which should have been carried out to protect both the railway employees and members of the public.[7]

The LSWR did not escape completely, for the railway had to apply to Parliament to raise the capital for the scheme. It was then easy for the National Physics Laboratory to press for protection from possible disturbance and this resulted in the London and South Western Act of 1913 setting out the stipulation that the electrification must not cause a disturbance of the horizontal magnetic field at the National Physics Laboratory exceeding 20 per cent beyond any existing disturbances.[8] This sounds reasonable – if complicated – but later, proving that any detriment was due to the railway and not some other source was to be highly problematic with the passing of time after the railway was electrified.

Work soon began on the electrification scheme. The new generating station at Durnsford Road was a major civil, mechanical and electrical engineering project. It was considered expensive at the time costing £341,815 – about £27 million in 2020 – thereby justifying Herbert Jones' concerns about the railway embarking on such a scheme. Mowlem constructed the foundations for the boiler house which was 237 feet long and 130 feet wide (72 metres by 40 metres). It contained 16 Babcock and Wilcox water tube boilers with operating pressures of 200 pounds per square inch.[9] The water for the boilers was obtained from the public supply which was first softened and purified. A water tank with a capacity for 110,000 gallons was installed for the feed water. Some sources state that the cooling water used to condense the used steam back into water was drawn direct from the adjacent River Wandle, but Alfred Raworth's notebook records that one contractor, C. Islee and Company, was employed to sink an artesian well on the site. Set behind the main building were three rather ugly cooling towers which were rectangular with external bracing, much less attractive than the elegant concrete structures that we are all used to seeing today at modern power station sites.

In between the two tall and thin chimneys there was a siding on a ramp from which the loaded coal trucks were discharged directly into the boiler house's overhead bunkers. This ramp was electrified to enable the LSWR's only electric locomotive to shunt the coal wagons. This small steeple-cab locomotive, having been relieved of its shunting duties in the depths of the Waterloo and City Railway, was brought out into the sunlight and survived until 1959, numbered by British Railways DS74. A suction plant was used to dispose of the ashes from the boilers – this was a new and innovative feature recently devised by Babcock and Wilcox.

The turbine house was 275 feet long by 55 feet wide (84 metres by 17 metres) and contained five Dick, Kerr and Company 5,000kW alternators driven by steam turbines operating at 1,500 revolutions per minute to produce the 11,000V three-phase alternating current supply at the railway frequency of 25 cycles per second. There were also three smaller auxiliary dynamos to provide the 220V direct current required for use within the station for lighting and the various plant motors at the generating station.

The switchgear operating at 11,000V was installed within the building adjacent to the turbine house and was on three levels, each almost 13 feet (4 metres) high and was built to a principle that was later to become

Durnsford Road generating station. An LSWR G6 0-6-0 tank locomotive shunts coal wagons up the incline to the boiler house. This official LSWR photograph was probably staged and was intend to show also the new Wimbledon Depot and the electric trains. Perhaps the staff on the railway are installing conductor rail. (Railway Magazine Archive)

The turbine hall within the Durnsford Road generating station. (Railway Magazine Archive)

known as a 'cellular' arrangement, using uninsulated copper bars. Each individual item of plant was safely housed within its own moulded stone 'cell'. The switchgear was provided by the Rugby based British Thomson-Houston Company (BTH) – the British subsidiary of the American General Electric Company.

The bottom tier contained the cable boxes – the transition between the insulated cables and the bare copper bars. These consisted of a cast iron 'boat' with the three-phase cable entering at the bottom of the casting. The cast iron container would splay out as the individual cores from the cable separated away from each other. The boat was filled with a bitumen-based material and the three live phase cores emanating from this hard, black lump would be connected to the copper bars dropping down from the tier above. The cables either came direct from the generators or were outgoing feeder cables to the trackside substations.

The middle tier on the switch house first floor contained the circuit breakers, one for each of the generators or feeder cables. Strangely, contemporary accounts refer to these as 'oil switches' – a completely different modern connotation. A circuit breaker is a device that has the capability of not only being able to switch on or off an electrical load but can also safely interrupt the massive short-circuit current that will flow in a fault situation. The 'oil' in the description refers to the thin mineral oil (like that in transformers) used to quench the ferocious electrical arc that is drawn between the opening contacts when the sensing ('protection') equipment detects an overload and the breaker automatically opens. Opening and closing of these switches was carried out remotely, the controls being in a room in the gallery overlooking the switch house first floor. Breaking the current within these switches was achieved by the opening of two sets of contacts connected in series and these were housed in steel tanks immersed in the mineral oil. Each switch had three separate tanks – one for each phase – and each of these circuit breakers was within its own cell with a locked steel front panel to prevent unauthorised access. On the same floor level as the circuit breakers were various offices and stores.

The top tier and switch house second floor housed the busbars. The term 'busbar' has the same Latin root as 'omnibus' (bus), and implies 'belonging to all', and each circuit – either outgoing feeder or generator cable – had its circuit breaker to connect it to the busbars which consisted of chunky copper bars standing on porcelain insulators. The busbars were arranged in three sections with interconnecting switchgear. For security, each of three sections was separated by solid firewalls. Each section of busbar was designed to take two generator circuits and four feeder circuits, thus at the outset there were some spare positions for any additional circuits that might be required in the future. These were the main busbar sections, but a reserve busbar ran the length of the third tier for emergency use should any section of the main busbar be out of service due to either planned maintenance or a breakdown. To facilitate the changeover to this reserve busbar and to safely isolate the circuit breakers for maintenance, additional selector switches were provided to connect the circuit breakers to either – or neither – busbar. These were probably simple knife switches operated manually using long insulated wooden rods, and these switches, as for the circuit breakers, were housed in their own individual steel panel fronted stone cell. It was safe to manually operate these knife switches 'off load' when the associated circuit breaker was in the open position and thus there was no load current to break.

Jones had stipulated in his report that substations – or the 'feeding points' as he referred to them – would be required at distances varying between three and a half and eight and a half miles. From the Durnsford Road generating station 11,000V cables were installed alongside the track. To ensure security of the electrical supply these cables mainly ran in pairs between the substations and they also formed a ring configuration. The original high voltage ring from Durnsford Road generating station was two cables alongside the LSWR main line to Raynes Park substation; then beside the Kingston loop to reach Kingston substation, followed by Twickenham substation and then (on this section using a single cable only) to Barnes substation before returning to Durnsford Road. Running north, again beside the main line, were four cables installed to Clapham Junction substation and then three only on to the 'new' Waterloo substation. Not on the main ring were duplicated cables to Sunbury substation on the Shepperton branch; to Hampton Court Junction; and to Isleworth substation on the Hounslow loop.

The 11,000V three-phase cables were manufactured by Siemens Brothers at Woolwich, the one-time employer of both Alfred Raworth's father and uncle. Each cable contained three copper phase cores insulated with oil-impregnated paper and enclosed within a lead sheath to prevent the ingress of moisture. Above the lead sheath were wrapped steel wires as 'armouring' to protect the cable from mechanical damage. These cables were placed above ground and hung in vertical formation on cast iron brackets fixed to creosoted wooden posts about 5.5 feet (1.7 metres) apart.

The new substations followed the pattern observed by Jones in New York. They contained the step-down transformers with the rotary convertors supplied by BTH and switchboards to control the 600V cables feeding out onto the track. These architecturally very pleasing brick-built buildings were about 46 feet (14 metres) wide with the length depending on the number of transformers with rotary convertor sets installed. Many of these buildings survive today – for example at Twickenham – grubby and graffiti besmirched and now being used for a variety of things by the railway.

The 11,000V switchgear at the substations was like the three-tier arrangement installed at Durnsford Road except that only a single busbar was provided connecting each of the incoming feeder cables to the local rectification sets. Space was reserved for additional feeders and extra transformers with rotary convertors. The rotary convertors sets were rated at either 1,500kW or 1,000kW depending on the individual site. For example, Sunbury substation had two 1,000kW rotary convertors, while Raynes Park substation had three 1,500kW sets and at Twickenham substation there was four 1,500kW sets. The transformers were single-phase units interconnected externally.

To control the outgoing direct current traction supplies, distribution switchboards were installed. These consisted of vertical slate panels with all the bare copper connections mounted on slate insulators out of sight behind the panel. The switches and instruments were mounted on the front of the panel. Each 'live' connection to the conductor rails was controlled by a simple manually operated air-break circuit breaker which had to be used in conjunction with the negative 'return' switch, which was a simple knife switch mounted below. Each slate panel accommodated switches and instruments for two outgoing direct current circuits. The instrumentation was simple: an ammeter to measure the load current and indicator lamps to show that the cable was 'live' – this was needed as the conductor rails ran between the substations and could be energised from either end; therefore an open switch did not necessarily mean that the cable was fully disconnected and safe to work on.

The conductor rails, mounted on porcelain insulators, were positioned three inches (76 millimetres) above the running rail and 18 inches (457 millimetres) away from the rail. They were made from special high conductivity steel having high carbon content and an electrical resistance only six and a half times greater than pure copper. They were supplied in 50-foot (15 metre) long sections. This conductor rail was wherever possible positioned on the

A twenty-first century view of Waterloo substation. Note the 'new' transformers outside, these would have been at least the third set installed since the substation was built for the original LSWR electrification. (Peter Winchester)

Interior of Waterloo substation. On the left of this view is the direct current switchboard and, in the centre, three of the rotary converters. At the right hand ends of the converter shafts can be seen the the small synchronous motors used to start the machines, and in the centre there are the six slip rings. Just visible is one of the substations transformers. (Railway Magazine Archive)

Twickenham substation building in 2020 – modern replacement plant has been installed outside or in the graffiti-stained steel cubicle next to the original building. (Author)

trackside away from any platform edge. For safety reasons, wooden boards were placed either side of the live rail at certain locations at stations.[10]

The LSWR originally electrified 12 platform lines at Waterloo station, six in the 'main' station and six on the 'Windsor' side of the station. Between Clapham Junction and Waterloo eight tracks were electrified and on the main line to Hampton Court Junction and the Windsor lines as far as Wandsworth all four tracks were electrified. Both tracks were electrified on the remaining double track sections.[11]

With the installation of the new substation at Waterloo it was possible to close the old generating station built for the Waterloo and City Railway and to supply the electricity for the LSWR's tube railway from Durnsford Road. The existing Waterloo generating equipment was removed except for two sets to provide electricity for lighting and the lifts at Waterloo station.[12] Five small boilers were installed for heating the company's buildings at Waterloo[13] thereby keeping Herbert Walker warm in winter.

To run on this new electric railway the LSWR needed new electric trains, and Jones had made his proposals for these in his report.[14] He emphatically recommended multiple units in preference to locomotive-hauled trains. These, he claimed, would achieve, depending on train length, a 12 per cent to 45 per cent reduction in energy consumption compared to locomotive-hauled trains and would reduce lie-over times and simplify the operation of trains at terminal stations. This view would not have ruled out using the electric locomotives on the fast trains made up of 'steam' stock to Woking if this proposal had been accepted, as this would have been an extra to the suburban scheme. Eschewing the American electric train arrangement Jones, based on his own observations, claimed that trains with the then British design of compartments with side doors could be emptied or filled in one third of the time compared to open carriages with end doors. The time allowed for station stops was only 20 seconds, but these were the days when passengers could be trusted to open and close the doors themselves. Jones' recommendation was for the trains to be made up of two-car sets. Each two-car set was to have a motor coach with two power bogies with four motors – one per axle – and a trailer coach.[15] These sets would have a driving compartment at each end and with a stylish 'V'-shaped front – often referred to as a 'torpedo' end with two windows either side of the route indictor board providing a 'streamlined' feature to reduce wind resistance. To extend these two-coach trains to three coaches at busy times, an additional driving trailer was proposed – complete with streamlined front end. The new motor coaches would have been 56 feet (17 metres) long and contained a guard's compartment, three first class compartments and four third class compartments. The trailer units would have been converted from 51-foot (15.5 metre) ex-'block' set brake thirds and contained a guard's compartment and seven third class compartments. The strengthening driving trailers would have been converted from the 51-foot tri-composites (first, second and third-class compartments) and contain three third class compartments and four third class compartments.[16]

But such decisions as to train formations were not within the remit of the railway's electrical engineers. This was the business of the traffic operators and the chief mechanical engineer. At the LSWR officers' conference held in June 1913 it was reported that three-coach sets were to be used to form the electric trains. No completely new carriages were to be built as all the electric trains were to be converted from the 'block' stock. Eighty-four three-coach sets were to be built for the first stage by converting these existing steam train coaches. Each three-car set was to comprise of two motor coaches with driver and guard's compartments with third class passenger accommodation in traditional compartments with a third, trailer vehicle, placed in between.[17] The unpowered trailer coaches would have first and third class accommodation. The proposed V-shaped driving cab front end was however retained. (See plate 2 (bottom)) There was originally a very high ratio of first class seats to third class seats because the LSWR expected that its second class clientele, when robbed of their usual accommodation, would trade up to first class rather than demean themselves in third class, but the exact opposite was to be the case and looking ahead to 1919 it was found necessary, due to overcrowding, to construct several two-car trailer sets to be marshalled between pairs of three-car sets to make up eight-coach trains. This rather clumsy arrangement – likely to result in dangerous situations during shunting operations, was superseded by the building of four-car sets for suburban services in later years.[18] The new trains were given a new livery. Gone was the salmon and umber to be replaced by a bright green lined with yellow.

The trains were fitted with the standard Westinghouse air brakes,[19] and it was to be the throbbing sound that these compressors made before departure that was to become so familiar with travellers on the Southern Railway electric trains. Alfred Raworth had carried out an investigation as to which braking system should be used – either the vacuum brake then standard on the LSWR – or the alternative air brake – used by some railways, for example the LBSCR. He had made enquiries with the MDR, the LBSCR and with his former colleagues at Brush and concluded that the cost of providing brakes for each three-coach set was £209 if air brakes and £305 for vacuum.[20] Jones in his report had recommended the air braking system for, as well as being cheaper, it allowed 10 stops after failure of the compressor.

The conversions were supervised by the chief mechanical engineer's assistant for carriages and wagon design, the grandly named Surrey Warner, who impressed Herbert Walker by his economical approach

Original LSWR three car set built using carriages from the 'block sets' Set 1201, now part of the Southern Railway fleet is standing at Clapham Junction on its way to Waterloo c1924 (John Scott-Morgan collection)

to the rebuilding programme.[21] Given that they were conversions from 1902 carriages, these 'new' trains had wooden bodies with the usual attractive panelling common to all early railway stock. The new motor bogies, one for each power car, were provided with two 275 horsepower (205kW) electric motors. The motors ran at speeds of up to 590 revolutions per minute. The gear wheels transferring the motion to the axles had a diameter of 31 inches (0.8 metres) – the largest in use on electric trains at the time.[22] A wooden beam was positioned either side of each motor bogie to support the collector shoes.

All the electrical equipment was installed in an area between the driver's and the guard's compartments in the motor coaches. Alfred Raworth may have been an expert on the control of electric vehicles, but he had no direct design input since the control system on these trains for multiple unit working was the Westinghouse Company's 'all-electric' system.[23] In this system the traction current was increased or decreased by the operation of a number of contactors mounted in the control compartment and connected through the train by control wires. The driver operated the master controller – described as being of the 'drum' type – to the desired 'notch' and the contactors automatically applied the setting for all the motors in the train. Naturally a 'dead man's handle' feature was used which required the driver to keep a continuous pressure on the operating handle, otherwise all the contactors would operate to shut off the power and apply the brakes. There was to be no attempt at 'regenerative control' or 'regenerative braking' – such technical virtuosity did not impress Herbert Walker. His solution to the use of the stored kinetic energy in a heavy moving train was to use it all up by dropping off the power early and coasting the trains into the stations. Consequently, he ordered trackside marker boards to be set up to show the driver when he should close off the power before each stop.

The whole project was well underway in August 1914 when the world turned upside down. The previous June an amateurish assassination of an archduke who was the heir to the Austrian throne and his wife by a gang of Serbian hotheads had nudged the whole of Europe towards a terrible war. The Austro-Hungarian Empire, with its German allies standing behind and edging it on, wanted to retaliate and punish the Serbs, but knew that this would drag in Serbia's old friend Russia. The Russians had a treaty with the French, so while the Russians assembled their army the German High Command took out of a drawer a plan to invade France. A simplistic plan: you knock out France with an unexpected lightning attack and then deal immediately afterwards with the still unprepared Russians who would then be standing alone. The result was that, after first asking the Belgians nicely if they could crash through their peaceful country and getting only a blunt refusal, the Germans threatened force which brought in the British to line up with the French to 'protect Belgium's neutrality'. Germany then brusquely rejected any diplomatic solution and did more than just posture and mobilise her troops as the other nations were doing at the time, choosing instead to go one step further and move on to the actual fighting stage by invading Belgium. For Germany, the outcome was not the swift victory they had planned but four years of virtual military stalemate when their rash plan to reach Paris failed. This catastrophic example of what has been described in modern times as 'chaos theory' was in reality a war waiting to happen, for Germany had designs to become a world power and had been building up its army and navy in preparation.

The railways of Britain were of strategic importance and were immediately taken into a form of government control at the outbreak of war as prescribed by the Regulation of the Forces Act 1871 but remained 'private' companies. Herbert Walker had a second job at the time – deputy chairman of the Railway Executive Committee (REC) – the body that was to take charge of the railways during any conflict. The REC was set up by the Board of Trade in 1912 and was made up of railway general managers with the intention that in times of crisis this body would work with the armed forces to secure the nation's main transport links. The nominal chairman for such an important body was the president of the Board of Trade himself, but in practice the de facto chairman was elected from one of the general managers and designated deputy chairman. In June, just before the outbreak of war, the young and ambitious Herbert Walker was elected to the post of deputy chairman. It was therefore Walker who issued formal notice to all the railway companies stating that the railways had been taken into government control to ensure that all their resources could be used as one unit to best facilitate the movement of troops, stores and food supplies. Walker took charge of the REC, which included the general managers of eight other British railway companies. What appeared to be government control was in fact decision making by professional railwaymen in the national, not shareholder, interest.

One can only speculate what the directors representing the shareholders would have said if they could have seen into the future, for far from the 'it will all be over by Christmas', of popular opinion, these august gentlemen on the boards of the great railway companies would never again have full control of their railways' destiny. A rather poor financial arrangement was made between the government and the REC, and the acceptance of the arrangement is perhaps the only major failing that has been attributed to Herbert Walker during his illustrious management career. Despite the prospects of a massive wartime increase in traffic and hence revenue, initially dividends were to be paid by the government at 1914 levels; but because 1914 was deemed to be an exceptional year, this was changed to payments based on 1913 levels. This arrangement, far from a temporary expedient as expected, was to result later in the government regulation of the private companies' earning ability by basing rates and earnings on 1913 levels, and leading in the years after the war to the standardisation of rates, in turn setting the railways at a distinct disadvantage to the new road haulage companies. This eventually led to the railways' 'Square Deal' campaign in the 1930s which was never resolved before the outbreak of the next great conflict in 1939. Also, following the Great War, payments to cover such matters as capital expenditure to replace overworked assets were initially to be reneged on by the post-war government as it procrastinated as to the future ownership of the nation's railway network. Among all the impending turmoil, which was completely unexpected in the heady patriotic atmosphere in 1914, it was the high capital cost projects such as railway electrification that were the most likely to be influenced by the hand of any remote government.

On outbreak of war all contracts relating to the LSWR electrification were suspended. These had been issued for all manner of goods and services including orders for the 11,000V switchgear and the construction of the cooling towers at Durnsford Road; cables for the 600V direct current supplies; the purchase of a traverser for the carriage sheds; and the construction of the substations at Isleworth, Sunbury and Hampton Court Junction.[24] But on 29 October 1914, the members of the Locomotive, Carriage and Estates subcommittee were informed that it had been arranged that work included in the electrification contracts was to be resumed. With Herbert Walker in the background, who was then in a position of considerable power and influence as head of the REC, when one reads the word *arranged* it must be speculated as to just how much string pulling and lobbying was going on behind the scenes and just how close the LSWR came to be forced to cancel the whole project.

Alfred Raworth had played his part in the planning and design work for the LSWR electrification, but he did not remain with the LSWR long enough to see the first electric train in service.

CHAPTER 6
RAWORTH RN

The first scheduled LSWR electric train service commenced on Monday 25 October 1915.[1] The initial route was from Wimbledon to Waterloo via Putney, the new trains sharing the tracks with the existing MDR trains between Wimbledon and Putney. Trains ran every 20 minutes from Wimbledon to Waterloo, from where passengers would find trains running every few minutes to the City using the LSWR's own Waterloo and City tube line, or alternatively they could travel to the West End using the Bakerloo tube railway. The technical press, such as the 18 November 1915 edition of *Tramway and Railway World*, delivered warm appreciation of the LSWR's electrification project and congratulated the company on completing the scheme with minimal delay despite the exigencies of war. One practical wartime issue was a shortage of labour due to the absence of many railwaymen who joined the British Expeditionary Force (BEF) in France, so women had been recruited for many tasks including electric train cleaning duties at the new Wimbledon depot.

But nearly three weeks before the momentous event when public services began, Alfred Raworth left the railway service to perform his patriotic duty by joining the 'Senior Service', the Royal Navy. His main role in the LSWR electrification had been as a design and planning engineer, so with all such work complete there was moral pressure upon him to 'do his bit' for the nation in wartime, even though conscription for married men between the ages of 18 and 41 did not come into force until May 1916. His brother Basil (Arthur) had in 1908 joined the reserve army as a 2nd Lieutenant in the 7th Battalion of the Manchester Regiment but due to his taking up a post with General Electric in Hong Kong he saw out the war as a lieutenant in the Hong Kong Defence Corps.[2] Alfred obtained a temporary commission as a lieutenant in the Royal Naval Volunteer Reserve (RNVR). It was very much part of the 'public school' ethos for young men to enlist in the forces as part-time reservists just as his brother had done, but in Alfred's case this was his first encounter with the military.

On 6 October 1915 Alfred received his first and only naval posting to the Royal Navy shore base, or 'stone frigate' – HMS *President II*.[3] These shore bases were often no more than accounting centres, somewhere to hold sailors between postings, facilitating the allocation of their pay and rations, etc. *President II* was the accounting base for London, Chatham and Shrewsbury and as such was the home to many Admiralty staff members. Occasionally old ships would be moored on the Thames to provide additional accommodation for this shore establishment.[4] Alfred was assigned to the new Royal Naval Air Service (RNAS) for what was termed 'inspectional duty'. The RNAS was set up by Winston Churchill as part of his naval reforms after his appointment as First Lord of the Admiralty in 1911. Churchill, a man of whom it must be said was excited and driven by war and combat, saw that there was a huge potential for aircraft to be used as an effective weapon against the enemy.[5] The RNAS did not restrict itself to seaplanes or launching aircraft from ships, for early in the war Churchill's RNAS had more aircraft than the army's Royal Flying Corps (RFC) and was to form many front-line land-based fighter and bomber squadrons for use in France.

So forceful was Churchill's influence that the RNAS was often using the latest fighter aircraft while the RFC, which should have been fighting over the land, was having to use second best, obsolete craft.[6] But on his 'inspectional duty' and based in London, there was to be no excitement flying over the German front lines for Alfred Raworth. Probably his duties were no more exciting than checking equipment as it was delivered from the manufacturers, a rather mundane task but an essential one, and given his previous experience a task that he was well suited for.

On 24 March 1917 Alfred's father died at his home in Streatham. He had been in poor health for some time and had relinquished most of his interests in activities associated with the various institutions of which he had been a member. He had continued to engage in some consulting work and had remained a director of British Electric Traction.[7] There were many glowing obituaries written, painting a picture of a man who was not only a brilliant engineer but also a popular and well-known figure renowned for his powers of oratory who used his vast experience and knowledge to good effect during shareholders' meetings when he was either a board chairman, a board member or simply a critical shareholder. He was also praised for his contributions to the technical debates held at the various institutions to which he was a keen and active participant, giving his greatest service to the Institution of Electrical Engineers. He was praised for his common-sense approach to professional and social problems, particularly in the area that we would now call 'industrial relations'.[8] Many sources today refer to him as 'the inventor of regenerative braking', but as was previously explained this is not a completely true attribute, but surely it is better that a man of his talent is at least remembered for *something*, even if his main work in the textiles industry, electricity supply and tramway promotion has long been forgotten.

On 1 May 1917 Lieutenant Raworth was 'lent' by the Admiralty to the all-important Ministry of Munitions.[9] This ministry had been set up in 1915 after a public outcry due to the lack of shells being made available to the troops fighting in France. Engaged at the Ministry were the cream of British businessmen who were loaned from their companies to assist with the war effort. This was therefore a prestigious appointment for Alfred Raworth, whose organisational abilities had obviously been recognised. Also included within this elite band of talented men was Sir Philip Dawson.[10] The manufacture of munitions had been completely taken over by the Ministry, the function of which was to coordinate the needs of big business with the needs of the state at war and to reach a compromise on price and profit that was acceptable to both sides. Additionally, Ministry of Munitions agents would buy essential supplies from abroad, and once bought, the Ministry would control their distribution to prevent speculative price rises and enable normal marketing to continue.[11]

The first Minister of Munitions was David Lloyd George, a man who was to make a significant impact on post-First World War Britain. Lloyd George was born in 1863 in Manchester but was of Welsh stock and spent his formative years in Wales speaking Welsh as his first language. He trained as a solicitor with a firm in Porthmadog and after qualifying he set up his own practice but was soon attracted into politics, becoming a Liberal MP for Caernarvon Boroughs in 1890. Before the First World War he served as Chancellor of the Exchequer and was instrumental in introducing many welfare reforms such as his National Insurance Act that made provisions for poor workers faced with sickness or invalidity. Lloyd George was appointed Minister of Munitions on 25 May 1915, a post he held until July 1916.[12] He was keen to recruit men from outside the usual government sphere of activity, men who could organise and motivate and could be relied upon to achieve his goals.

Meanwhile, new electric trains were duly introduced onto the 'Kingston Roundabout' and the Shepperton branch on 30 January 1916. The introduction of these electric services had been postponed from 6 December 1915 due to a dispute with the General Post Office (GPO) that had been claiming the new installation was interfering with telephone circuits in the vicinity. The LSWR denied this, but some remedial measures were put in place (most likely by improving the electrical bonding at rail joins) and subsequently the full service commenced.

The timetable for the new electric services was based on Herbert Walker's commercial philosophy that departure and running times should be standardised and be easily remembered by the travelling public. This is often termed a 'clock face' approach to timetabling. He also believed that the longest wait that could be tolerated by a potential passenger who had not looked up the times would be between 3 and 10 minutes and at the most 15 minutes.[13] As proposed by Jones and Raworth the timetable consisted of four trains an hour in each direction around the 'roundabout' and two trains an hour from Waterloo to Shepperton via Kingston. Electric trains on the Hounslow loop began running on 12 March 1916, the service comprising two trains an hour in each direction. The services to Hampton

LSWR poster advertising the new electric services on the 'Kingston roundabout'. (Author's Collection)

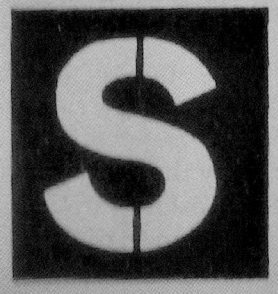

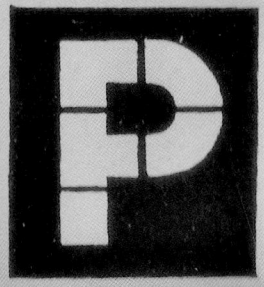

Court – three trains an hour – commenced running on 18 June 1916. The LSWR made a rather late decision to electrify as part of Stage One the short section of the New Line from Hampton Court Junction on the main line as far as the first station on the route, Claygate. This rather odd move resulted in Claygate becoming a changeover point where passengers who were continuing to Guildford had to transfer to steam-operated pull-push trains. The Waterloo to Claygate service commenced on 20 November 1916 with a half-hourly interval service provided. Unfortunately, shortage of available electric trains forced the cessation of electric service on the Waterloo to Claygate service and Claygate had to await the implementation of Stage Two of the LSWR plan by the Southern Railway in 1925.[14]

By using the third-rail earth return system the LSWR, in addition to sparring with the GPO in the Kingston locality, continued to conflict with the National Physics Laboratory at Teddington. In 1916 after the commencement of electric trains running through Teddington on the out and back route to and from Waterloo, the laboratory staff were in no doubt that there was a considerable increase in the disturbance to their sensitive magnetic equipment due to stray electrical currents. What could not be easily proved was the origin of these stray currents. Herbert Jones collaborated with the laboratory conducting tests to determine if the fluctuating increase in magnetic disturbance was coincident with the increase in current when trains were moving off from the local stations. The results were inconclusive, for some increase in magnetic disturbance was observed with some of these fluctuating events possibly in time with trains starting, but some not. It must be remembered that there were other potential sources of stray electric current such as the public electricity supply generating stations and more trams operating locally.

It was agreed to set up an independent arbitrator. Unfortunately, this story, traced through correspondence in the former Board of Trade files at the National Archives at Kew, does not continue beyond the agreement by both parties to move to arbitration. What is in the file is Herbert Walker probably playing his trump card.[15] He announced that since the railways were in government control and that the laboratory was a government owned facility, it should be the Treasury not the LSWR that should meet the cost of any extensive remedial work of say, improving the conductivity of the running rails with special steel, insulating the running rails or providing additional bonding. This tack probably closed the issue, even if it left bad feeling to be dealt with after the war.

Nonetheless, Walker had every reason to be pleased with *his* electrification scheme. Once the railway had a full year's set of traffic figures, the dramatic increase in passenger numbers was appreciatively noted by Walker and the directors. The new electric services were proving so popular that it was difficult to find living accommodation in the localities served by the electric railway.[16] This was one of the earliest instances of what has been termed the 'sparks effect' which is now the expectation of what will happen when a railway is electrified, enabling the public to ride on the faster, more frequent, cleaner and quieter trains which make more stops at stations due to their improved acceleration and braking, so as to serve more communities without any detriment to the overall journey times. The LSWR had not just been stemming the haemorrhage of passengers away from the railway and onto trams – it was positively attracting new business. These excellent results do need to be considered in the context that the previously burgeoning motor bus traffic had been severely curtailed, as the new operators' vehicles had been commandeered by the army for troop transport along the Western Front; for it appears that in the early part of the twentieth century the London suburban railways' managers had something of a blind spot in respect of this new form of competition.

But the success of the electrification was good news for the railway industry and was noted by another railway company in southern England which, on its suburban lines, was also encountering similar problems to those of the LSWR. Alfred Raworth, immersed in his work at the Ministry of Munitions, must have been surprised when the call came for him to leave his important war work and return to the railway industry. It must have been an even bigger surprise when he found that it was not the LSWR that was making the call. Only the highest authority could have released Raworth from his important war efforts at the Ministry of Munitions. His Royal Navy service record shows that on 20 February 1918 his temporary commission was to be cancelled. For his important yet hidden war service, Raworth was later to receive the British War Medal as he had been mobilised for a period of more than 28 days.[17] Raworth's naval service record reveals (incorrectly) that he *returned* to the South Eastern and Chatham Railway.[18]

CHAPTER 7

A RAILWAY UNDER STRESS

Alfred Raworth's new employer was to be the SECR which had been formed by a rather reluctant marriage between two railway companies, both of which after decades of bitter and, at times, unnecessary rivalry, were in poor financial circumstances. The SER and the London, Chatham and Dover Railway (LCDR) had fought hard to win traffic on their duplicated lines from their several London termini to destinations throughout Kent. From their Channel ports at Folkestone and Dover and their ports on the Thames Estuary, rival cross-Channel ferry services operated to France and Belgium. The behaviour of these two belligerent companies would make a case study to demonstrate everything that was wrong with the way the railways in the nineteenth century had been allowed to expand throughout Britain without any overall strategic plan – in many cases this rash expansion had been driven by the sheer greed of the promoters. Both companies attracted criticism for the construction of unnecessary and uneconomical lines across Kent. But despite this generally held view, over 90 per cent of these two railways' lines remain in use in the twenty-first century – a commendable survival rate. This corrosive relationship between the two has often been viewed as the consequence of the bitter personal animosity between the two chairmen – namely Sir Edward Watkin of the SER and James Staart Forbes of the LCDR.[1] Both men have been accused of constructing expensive and unnecessary duplicate railway lines almost as if to spite the other. This might be an unfair appraisal, for both companies needed to expand to attract business and neither could afford to sacrifice a potentially lucrative geographical area to the other.

The SER reached Dover in 1844 having constructed its line from London Bridge via Redhill, Tonbridge and through Folkestone Warren below Shakespeare Cliffs. This early roundabout route was due to Parliament's insistence that the SER share the route to Redhill with the main line to Brighton. By 1845, as described earlier, the Reading, Guildford and Reigate Railway had constructed its sinuous 46-mile line west from Redhill as far as Reading where it connected with the GWR having utilised running powers over the short stretch of the LSWR through the Guildford area. This line was bought by the SER in 1852, with the company now operating much further west than was decent for a railway with the word 'eastern' as part of its title. In 1868, the SER did manage to construct its own line from London Bridge through Orpington and Sevenoaks to join the original route at Tonbridge, thereby being able to cut the corner off the original route via Redhill.[2] The territory was filled in with branches to locations such as Hayes. Further expansion took place including a new main line from Tonbridge to Hastings on the south coast. The SER had started off being quite adventurous by not only managing to reach Reading in the west, but very nearly making it to Portsmouth in the south. In later years it was to become geographically restricted due to territory agreements it had made with the neighbouring LBSCR inhibiting further expansion to the west; and by the geographical boundaries of the Thames Estuary and the English Channel in the north, east and south.

The SER's London base was at London Bridge, the terminus originally built in 1833 by the London and Greenwich Railway. Other railways soon joined the SER, namely the London and

Croydon Railway and the London and Brighton Railway, both to become part of the LBSCR. In 1857 the SER promised Parliament that it would provide a 'West End' terminus for its Dover services[3] and this it did in 1864 by extending its line through London Bridge station and crossing the River Thames to a new terminus at Charing Cross. Incidentally, this extended line passed by the LSWR terminus at Waterloo, and the SER built a station which was eventually given its current name, Waterloo East, but was then named Waterloo Junction. The 'junction' appellation was earned because a line was built between Waterloo's platforms 1 and 2 across the station concourse to make connection with the SER line. This connection was used only briefly by the SER for its local cross-London service to Kensington (Addiscombe Road – now Olympia) and by the LNWR for its trains from Willesden. In later years, the line was used only for empty stock movements and by royal trains taking Queen Victoria from Windsor to the Channel ports.[4] The connection was removed in 1910 to make way for the LSWR's rebuilding of Waterloo station.

Meanwhile, the railway that had originally been a local line in Kent – the East Kent Railway, now renamed the LCDR – reached Victoria in 1858. The route taken from Rochester crossed over the SER Maidstone line and then headed due west through Swanley, crossing over the SER's line to Tonbridge near St Mary Cray, and then on through Bromley, Dulwich, Herne Hill and Brixton to reach Victoria station. Victoria had become the West End terminus for the LBSCR whose other main line to Brighton now ran from Victoria through Clapham Junction, Balham and Streatham to link with its original London to Brighton line from London Bridge station at East Croydon. The LCDR and LBSCR became joint occupiers of Victoria station which effectively became two conjoined termini. The LCDR extension into London was achieved by stringing together several dubiously financed small schemes which had, at the time of the completion of the line from Victoria to Dover, led directly to the spectacular financial crash of the

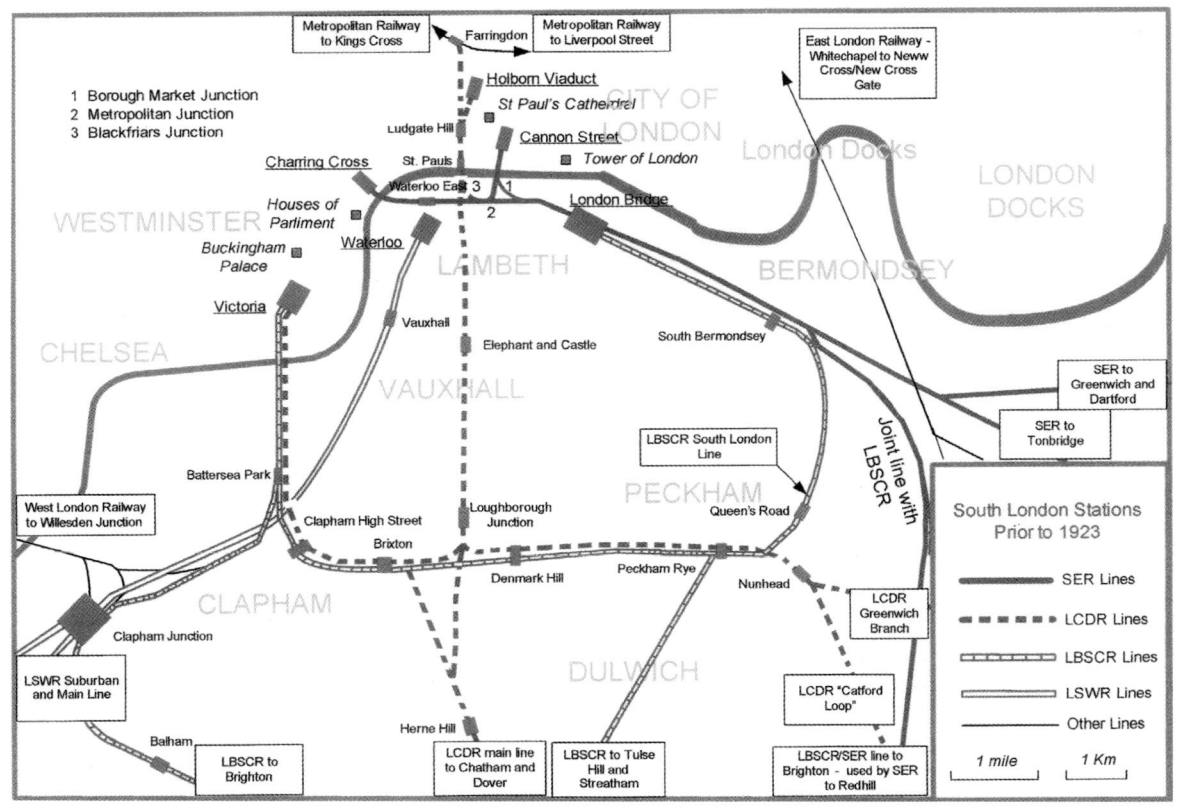

Simplified map showing the main termini for railway routes to the south of the River Thames. (Author)

bankers Overend, Gurney. This resulted in the bankruptcy of both the contractors and the LCDR, leaving the railway in receivership for many years.

Neither the SER nor the LCDR had a station in the City of London. The SER at first considered that its London Bridge to Charing Cross line could adequately serve the City from new stations it proposed to provide at the south end of Southwick Bridge and at Blackfriars Road.[5] However, the LCDR had a much better idea – it constructed a new line from Herne Hill on its new London extension, across the River Thames and then through the very heart of the City, making a junction with the Metropolitan Railway at Farringdon with 'City' stations at St Paul's and at Ludgate Hill which opened in 1865. A short, 264-yard (250 metre) line was constructed from a junction just north of Ludgate Hill to a small terminus at Holborn Viaduct which was opened in 1874. The short platforms of this tiny station were intended only for 'half trains', the main line expresses from the Channel ports would be divided at Herne Hill and run as two separate portions to Victoria for the West End and to Holborn Viaduct for the City.[6]

The competitive threat of the LCDR line spurred the SER to finance its own line directly into the City by, once again, crossing the River Thames to another new terminus at Cannon Street which was opened in 1866. The line to Cannon Street made a triangular junction with the original line above the famous Borough Market with curves facing towards both Charing Cross and London Bridge.

From the 1860s through to the 1890s several attempts were made to unite not only the SER with the LCDR but also to include the LBSCR and even the LSWR. Various deals were proposed with different permutations. All these proposals advocated the simple and sensible outcome that the railways should forget their wasteful rivalries, pool resources (including allowing each of the companies access to all the London stations) and balance the provision of resources to the actual traffic demand.[7] Eventually Watkin, in poor health, resigned from the SER board chairmanship in 1897 and his place was taken by H. Cosmo O. Bonsor.

H. Cosmo O. Bonsor (the 'H' was for Henry and the 'O' for Orme) was born on 2 September 1848 and was educated at Eton before he joined his father, Joseph Bonsor, in the family brewing business. He had many other interests besides the SECR; by 1922 he was the chairman of the brewing company Watney, Combe, Reid and Company; was a governor of Guy's Hospital and a director of the Bank of England. He had also been the Conservative MP for Wimbledon from

H. Cosmo O. Bonsor. (Author's Collection)

1885 to 1900. In 1907 he was offered the prestigious post of Governor of the Bank of England but turned it down as this would have required him to relinquish his interest in the SECR.[8] He had an estate at Kingswood Warren in Surrey and a London address at 38 Belgravia Square, where he lived with his wife, his two daughters and his son who were all attended to by a butler, three footmen, a manservant, a housekeeper, a lady's maid, a cook and nine more maids.[9] H. Cosmo O. Bonsor was apparently a great public benefactor and was loved locally for his kindness and generosity, seen by many as a benign, grandfatherly like figure.[10]

With Watkin stepping down, peace broke out between the two Kent companies. In 1899 they joined forces to become the SECR. The full title was the 'South Eastern and Chatham Railway Joint Management Committee', being a committee comprising of selected directors from both the SER and LCDR boards with H. Cosmo O. Bonsor as chairman. James Staats Forbes, outwardly at least, took a back seat, but remained chairman of the LCDR board. The two companies agreed to share the receipts, the split being 59 per cent to the SER and 41 per cent to the LCDR. There was never to be a formal amalgamation despite attempts to do so as late as 1921. Both railways had been very unpopular with the public – on one occasion in Parliament the MP for Battersea commented that the amalgamation had brought together two 'undesirables' to produce the worst railway in the UK.

From its shaky beginnings the new SECR was at last slowly able to make some movement forward and many improvements were made to integrate the two networks. For example, where the LCDR route to Dover crossed above the SER route also to Dover, at St Mary Cray near Bickley, the SECR installed various interconnecting spurs between 1901 and 1902 to permit running between the old SER and LCDR lines. Victoria station then became the 'Gateway to the Continent'. Generally the boat trains departed from the former LCDR side of Victoria but ran via the SER route to the coast, making use of the new St Mary Cray connection. Little attempt was made to close duplicate facilities; Bonsor always maintained that the working agreement between the SER and the LCDR prohibited this.[11]

If the locomotives were anything to go by the new railway certainly looked good. The engines were painted a rich dark green with deep red below the running plate, were sumptuously lined out in black and gold and there was plenty of polished brass on the domes and other fittings. The crest of the 'SECR Joint Management Committee' was proudly displayed on the tenders – a splendidly elaborate piece of company heraldry. The grandeur of this embellishment belied the fact that both companies had been almost bankrupt. Their new passenger carriages were also splendid, painted a dark red as rich as the dark green used on the locomotives, and again fully lined out. Even the Pullman carriages that the SECR began to use on the continental boat trains in 1910 (the forerunner of the *Golden Arrow*) were painted in SECR red and not in the standard Pullman umber and cream.

Despite the Edwardian elegance of its locomotives the SECR was tainted with the poor reputation inherited from its two parent companies. Any regular commuter taking his daily journey on the SECR from his home in the suburbs into the City or West End was likely to endure spartan accommodation in ancient six-wheeled coaches likely to have been built in the 1860s, some even being built in the 1840s.[12] While these old carriages may have been unsatisfactory as far as the passengers were concerned, restorers at the Bluebell Railway in Sussex have commented on the high standard of workmanship and quality of the building materials that were used in the LCDR coaches despite the poverty of the owning company. Many of these ancient vehicles were much later to be combined and rebuilt on modern eight-wheel underframes by the Southern Railway to form 'new' electric trains. While he may

A RAILWAY UNDER STRESS • 77

not have been so candid in the early years, Bonsor in his final speech to shareholders stated that it was no exaggeration that the SECR had been very unpopular due to its poor service. Many complaints had been received, he said, both public and private, concerning the punctuality of the train services. He went on to admit that his railway had been in 1900 an easy target for comedians in the musical halls and the clowns performing in the pantomimes.[13]

Funding any desirable improvements to rolling stock, stations and track – and most definitely the massive cost of any electrification scheme – was problematic since the SECR, along with all railway companies, was having difficulty raising money through the usual financial markets. The railways' income was healthy, but their profitability was not. The earliest railways (such as the Stockton and Darlington) had been built to easily transfer bulk goods such as coal short distances from the mine to the canal or sea wharf. But the first tranche of public railways following the completion of the Manchester and Liverpool Railway had been built to carry more valuable commodities and wealthy passengers who earlier would have travelled on the expensive road coaches. In the latter part of the nineteenth century the railways had expanded, revealing the previously unsuspected desire of the public to travel and became in the post-Industrial Revolution era, an essential service used by all. Merchants using the railways to carry their goods saw the charges paid to the monopolistic railway companies as a tax on their businesses. This led to Parliament stepping in to control the rates charged, adding to rising pressure for better pay for their employees and periods of high fuel costs. Railway profits were diminished to an extent that whenever there was a share issue offer it would be undersubscribed, as potential investors were turning to other opportunities such as foreign railways for a better return on their capital.

Its public image and lack of capital were not the SECR's only problems. Just like the LSWR and the LBSCR, it also faced potential competition in the form of proposals to extend the newly electrified London County Council (LCC) tramways, and the proposed deep-level tube railways which were likely to take away the SECR's natural commuter base. At a meeting of the SECR Joint Management Committee held on 2 May 1902, there was a discussion as to how to combat these new threats. In what appears to have been, what we might now describe as a 'brainstorming' session, the directors on the committee discussed the possibility of obstructing these schemes by formal objections when the private members' bills to extend the electric tramways came before Parliament. This would have been problematic as the new trams and tubes were popular with the public and were perceived as the way to reduce the chronic traffic congestion in the streets of London. Moreover, given the unpopularity of the SECR the railway could hardly expect a sympathetic hearing. More positive, if impractical suggestions, based on the 'if you can't beat them join them' principle, were raised by several directors who were present. One of these was that tube railways could be constructed directly below the existing SECR railway lines. The adventurous and obviously expensive nature of this proposal clearly sent shivers through the backs of some of the directors and it was agreed the subject would have to be discussed with the parent companies.[14]

The sole representative of the LCDR present at this meeting stated that he would have to refer the matter 'to Mr Forbes'. It seems that James Staats Forbes, while apparently no longer having his hands directly on the levers of power, was still a force to be reckoned with. Forbes may have had a reputation for looking after his own interests before those of the companies he worked for, but he was an astute businessman and in his role as chairman of the MDR (the District line on today's London Underground) he had recently been involved in the controversial decision to abandon the railway electrification proposal using the impractical three-phase Ganz system on both the Metropolitan and Metropolitan District Railway in favour of the American financiers' preferred direct current system – now the London Underground standard.

At the next committee meeting it was obvious that further informal discussions had already taken place outside of the London Bridge boardroom. The SECR Joint Management Committee had arrived at the obvious conclusion – that rather than promote the SECR's own new tube railways, as had been rashly suggested earlier, the SECR should seek parliamentary approval to electrify its existing lines. By so doing it might successfully compete with the tramways and would complement the new tube services which would take passengers out from the London termini to their final destinations throughout the metropolis. At the following meeting held on 12 December 1902, the secretary reported that a formal notification had already been published in the *London Gazette* announcing that the SECR would be applying for powers to work its railways by electric power.[15]

As for the LSWR, the SER and LCDR had been established by Acts of Parliament, so parliamentary permission was required for many of its business activities. The regular railway bills were portmanteau devices encapsulating many proposals for which the companies required parliamentary approval. The South Eastern and London, Chatham and Dover Companies Bill of 1903 was such a Bill and was needed for several other reasons in any event. The most significant item on the SECR's wish-list in the 1902 Bill might have appeared to be electrification, but it was wrapped up with several other more mundane clauses appertaining to such matters as land purchases, agreements with other railways and pension arrangements for staff.[16] The notification in the *London Gazette* stated that the Bill proposed to permit the SECR to operate all or any part of its railway by electric power. Approval was sought to construct any necessary generating stations, possible locations being at Camberwell, New Brompton, Hither Green, Redhill and Ashford. In addition, Section 8 of the Bill would permit the SECR to build its generating station at any other undetermined location, should this be expedient.[17] The Bill also sought permission to install distribution cables and substations and to make any necessary alterations to stations associated with the electrification scheme. As well as requesting powers to generate its own electricity the SECR also sought parliamentary approval to enter into agreements with any third party for the supply of electricity.[18]

The publication of the 1902 Bill provoked a few formal objections which were all withdrawn, probably due to private reassurances being made. The most interesting came from the City of London which complained that an electrified service would create overcrowding at Ludgate Hill station.[19] This was an issue because Ludgate Hill on the LCDR City extension already had a reputation for being overcrowded and dangerous with its cramped booking hall and its narrow staircases leading up to very narrow platforms. This station seemed to epitomise the cheap minimalist approach to facilities provided in the Forbes era and it was unfortunate for the railway that this station was frequented by Fleet Street journalists with the obvious bad publicity implications. The SECR agreed to put proposals for improvements to the station in its next Bill and ultimately, in 1910, Ludgate Hill was rebuilt with the 'main line' island platform removed allowing more space for a wider, and much safer from the waiting passenger's point of view, single island platform for the local services only.[20] A committee from the House of Lords allocated only one day to review the unopposed 1902 Bill[21] which received Royal Assent on 11 August 1903. One important section of the resulting Act that was to become relevant later was clause 12: 'Provision as to the Use of Electric Power'. This clearly stated the requirement to protect government buildings, particularly scientific establishments, from the effects of any 'stray' current returning to the railway's substations through the general mass of the Earth.[22]

The question must be asked, given the suddenness of the decision and the lack of any technical report or recorded board-level discussion, as to how serious the SECR really was about electrification in 1902. Bonsor was later to be candid enough to admit that the SECR Joint Management Committee had never been seriously interested in electrification, and for a least a decade this appears to have been the case. It must therefore be assumed that the acquisition of powers was a purely defensive measure, deterring rival schemes proposed by others, such as tramway operators, and allowing the SECR to make 'legitimate' objections to any of these schemes that would affect the SECR's laudable and apparently reasonable intentions to electrify its own suburban lines. Given that the South Eastern and London, Chatham and Dover Companies Act of 1903 was necessary for all the other non-electrification reasons mentioned before, the addition of a few extra 'electrification' clauses would have cost the SECR little in additional professional fees. For this subterfuge to work the SECR needed to make its spurious intentions public and this was of course achieved using the Act of Parliament. It is probable that this deception was proposed by the wily James Staats Forbes since the lone LCDR representative on the management committee had said that he would have to 'consult Mr Forbes'. With the passing of time the SECR's alleged intentions would have become more suspect but time had been bought. Had the SECR been serious in its intentions and been able to proceed immediately with electrification, it might easily have taken the same approach as the LSWR and others and maintained that it did not require Parliament's permission, as the original railway Acts had not specified what form of propulsion was to be used.

The SECR did however embark on several large capital expenditure schemes that it deemed essential to improve its operations. Every department of the SECR was having to carry the burden of operating with old, worn out and obsolete equipment. As well as track alterations to make new connections at locations such as at St Mary Cray, the SECR embarked on a massive programme of track and bridge renewals that were essential due to the initial inadequate construction followed by years of neglect, particularly on the former LCDR routes. Henry Wainwright, who had worked for the SER, became the locomotive superintendent based at Ashford works. He set to updating and transforming the railway's locomotive stock from what had been a rather mediocre collection of rather asthmatic contraptions into a fleet of reliable and economic machines. This he achieved by both the rebuilding of existing locomotives and carrying out some new construction. New modern passenger coaching stock was built, including the three-coach sets known to all as the 'birdcage stock', so named due to the appearance of the ends of the brake vans which had raised roof sections with glazing for the guard's look-out. The coaches in these sets were 60 feet (18.23 metres) long and carried on two four-wheeled bogies. The former LCDR works at Longhedge, Battersea, was closed in 1910 to rationalise the locomotive works, with all new construction and repairs at the former SER establishment at Ashford. This move cost the SECR £85,000.[23]

Besides improvements at various stations such as Ludgate Hill, the SECR embarked on a magnificent scheme to turn the notoriously inadequate LCDR Admiralty Pier at Dover into the splendid Dover Marine station. In charge of this huge project was the SECR chief engineer, P.C. Tempest. Percy Crosland Tempest was born in 1860 in Leeds. His father, Charles Tempest, was a solicitor. Percy was educated at Leeds Grammar School and Leeds University and took employment with the LNWR. He was a member of the Institution of Civil Engineers. He joined the SECR as a permanent way engineer and was later promoted to chief engineer.[24] The Dover Marine project involved widening and extending the ancient Admiralty Pier which first entailed reclaiming 11.75 acres of land from the sea by filling in the harbour with chalk. The original Admiralty Pier had never been intended for train to ferry interchange before the LCDR arrived on the scene, and where once there had been a single windswept platform, four platforms were now to be provided with four steamer berths. The berths comprised of a chief landing stage, a second landing stage, an emergency landing stage and a space to berth a ship between sailings.[25] Just how much all this expenditure on track, rolling stock and stations affected the railway finances was later explained by Bonsor to the shareholders when he reminded them that the inevitable consequences of the expenditure needed to strengthen the track and bridges during the previous years had meant the payment of much lower dividends.[26]

Meanwhile, unsurprisingly, there had not been any progress on electrification, but in 1911 the SECR had a new general manager, Francis H. Dent. He was the son of Admiral C.B.C. Dent, the marine superintendent of the LNWR. With this good connection with the railway, Francis had joined the LNWR at the age of 18 in 1884. He had several managerial posts in the goods traffic part of the business before moving on to the LSWR and then the LCDR. He was goods manager and assistant general manager of the SECR prior to his promotion to general manager.[27] As general manager, he received an annual salary of £4,000 taking charge of 18,000 employees.[28] Dent seems to have been keen to make his presence felt and shake the stolid SECR directors out of their complacency, for, among other things, he wrote a report proposing the reorganisation of the locomotive, carriage and wagon department, and while this was not carried out, his criticisms appear to have led eventually to the kindly Cosmo Bonsor offering the locomotive superintendent, Henry Wainwright – who had been a good servant to the SECR over the years – early retirement on the grounds of ill health.[29] This led to the appointment of Richard Maunsell as chief mechanical engineer, a man who was eventually to hold the same post with the Southern Railway.

During his tenure as the SECR general manager, Francis Dent was an advocate for electrification, but he needed to convince committee members who had showed little enthusiasm. He produced another report, this time advocating electrification, and presented it to the committee at a meeting held on 24 January 1912 – which was the same month that Herbert Walker arrived at Waterloo to set in progress the LSWR electrification. While Dent was an enthusiastic supporter of railway electrification, he seems to have had to tread very gently when addressing the sceptical committee, stating that while he believed that steam traction was more economical for most of the SECR traffic, his main concern was the chronic congestion at Cannon Street station, where too much track occupation was being taken by light engine and empty carriage stock workings. The extensive use of trains formed of electric multiple unit stock would, of course, remove most of this unnecessary track usage.[30]

Dent had a plan. It is probable that he had already paid a surreptitious visit to Merz and McLellan's London office in Victoria Street to seek their advice before approaching the SECR committee. At the time Charles Merz was vigorously promoting 1,500V direct current overhead electrification – a system to be dismissed by Herbert Jones as being unsuitable for a densely trafficked suburban railway. After discussions with Merz, Dent formally approached the committee urging them to fund a report from Merz and McLellan to investigate the possibility of electrifying some of the SECR lines. He soothed the reluctant directors by assuring them that any report would not commit them to any expenditure other than that for the report.

The consultant, Dent advised, would not only report on costs of any proposed electrification, but in doing so decide as to which system of electrification to use. Looking to its near neighbours, the LBSCR and the London 'underground' railways, it could be seen that two quite different schemes of electrification were in use locally – the LBSCR's overhead 6,700V alternating current scheme and the 600V direct current electric railways. Dent offered a view on this, stating that he thought that the overhead system being used by the LBSCR would best meet the railway's needs. This was clearly a working and profitable electrification that the directors could witness for themselves if they looked outside the SECR offices towards the 'Brighton' side of London Bridge station. With some difficulty Dent managed to persuade the nervous directors to authorise expenditure of just over £1,000 to instruct Merz and McLellan to produce a report for the whole SECR system.

By requesting a 'whole system' solution Dent was directing Merz and McLellan towards the Brighton system for eventual main line electrification to the Channel ports. He would have been aware that the generating station owned by the LESC at Deptford was not only conveniently located close to the SECR lines, but the corporation also had the expertise and experience to provide a reliable 6,700V alternating current supply and had shown by its relationship with the LBSCR that it was prepared to invest in additional plant to expand its business. But the relationships between the SECR and both the LBSCR and the local electricity supply company were not cordial. The friction between the two railways was in part due to the difficulties they were having in sharing so much railway track in South London – a legacy of the parliamentary decision in 1836. As far as the LESC was concerned, for many years, the power company had run its cables attached to the walls and arches of the SER until the railway, rather unreasonably, deemed that they were a fire risk and insisted they be removed. It must be suspected that, while there is no record of any such instruction, before Charles Merz had been given the consultation work he may have been clearly told that cooperation with the LBSCR and the LESC was not an option.

Whatever Francis Dent had told the committee however, his early informal discussions with Charles Merz would have soon indicated that the only likely electrification option he would recommend would be the 1,500V direct current overhead wire system – the system of electrification that was soon to become almost universal across the British Empire. Merz believed, like all other engineers, that a voltage as low as 600 would be insufficient for main line use. As well as designing the NER's 1,500V system four years earlier he had prepared a report proposing a similar electrification of the Melbourne suburban lines in Australia. So impressed was the Victoria government that it requested a second report covering an even larger area and this report for Australian railway electrification was delivered in 1912 and was therefore contemporary with Merz's work for the SECR. The result of the antipodean proposal was to be the extensive Melbourne 1,500V direct current suburban electrification which is still in use today.[31]

Merz and McLellan duly set about their work, but it was a long time before their completed report was to be formally presented to the committee at a meeting held on 27 June 1913, when Chairman Bonsor briefly outlined the recommendations.[32] The scheme, to electrify effectively all the SECR suburban line and the main lines would have cost, according to Merz and McLellan, £5,599,000.[33] Unsurprisingly 1,500V direct current was proposed – but there was a snag. Much of the SECR lines were jointly operated with the LBSCR which was planning to extend its electrified network and, of course, the two overhead systems could not be used together. To overcome this difficulty, Charles Merz proposed a mixed installation comprising both 1,500V overhead and 1,500V ground-mounted conductor rails. A conductor rail voltage as high as 1,500V had not been used in Britain and Merz proposed a shielded conductor rail with below-rail contact with the trains' collecting shoes rather than top contact to minimise accidental contact by personnel.[34]

Despite Merz's publicly known vision for a standardised and widespread public electricity supply and his later condemnation of the railways for building large power stations sited in inappropriate locations, the recommendation in his report was that the SECR should build its own generating station at Angerstein's Wharf, Charlton – a site *not* included in the 1903 Act. Merz, like Herbert Jones, was aware that the many small, disparate local supplies around London would never be able to supply the railway's needs. The only company that was ever likely to be able to do so, and then probably only for the early stages without significant extension to its generating station, was the power company embedded in SECR territory with its Deptford power station: the LESC. Merz's integrity as an engineer cannot be questioned if he had been informally told not to approach the LESC; he would have had to follow his client's instructions.

Why did this relatively inexpensive report apparently take so long to materialise? Probably it was completed months previously and circulated to officers and some directors, but they were unable to come to a decision. The committee meeting on 27 June was to be just a formality in order to close down the proposal. Cosmo Bonsor, with his copy of the £1,000 report already in the chairman's wastepaper basket, while speaking at a shareholders' meeting a month later, admitted that electrification would have to wait as it was not the right time to ask the shareholders for powers to raise the large amount of capital that would be needed to electrify the SECR's lines with its many complex junctions and six London termini.[35] By claiming six London termini Bonsor was adding St Paul's as a starting and arriving destination to add to Victoria, Holborn Viaduct, Charing Cross, Cannon Street and London Bridge. He was later to make the truthful claim that in about 1912 suburban electrification had been considered and a report commissioned – but he blamed the outbreak of war in 1914 for the lack of progress, not financial restrictions and scheme complexity.[36]

The Great War however – despite being made the excuse for electrification inaction – can justifiably be described as the SECR's finest hour. As for the LSWR, the government in the form of Herbert Walker's REC took control of the SECR as prescribed by the Regulation of the Forces Act 1871. The SECR was soon in the fray as the German army swept through Belgium, hindered only briefly by the Belgian military who, besides putting up some resistance, did have the foresight to sabotage the railway system. The British may have supposedly entered the war to 'protect Belgium's neutrality', but the army, despite having well-prepared plans to mobilise using the railways, had no military plan prepared when the international situation quickly deteriorated, other than to send the small BEF across the Channel to line up with the French army along the Belgian border with France. The suddenness of the German invasion when it came meant the Allies were too slow off the mark to enter Belgium and offer substantial assistance. Consequently, as the British and French were unable to effectively confront the invaders in Belgium itself, this left the German forces a clear run to the Channel coast at Antwerp and Ostend.[37] With the Germans on their way, thousands of refugees fearing that atrocities would be committed by the invading forces were collected from the Belgian ports and delivered safely to the SECR's ferry terminals at Dover and Folkestone.[38] After the fall of Antwerp, on a single day, a maximum of 6,000 Belgian refugees arrived at Dover, and in one week 26,000 in total. Later, with the invaders approaching Ostend, 35,000 refugees were received. The total number passing through Folkestone was 120,000. All these destitute people were then taken on by train to various locations around the country. Besides the human cargo, under conditions of strict secrecy, the Belgian crown jewels, and other national treasures were transhipped into an SECR train for conveyance to the safety of the Bank of England's vaults in London.[39]

This was just an overture to the main event. Percy Tempest's new Dover Marine station was not finished at the beginning of the Great War, but construction was sufficiently advanced for the station to be made operational for military traffic. Due to the importance of the station to the war effort, the construction work was continued to complete the overall roof to protect from the elements the wounded that were to be transferred into the ambulance trains. With its harbour stations at Dover and Folkestone and its London termini the SECR took on the Herculean task of transporting men and materials to support the BEF in France. Ambulance trains were routed from Dover Marine to Charing Cross. These special trains were provided by several railways and the SECR contributed two which were converted from existing stock at Ashford works. During the war period (up to February 1919) over 4,000 hospital ships arrived at Dover carrying over a quarter of a million wounded men for transfer into nearly 8,000 ambulance trains. Mail trains also ran from Dover to Victoria, and over 10.5 million sacks of military mail were conveyed.[40] Leave trains ran between Dover and Victoria. The SECR ran special trains to provide over 6.5 million troop journeys. On arrival at Victoria the troops had to change their French francs to pounds sterling – the Bureau de Change at Victoria changed over £10 million during the war period. In total over 12 million troop journeys were carried by the railway.[41]

Freight, however, was the biggest challenge. The SECR, as for the LSWR, was predominately a passenger line; its goods traffic in 1913 was only 25 per cent of the total traffic. The advent of war brought a massive increase in all goods traffic, both locally generated and through trains from other parts of the country. This necessitated the installation of new sidings at many locations on the network. The SECR's western arm to Reading was used for traffic between the West Midlands and South Wales, the trains travelling via Guildford, Redhill and Tonbridge to reach either Dover or Folkestone. Traffic from the North and Midlands travelled from Willesden on the LNWR main line along the West London Railway to reach the original LCDR main line. Other traffic from the North and East was routed from King's Cross and the Metropolitan Railway through the LCDR City extension line to gain access to the SECR network. To handle goods traffic the SECR closed Cannon Street station to passengers during the middle of the working day, on Saturday afternoons and on Sundays.[42] As a result, the SECR goods traffic went up by 150 per cent and the average train load went up by 65 per cent.[43] Percy Tempest assisted the War Department by supervising the construction of a military port at Boulogne and Francis Dent served on the REC under Herbert Walker.

While the SECR struggled valiantly on with transport, the battlefield war was not going well. Stalemate had been reached, with both sides staring at each other across no man's land with periods of inactivity interspersed with ghastly battles and huge loss of life. Because of the poor military performance, a new man was to step forward to take over the reins of government. In 1916, Field Marshall Kitchener, the secretary of state for war, was killed when travelling to Russia, when the cruiser *Hampshire* struck a German mine off the Orkneys.[44] Kitchener, whose fiercely pointing finger and '*your country needs you*' slogan on the famous and often parodied recruitment poster, was ironically the only member of the high command to lose his own life in the Great War. Lloyd George was 'promoted' from the Ministry of Munitions to become secretary of state for war, but immediately discovered that the other politicians and military commanders had manoeuvred the out of favour Kitchener away from the main decision-making process – in fact the trip to Russia was a ruse to 'get him out of the way'. The war was not going well in 1916, and Parliament wanted a more robust attitude to ending the conflict. Lloyd George had hoped he would be able to improve the situation from his new role but, as things stood, he was powerless. A coup was required! Assisted by the Conservative Party and most Liberals, and by enlisting the backing of the then small Labour Party, Lloyd George formed his own coalition government. With the Labour Party with him, Lloyd George had secured the support of trade unions and the factory workers. He ousted from office his former Liberal colleague Asquith to become prime minister himself and he put together a streamlined Cabinet of only five members, leaving Asquith and many former Liberal Cabinet members to form the first official opposition since May 1915.[45] With Lloyd George at the helm the conduct of the war may have changed up a gear, but he was also in a prime position to set about preparing the social and economic changes that he wished to see ready for the time after the war. Lloyd George was a passionate reformer; he could now make real changes that would attempt to transform Great Britain's social and economic structures.

One of the changes that Lloyd George was actively considering was drastic reorganisation of the railways. Early in 1918 he received at 10 Downing Street a delegation of trade union officials, a group from whom he required full support in the war effort. The representatives of the nation's labour force requested that consideration be given to the state taking control of the nation's railways, canals and waterways. The delegation was well received, and the visitors were told that a committee – the Reconstruction Committee – had been set up to investigate the matter. Furthermore, Lloyd George told them that the government would not necessarily await the end of the war before making decisions or taking definite steps towards enacting any recommendations. The delegation left feeling satisfied with the information that they had been privileged to have received. News of the meeting reached the press and a spokesman from the government would not deny the substance of what the trade unionists had reported.[46] The *Railway Gazette*, while insisting that it never took a political standpoint, nonetheless offered a 'political' opinion, one surely echoed by all the railway directors across the land, that it was the railway companies and their owners that knew how to run the railways – not the politicians.[47]

The SECR directors were fully aware of a government committee set up by the Board of Trade to consider options for railway governance.[48] As well as politicians the committee had representatives from the railways including the recently knighted Sir Herbert Walker who was, of course, still the general manager of the LSWR and head of the REC; Sir Gilbert Claughton, who was a director of London Underground Railways; and Mr J. Potter, who was the general manager of the GWR. The SECR board members were dismayed that there was no official representative from their 'club', that is the Railway Companies Association, and this must have spawned the suspicion, probably justified, that they were being sidelined and that the wily Lloyd George, who had no mandate from the electors of the country to make changes to railway ownership or organisation, was nonetheless to instigate dramatic changes. The Reconstruction Committee was also considering railway electrification and in November 1918 a subcommittee dealing with electrical services asked the railways to provide details of any of their railway electrification proposals.[49] Furthermore, in June, a Board of Trade departmental committee had publicised the view that any future railway electrification should be dealt with as part of an overall national plan and that any power stations must be situated with due regard to both civil and military requirements.[50]

Lloyd George was a man in a hurry who wanted to achieve change. He was a man of action, not prevarication, and wanted to maintain the benefits of cooperation that had been achieved by national agencies during the Great War, particularly by the Ministry of Munitions. He also wanted to wrest away the stranglehold on industrial power held by the aristocracy and super-rich by, among other things, doing what was unthinkable to company directors in 1918: ensuring that there would be workers' representatives on the railway management boards. What is surprising is that this war leader was giving so much attention to what should happen after the war when the military situation was so very grave. Lloyd George seemed to be devoting as much energy to winning the coming peace as he was to the war itself.

Francis Dent too, despite serving as a lieutenant colonel in the Engineer and Railway Staff Corps and simultaneously carrying out his duties as a member of the REC, also seems to have been considering what to do 'after the war'. Both he and William Forbes of the LBSCR, in respect of their service with the territorial reserve forces during the war, received knighthoods for their efforts. To celebrate both men's recognition by the king, Sir Herbert Walker organised a celebratory dinner at the Euston Hotel to honour both Sir Francis and Sir William.[51] Such conviviality implies a good personal relationship between Sir Herbert and Sir Francis, both of whom were then working under the stressful conditions of not only being general managers of their respective railway companies, but also having to carry out the work of the REC during a time of national crisis. The controversial questions around railway electrification would have been well understood and discussed with Sir Herbert, due to his practical experience of electrification with the LSWR. Sir Francis would have expressed to Sir Herbert his desire to electrify much of the SECR and they would also have discussed the SECR's electrification proposals, or perhaps more accurately, the lack of any plans for his railway. Sir Herbert was very keen to explain the success of his railway's electrification and thus Sir Francis heard at first-hand that the prediction of increased passenger numbers with the associated increase in revenue had been realised. Sir Herbert's views too were beginning to shift ever so slightly. Previously he had been in the 'electric trains are more expensive to run than steam, but revenue is increased' camp along with other electrification protagonists such as Sir John Aspinall, but after the rising costs of locomotive coal and the government imposed increase in wages for the two men on the footplate of a steam locomotive, he would now argue that the cost comparison between steam and electric was being turned around and was about to be reversed.[52]

With the Americans now joining the war, the prolonged stalemate would surely not last much longer. The question of railway electrification was at last to be seriously considered by the SECR Joint Management Committee, bolstered by the glowing recommendations passed on by Sir Francis Dent from Sir Herbert Walker. Also, the severe congestion between London Bridge and Charing Cross was troubling the directors, and this could only be solved by widening the tracks and bridges and expanding the main station – or by electrification. The SECR could see that the results from the LSWR and others predicted a massive increase in receipts, so there was now a strong and compelling business case to proceed with electrification on busy suburban lines.[53] Whatever was decided the SECR still had to finance this huge capital project, but now electrification was promising reduced costs and a better return on capital. The railways may have initially struck a bad financial arrangement with the government, but nonetheless some compensation was anticipated at the end of the war as the railways had made large wartime profits for the government due to the massive increase in traffic. Unfortunately, the railways were due to be disappointed in this respect.[54]

So, despite the war still raging within earshot of the SECR's Dover Marine station, the Joint Management Committee, guided by the general manager, began at last to seriously consider the electrification of the railway. There was the question as to whether the SECR only wanted to electrify the London suburban routes or go for full main line electrification. 'Main line' in SECR terms meant continental traffic to the Channel ports, and potentially France through the Channel tunnel, for there was at the time the strong possibility that a tunnel would soon be built. In 1913, a House of Commons Channel Tunnel Committee[55] was set up whose chairman, Sir Arthur Fell MP, was in 1918 busy giving lectures describing all the technical details and outlining the commercial advantages of the proposed scheme.[56] Previously, in November 1916, the SECR had received a letter from 'The Channel Tunnel Co. Ltd' which was a near neighbour of the SECR at 84 Tooley Street, London Bridge. In the letter the company formally requested the services of Percy Tempest to act as its engineer. Tempest was to work with Sir Douglas Fox and Partners,[57] a firm of consulting engineers that had previously been involved in the construction of the early tube railways in London so would be qualified to take on the task of boring beneath the Channel. The management committee agreed to this request and Tempest covered both posts concurrently.

To resolve the complex technical issues, Sir Francis Dent proposed at a Joint Management Committee meeting held on 12 December 1917 that the SECR should follow the LSWR example and appoint its own electrical engineer to investigate and report on the various electrification options. As an SECR employee, not a consultant, this officer would provide an unbiased recommendation as to how the SECR should electrify its railway.[58] The consequence of this proposal was that Alfred Raworth was to be appointed as the SECR electrical engineer at a salary of £600 per year.[59]

As for Alfred's original appointment at the LSWR, the exact sequence of events leading to his selection are shrouded in confidentiality. It is possible however to hypothesise. To overcome the financial woes of the railways during the early years of the twentieth century in the years leading up to the Great War there were attempts to amalgamate the larger companies into even bigger units to reduce wasteful competition and overall operating costs. Sir Herbert Walker and Sir Francis Dent must have foreseen the possibility of the main railways south of the Thames joining forces. Given Sir Herbert's enthusiasm for electrification he would have wished that other railways that might join the LSWR to proceed along similar lines to his own. So, when the SECR was seriously contemplating electrification and needing technical support, who better to assist them than *his* engineer, Alfred Raworth, a man who might be expected to propose something every bit as good as, and compatible with, the successful LSWR electrification?

But Alfred Raworth was now a serving naval officer in wartime, so it was necessary to make a release request to the Admiralty. This was probably made by Sir Herbert as head of the REC on the SECR's behalf. This was a relatively easy task for Sir Herbert because the First Lord of the Admiralty – a political not a service appointment – was at the time a seconded railwayman, the deputy general manager of the NER, Eric Geddes, who now rejoiced in the full title of Sir Eric Campbell Geddes having also received a knighthood for his wartime efforts.[60]

Geddes was not a career politician having been brought into the wartime administration in 1915 as one of many who Lloyd George perceived as having the qualities of vigour, courage and the ability to improvise.[61] He was a successful administrator, but he had to admit that he did not like politics.[62] He had joined the NER in 1904 having been in railway employment abroad[63] and rose to deputy general manager in 1911.[64] That he achieved this post after only seven years is a striking acknowledgement of his management skills.[65] In 1915 when Lloyd George took charge of the Ministry of Munitions, he seconded Geddes as his deputy director general of supply and later he was to be the successful director general of military railways in France. He subsequently entered politics as First Lord of the Admiralty in 1917 when, sponsored by the Unionist (Conservative) party, he was elected unopposed at a Cambridge byelection. As an NER man Geddes would have been sympathetic to electrification proposals and, since at the time Lieutenant Raworth was seconded to the Ministry of Munitions, Geddes would have been able to pull strings with the Ministry, having first served there himself when brought into government by Lloyd George.

The importation of an outsider must have ruffled a few well-established feathers within the SECR. It was made clear that Raworth was to report directly to Percy Tempest. The existing head of the electrical branch of the engineering department, Mr G.F. Glover, despite being effectively demoted, was given a pay increase from £375 to £425 commencing on the date of Raworth's appointment which was of course some compensation for him. Glover would have been responsible for the small SECR-owned power stations such as at Slade Green and Queensborough, which were then used for internal electricity requirements that might otherwise have been provided from the local public mains supply had this been available at the time.

It is also an enormous compliment to Alfred Raworth who at 36 years old was still a young man. The recommendation from Sir Herbert Walker was enough for the Joint Management Committee to accept Francis Dent's request to appoint him as a salaried officer, rather than, yet again, involving consultants or making a temporary appointment. Raworth's arrival at the SECR was delayed until the end of February 1918, as he was not released from the RNAS until 21 February. When you are appointed in 1918 at the princely salary of £600 a year on the recommendation of arguably the most senior railwayman in the land, you would not be expected to 'find a desk somewhere'; the SECR rented the first-floor rooms of 19 Railway Approach, London Bridge, as temporary offices for Raworth and his staff on a three-year lease at £60 per year.[66]

It might have been expected that Raworth first dusted off the Merz and McLellan report from 1912 as a starting point for his endeavours. But, of course, he and Jones had rejected the 1,500V system in 1912, and when it was much later put to him by no less a person than Sir John Snell, the chairman of the Electricity Commissioners, that the Merz and McLellan report had suggested 1,500V as the traction voltage for the SECR, he tersely replied: 'I believe so.'[67]

If Raworth and Tempest wanted to put forward any scheme utilising the very latest technology, they would expect to have to look either to the USA or to Germany.[68] But Germany was still at war with Great Britain and was to be a defeated nation in 1919 when the report was completed. Meanwhile in the USA there had been considerable development in railway electrification with the use not only of high voltage direct current, but also alternating current both single-phase and three-phase. Improvements in electric motors and other plant had led to improvements in overall efficiency.[69] So, as was the case for many others before him, it was to America that Raworth decided to look to obtain his knowledge of the very latest expertise for the electric railway he was envisaging. Consequently, on 9 October 1918 the Joint Management Committee approved Sir Francis' request that Raworth should cross the Atlantic to meet and have discussions with railroad managers and electrical engineers in the USA and then compile a report for the directors.[70]

But Raworth had to wait a while before making his journey for the war was still raging on. Earlier in the year, during March and April 1918 the German forces had counter-attacked gaining more ground than at any other time during the war. Paris was under serious threat from the German advance and being shelled by a massive gun, 'Big Bertha', from 71 miles away. Lloyd George mobilised all his available troops and requested more assistance from the Americans. In early 1918, when Lloyd George was entertaining the trade unions in Downing Street, the capital was being subjected to Zeppelin air raids, creating panic and hysteria in the population. The allies counter-attacked in September and by November the Germans had been virtually pushed out of France. The end came however, not so much from a military victory but from the German civilian politicians ruling that enough was enough despite the German military's over-optimistic predictions that they could still win the war.[71] The circumstances around this controversial capitulation by the Germans were used later by Adolf Hitler to drum up support with his Nazi propaganda – so history rolled on …

With the war over Cosmo Bonsor received a letter from General Haig thanking the railways in general but singling out the SECR for praise. He wrote that the bulk of the army had passed through the SECR system and that as they were nearest to France and Belgium the railway had been subjected to hostile attacks, but despite this the traffic for the armies in France had continued uninterrupted. This he concluded reflected great credit on the SECR and its employees. Some 5,222 SECR staff joined the forces and of these 556 lost their lives.[72] A war memorial to the SECR fallen was built, appropriately, at Tempest's Dover Marine station, consisting of a sculpture by W.C.H. King.

The war had taken its toll on the railway, and now the war weary SECR was in a new, more austere era. The progress since 1899 had been reversed as the assets had been worked to the extreme with less skilled staff. Gone was the previous opulence of the Edwardian period. The post-war age saw the brass domes on locomotives painted over – some locomotives were even painted 'wartime' grey. Lining on the rolling stock was simplified and the beautiful deep red that had previously adorned the carriages was replaced by a rather drab and subdued colour: Wellington brown.

CHAPTER 8
ALFRED RAWORTH'S ELECTRIC RAILWAY

Following the Armistice in November 1918, plans were made for Alfred Raworth to travel to the USA. On the 18 November £350 was made available to meet his expenses while there and the SECR arranged life insurance for him to the value of £11,000. So, on 30 November 1918[1] – only 19 days after the cessation of hostilities – Raworth, accompanied by neither family nor SECR colleagues, set off from Liverpool on the White Star Dominion Line ship, *Adriatic*. This vessel was not one of the line's largest ships (at just over 35,000 tons) and had been built for the Liverpool to New York route by Harland & Wolfe of Belfast in 1907. A winter crossing of the Atlantic was potentially a very bleak prospect in 1918, but naturally the SECR bought him a first-class ticket. The *Adriatic* docked in New York on 11 December. The immigration officials duly recorded Raworth's appearance noting that he was a relatively small man being only five feet seven inches tall with brown hair and grey eyes. They noted his facial scars and observed that he had a sallow complexion – that is, he was pale, yellow and unhealthy looking. Perhaps he was suffering from the very worst effects of an Atlantic crossing in the depths of winter – although this description was not applied to any of his fellow travellers.[2] After completing all the immigration formalities, Raworth proceeded to 30 Church Street, Lower Manhattan, the offices of the General Electric Company where Mr A.K. Taylor was to be his contact and host.[3]

There is no record of what, or who, Raworth saw while in America. Herbert Jones had made the same journey in 1912 and returned with a design for an electric railway that might be described as a 'clone' of the electric suburban railways serving New York and the surrounding area, so Raworth would learn little that was new by retracing Jones' steps. But, given what he was later to recommend both in his report to the SECR and in later reports for the Southern Railway, two schemes among the many he saw clearly impressed him.

General Electric would not have failed to make Raworth aware of the Chicago, Milwaukee and St Paul Railroad's 3,000V direct current overhead electrified lines running through the Rocky Mountains. By 1919 this railway was operating over 600 miles of single-track electrified railway on its steeply graded main line. The railway took a 100,000V supply from the Montana Power Company which owned hydroelectric power plants in the mountains. At the railway-owned substations, the supply was transformed down to 2,300V to drive the 3,000 or 2,000kW alternating current motors coupled to pairs of generators each with an output at 1,500V direct current. Connecting the outputs from pairs of generators in series produced the 3,000V traction supply.[4] This railroad, predominantly a freight line, operated massive 3,440 horsepower (2,500kW) locomotives. Raworth's reluctance to use overhead wires was reinforced because even at the high voltage of 3,000V, to be able to carry the high current drawn by these huge locomotives a pair of overhead conductors were often installed in place of the usual single contact wire.[5] The use of two 1,500V supplies to achieve 3,000V would however have been of interest to Raworth.

A scheme that also impressed Raworth was the rebuilding of Grand Central Station in New York which had been completed only six years before in 1913. This had been prompted by the city having effectively banned steam locomotives from the central area by introducing smoke abatement legislation, thus all the railways in Manhattan had been electrified using either third-rail direct current or higher voltage alternating current with overhead wires.[6] The work at Grand Central had been carried out under challenging conditions without any interruption to train services. The rebuilding provided for two levels of platforms enabling suburban and long-distance services to be segregated. Electric traction was key to this innovative design since the use of steam trains with the consequent steam and smoke emissions would not have been possible in the subterranean depths of the station. The former 'steam' station area was covered over to create Park Avenue and 'air rights' leases issued allowing more than 25 buildings to be constructed above the railway.[7]

Raworth returned home via Southampton, arriving on 8 March 1919 aboard the White Star Dominion line

A Chicago, Milwaukee and St Paul and 3,000V locomotive hauling a passenger train. (Author's Collection)

The main concourse area at New York Central station in November 2018. The station had been allowed to deteriorate over decades of use but has now been restored to its former glory as had been seen by Alfred Raworth in 1919. (Author)

ALFRED RAWORTH'S ELECTRIC RAILWAY • 91

ship *Olympic*,[8] which at over 46,000 tons was a larger ship than the *Adriatic* and at that time a regular on the New York–Cherbourg–Southampton run and the sister ship to the ill-fated *Titanic*. Previously he had telegraphed Percy Tempest requesting additional spending money and an extra £100 was made available to him.

Raworth's report, titled *SECR Management Committee, Electrical Engineer's Report on Electrification*[9] was completed six months later in September 1919 and it proclaimed that it was from the Electrical Engineer's Office, London Bridge Station, SE1. Raworth now had enough status to be able to sign the report 'Alfred Raworth', without any 'on behalf of' his superior, the SECR chief engineer, Percy Tempest. In his all-encompassing report Raworth set out what must be interpreted as his own personal vision for an electric railway. In the writing of this, his first report, he had his 'Brunel' moment. As Brunel had done when given the task of designing the London to Bristol railway, Raworth ignored all precedent and designed his electric railway from first principles and proposed something that was completely different to what had been done before. This, his patron Sir Herbert Walker – the champion of standardisation – had not anticipated.

To the shock and surprise of many engineers, Raworth proposed a 3,000V direct current scheme using two conductor rails. This was to be achieved not by connecting two 1,500V sources in series, but by having a positive polarity 1,500Volt conductor rail and a negative polarity 1,500Volt conductor rail. (See plate 3)

The traction current flow would be from the substation out along the 1,500V positive conductor rail to pickup shoes on a power bogie having two 750V, 200 horsepower motors connected in series. The return current from these motors would pass through the wheels onto the track running rails as for the conventional earth-return system. The three-car trains would however have a second 'negative' bogie at the other end of the three-car set which would collect the current off the running rails, use it to drive the motors, and then through pickup shoes return it to the substation along the 1,500V negative conductor rail.

As described, contemporary electrical engineers would immediately recognise such an arrangement as a *three-wire* system. This was commonly used for direct current public electricity supplies. Typically, there would be a positive conductor charged at around 200V positive and a negative conductor charged at minus 200V. The third wire would be termed the neutral. Electricity customers' installations would be connected to either positive and neutral or negative and neutral. Often the buried mains cables comprised of three lead-covered insulated wires (positive – neutral – negative) laid within wooden troughs filled with pitch. Just as for a three-phase supply, if each phase (or live wire) carries the same current there will be no current flowing back to the source in the neutral wire. Anything close to such a balance is difficult to obtain on a public electricity supply with individual consumers, but the SECR could build and operate trains with a perfect balance of positive and negative motors.

By using such a system, the overall resistance of the whole electrical circuit to and from the substation would be greatly reduced. But one of the principle advantages Raworth claimed for his design was that it eliminated the earth return feature with its attendant risk of stray currents affecting any third party's property. In doing so he asserted that his design fully met the conditions prescribed by the SECR's 1902 Act of Parliament requiring the railway to take measures to protect third parties, particularly government scientific establishments, from the effects of stray earth return currents.

In his report (for which the intended audience was the Joint Management Committee, not engineering colleagues) Raworth is not explicit as to the positioning of the two conductor rails on the track, stating only that they would be protected by timbers to minimise accidental contact. The, at first, obvious solution would be for the two conductor rails to be placed outside of each running rail, but this would create two enormous problems. Firstly the 'gapping' issue would have to be resolved. The LSWR system overcame this by arranging for the conductor rail to swap over to the other side of the track when the route passed over complex pointwork. Allowing the two rails to overlap slightly enabled the pickup shoes on either side of the power bogie to maintain contact during the transition. To connect the sections of conductor rails, short pieces of buried cable were used. The other difficulty that would need to be addressed when positive and negative rails were to be used was the question of polarity. If the positive rails were, say, always on the left and the negative rails always on the right, as the train left London what would happen at 'out and back' loops or at triangular junctions? A train would encounter a positive rail where a negative had been and vice versa. Anybody who has ever wired up a 'two-rail' model railway will understand this issue.

The SECR records are also very vague as to what was planned but the only practical solution (confirmed by referring to Ministry of Transport files)[10] would be to have one side conductor (which would change from side to side as necessary) to carry the positive supply, while a centre conductor would carry the negative supply. This would potentially involve frequent gapping of the negative rail – but the train could momentarily continue using 'earth return' on the positive rail. Stray currents from a negative-only earth return system would have had to be avoided as any such currents would flow outwards from any buried pipework towards the track, not from track to earthed pipework. Severe corrosion would result as the buried pipes would effectively become a positive 'anode' and as such subject to dissipation due to the electrolytic action. In America, several tramways had used negatively charged trolley wires and this had resulted in a massive detrimental impact on other parties' buried equipment.

With this three-wire arrangement the third wire – the running rails – would only ever carry current back to the substation momentarily while the bogie at either end of the set became temporarily 'gapped' (almost always the centre track negative rail) – an inevitable occurrence upon encountering any of the short breaks determined by track pointwork or at a boarded foot or vehicle crossing.

Most commentators have assumed that a side-contact arrangement such as used by the LYR would have had to be installed for greater safety when using such a high voltage. It can be argued however that, given the wording of his report, Raworth originally considered a top contact rail with simple timber side shields such as the LSWR were using at safety critical locations, such as at stations. This would avoid the added complexity resulting from the need for the shields around the central conductor rail to be bi-directional – difficult to imagine how this could be achieved, so duplicate pickup equipment would have to be provided to supply the negative power bogies.

Technical details are tantalisingly scarce in Raworth's report. He describes '1,500V' supplies from rotary convertors. He may have envisaged more than one machine connected in series to produce a 1,500V output since Merz and McLellan had used pairs of 750V machines in series as a 'set' for their first 1,500V electrification scheme in India.[11] Given Merz's preference for 50 cycles per second systems, he would have deemed this necessary as contemporary opinion was that to avoid flashovers on the commutators, the maximum frequency that could be used at 1,500V was 33 cycles.[12] Therefore at only 25 cycles per second Raworth probably specified single 1,500V converters. The 1,500 plus 1,500V positive and negative supply would have been obtained by connecting in series two 1,500V rotary convertor sets and providing an earthed neutral connection between them.

While no national or world standard had yet been agreed, a consensus among operators and manufacturers, led by Charles Merz, was that 1,500V direct current was the best system for main line electrification – it was being used by the NER and was to be introduced into France within the next decade. The SECR would potentially be able to source 1,500V equipment that was becoming available, since manufacturers were likely to have at least preliminary designs for switchgear, etc. What Raworth was proposing was a system that matched the third-rail system for simplicity of installation and was also suitable for main line use without the complexity of overhead catenary structures entailing significant alterations to bridges, tunnels, signal locations and station platform canopies.

Raworth's proposal had one distinct advantage over the LSWR system apart from eliminating the stray current problem. The use of a much higher voltage would have meant that more energy could be carried over a greater distance, thereby reducing the number of trackside substations required. He proposed that to maintain the required voltage only four substations would be required for the whole of the SECR suburban area network. The proposal to use conductor rails energised at 1,500V was controversial; the highest voltage to date had been installed in 1913 when Sir John Aspinall's LYR had electrified the Manchester Victoria to Bury line using a single 1,200V direct current conductor.[13] As part of Raworth's investigation, enquiries were made with the LYR to find out if this relatively high voltage system had caused any danger to railway staff working on the track, or if they had encountered any additional maintenance problems.[14] Given that this was not deemed to be a problem, Raworth felt that he could increase the conductor rail voltage up to what was likely to become the industry standard of 1,500V direct current, and by using two conductor rails achieve his 3,000V system.

Raworth in his report proposed three stages of electrification. Stage One would be the inner suburban lines of the SECR; Stage Two, more outer suburban destinations; and Stage Three would be to increase the power supply and to electrify sidings for goods traffic to enable electric locomotives to haul both the goods trains and the longer distance passenger trains that would pass through the electrified area. (See plate 4)

For Stage One Raworth proposed that the principal SECR London stations were to receive the conductor rails: Victoria, Charing Cross, Cannon Street and London Bridge – but not Holborn Viaduct as the first stage trains would terminate at St Paul's (now named Blackfriars).[15] The electrified routes from Charing Cross and Cannon Street were to be as far as Orpington on the SER's main line to Tonbridge and the branch to Bromley North and the SER's mid-Kent line from Lewisham to Addiscombe Road and Hayes. All three former SER routes to Dartford were included. From Victoria, the former LCDR main line and the Catford loop were to be electrified, but at Bickley electric trains would transfer to the SER main line for onward running to Orpington. The former LCDR branches to Crystal Palace High Level and to Greenwich Park were included, despite both being closed at the time due to wartime economy measures. The total length of running lines to be electrified at this stage was to be 198.35 miles. In addition, 11.65 miles of sidings were to be electrified for the use of the electric trains for overnight berthing and for layover between services.

Stage Two proposed electrifying the original SER line jointly operated with the LBSCR from London Bridge through Purley and Croydon to Redhill and then eastwards through Edenbridge to Tonbridge. The former SER branch lines from Purley to Tattenham Corner and Caterham were also to be included. On the SER's adventurous westward route from Redhill to Reading electrification was proposed but only as far as Dorking.

The Stage One electrification to Orpington was to be extended along the SER main line to Tonbridge and this section was to include the branch to Westerham. Raworth's inclusion of the delightful Westerham branch is interesting for it survived until 1961, remaining to the very end of steam operated on the 'pull-push' system using ancient SECR H class locomotives. While traffic was light on this rural byway, with Raworth's proposals requiring overall so few substations, all that would have been needed to electrify this branch would have been the installation of the conductor rails and track bonding. Sadly, the route of this line has now disappeared beneath the M25 motorway.

The former LCDR branch linking Swanley and Sevenoaks was to also be electrified as part of Stage Two. The eastward progress of the conductor rails was to be from St Mary Cray junction to Swanley, Chatham and Gillingham and, on the former SER line from Dartford, through Gravesend to Strood and hence onto the former East Kent Line that had become the LCDR to reach Chatham. The former LCDR branch line from Southfleet to Gravesend West was also included. With the completion of Stage Two, electric trains were to run from Holborn Viaduct and four tracks were to be electrified as far as Herne Hill. A further running line was to be electrified to make three in total between Victoria and Brixton as only two would have been electrified in the first stage. Similarly, between Cannon Street and Borough Market Junction, Charing Cross and Borough Market Junction and Hither Green to Orpington an additional two electrified tracks were to be installed to make a total of four at each of these locations. The total length of running lines to be electrified in this second stage was 320.34 miles with an additional 14.66 miles of sidings for berthing and for the layover of trains between services.

Stage Three did not require any additional running line electrification. Raworth proposed increasing generation and substation capacity to enable goods and passenger trains travelling through the electrified area to be hauled by electric locomotives. Sidings would be electrified using overhead line equipment rather than conductor rails, an obvious requirement to protect staff working in the goods and marshalling yards, but he did not provide any details. These overhead wires might possibly have been energised by the positive 1,500V supply to avoid the corrosion problems mentioned before, or alternatively using two overhead wires as later used by the road trolley bus systems. Electric locomotives would be equipped with additional overhead pickup gear to be able to work in these sidings. Raworth estimated that 36 passenger and 58 goods locomotives would be required at Stage Three. Given the later engineering problems concerning electric locomotives using conductor rails, it is disappointing that he gave no hint as to his intentions for the design of these locomotives. Based on the ratio of sidings lengths to main line lengths on the railway in 1919, Raworth's estimate was that 26.31 miles of overhead electrification equipment would be required.[16]

Part of Raworth's case for his three-wire system was that while only suburban and outer-suburban electrification was being proposed at this stage, his design could become a springboard for main line electrification to the Kent coast – and even through the Channel tunnel should the link to France ever be built. It is tempting to speculate that Raworth's system might have been used for the actual Channel Tunnel. Giving evidence to a parliamentary committee on 16 October 1918, Sir Francis Dent had stated that he preferred a tunnel to the alternative schemes for train ferries as he believed the tunnel would be cheaper to operate and would generate more traffic. He went on to say that, based on advice from Percy Tempest, the tunnel would take eight years to build and would cost £25 million. It would (and it is hard to imagine steam operating in the tunnel) be electrically operated. The cost of widening bridges, etc. to accommodate the wider and higher continental trains between Folkestone and London was calculated to be a further £10 million.[17] Given that the Raworth plan required so few substations, and his Stage Two proposal would have required feeder cables over 20 miles long between Lewisham and Strood and between Lewisham and Redhill, a substation at either end of a 30-mile-long Channel tunnel linked by a pair of 33,000V cables placed within the (most likely) twin bores, would easily provide a secure supply without the need for any submarine substations. In addition, using the conductor rails rather than the space-hungry overhead lines that were to be used later for the actual Channel Tunnel, the bore could have been much smaller. (See plate 5)

An important element of Raworth's plan was that the SECR would construct its own generating station with a capability of being extended to meet the requirements of any future electrification beyond the suburban area. The site chosen was at Angerstein's Wharf, the same location that had been recommended by Merz and McLellan in 1913. Raworth made it clear in his report that all the costs for his scheme had been calculated based on the assumption that the SECR generating station would be built at Angerstein's Wharf but added the caveat that there was a looming possibility that the SECR might not get permission to build the station due to the provisions of a parliamentary Bill which at that time was passing through the committee stage. Raworth advised in his report however that should the new legislation prohibit the SECR building its own power station, the new electricity commissioners might sanction a 'public' 'super-station' at Angerstein's Wharf three times the size of the MDR's Lots Road to feed local industries in the Wharf locality who used electricity at 25 cycles and could also meet the electrification needs of the SECR, LBSCR and the Great Eastern Railway (GER).

Raworth's planned Angerstein's Wharf generating station would, if progressed through to his third stage proposal, have a potential output of 75,000kW – three times that of the LSWR's Durnsford Road station. Even for the first stage Raworth proposed to install four 12,250kW generators eclipsing the LSWR's station output. At the second stage another 12,250kW generator would be installed and for Stage Three a sixth 12,250kW generator. The electricity would be generated with a frequency of 25 cycles per second to allow the use of rotary converters, but Raworth boldly stated that railway supplies had to be at that frequency. This caused one recipient of the professionally printed, hardback report to neatly pencil in the margin 'why?'

For the transmission of the electricity from the generating station to the remote substations, Raworth proposed that, unlike the LSWR which used 11,000V supplies, the SECR should install 33,000V cables. For the first stage the distance between the generating station and the first substation was small, but for Stage Two and Stage Three – and looking further ahead into the future – the higher voltage was necessary due to the distances involved. Given the Durnsford Road generating station was only four years old at the time and that Raworth had been involved with its design and construction, it can safely be assumed that the SECR station design would have been similar. The main difference was that while the electricity would be generated at 11,000V each generator set would have a step-up transformer enabling direct connection to the 33,000V switchgear.

For transmission to the substations three-phase armoured 33,000V cables were proposed. Raworth dismissed any suggestion of using overhead lines, for while he conceded that bare copper wires at such a high voltage would be cheaper to install, they would be vulnerable to damage and short lengths of cable would frequently have to be inserted when bridges and tunnels were encountered. For example, the four cables that were proposed for installation between

Angerstein's Wharf and the first substation at Lewisham would be laid through Blackheath Tunnel. From Lewisham, duplicate 33,000V cables would be installed to the three outlying substations at Redhill, Tonbridge and Rochester Bridge. As for the LSWR scheme, these cables would be laid in pairs to allow for either to be switched out for maintenance or repair. The substations would have followed the same design principles as the LSWR substations with 33,000V alternating current switchgear, step-down transformers, 1,500V plus 1,500V rotary convertors and 1,500V direct current switchgear for both the positive and negative conductor rails.

In his report Raworth argued that to make the fullest use of electrification, *new* trains must be built – the SECR must not resort to using existing carriages refurbished with electric motors and control equipment as his former employer the LSWR had done. He argued that the SECR should build the highest capacity carriages possible within the existing track constraints. Current stock was too narrow he maintained, so he proposed carriages with a maximum width (at 'the waist' as he described it) of 9 feet (2.7 metres). There was a pull-out from the report revealing a beautiful one-quarter of an inch to the foot drawing depicting the 'Raworth' three-car multiple unit set. His description of these trains was that they were motor trains for suburban services. The carriages are shown to be of the compartment type – no side corridors or lavatories. The width specified would permit six third class passengers abreast in a compartment just short of six feet long (that is 12 in each compartment). First class passengers could luxuriate four abreast in a nine-foot-long compartment (that is eight in each compartment). At the time, there was still a second class on the SECR but, as was the case on the LSWR, Raworth assumed that it was about to be abolished.

Within the three-car sets each coach was to be 57 feet (17 metres) long, slightly shorter than the SECR's new 'birdcage' stock as each of these coaches was 60 feet (18 metres) long. Due to tight clearances to bridges and platform edges on many curves on the former LCDR lines, Raworth was having to trade off length for additional width. Later, 9-foot (2.7 metre) wide electric trains were to be built with a length of 62 feet (19 metres), but their use on certain lines was restricted until work was carried out to ease awkward tight clearances at many locations. The outer vehicles of each of Raworth's three-car sets were to be powered and to have a small driving and guard's compartment at the outer ends and eight third class compartments. Sandwiched in between the powered cars a trailer coach with eight first class compartments was proposed. The whole three-car set would have provided 192 third class seats and 64 first class seats.

The obvious desirability of using multiple unit trains for suburban services was emphasised by Raworth. One additional advantage, pertinent to the three-wire electrification proposal, was that to avoid gapping, pickup shoes could be placed on any of the six bogies, powered or unpowered. This was particularly important for the central, negative rail pickups. Cabling within the set could link together several pickups to, hopefully, bridge any gap anticipated on the system.

Within the report were comprehensive details of the proposed passenger services. Raworth reiterated the now proven expectation that the more frequent services made economically viable with electric trains would always result in large increases in traffic on lines serving stations which could be reached with a less than 50-minute journey time.

The report might have been Raworth's, but to produce such a detailed proposal he would have had to consult the SECR's chief operating officer, Edwin Charles Cox. In later years, Cox, Raworth and Sir Herbert Walker were to work harmoniously together to develop what was to become known as the Southern Electric. In a manner well within the

'Walker' philosophy and as first developed by Jones and Raworth, a comprehensive regular interval service pattern was prepared for each route for each stage of the electrification. Typically, three trains an hour would be run. During the morning and evening commute times the new electric trains would comprise of two three-car sets, while outside these hours the trains would consist of a single three-car set. The services were almost all to be to and from the London termini, but there were a few others, namely Redhill to Tonbridge, the Gravesend West branch and the lowly Westerham branch which was to have a shuttle service from Dunton Green consisting of two trains an hour at peak times and one train an hour during the off-peak period, all with a single three-car set.

From these timetable assumptions, Raworth predicted that the annual train mileage during Stage One would be 5,875,000 miles a year, the three-car sets total mileage allowing for empty workings would be 8,667,000 miles. For this intensity of service 136 three-car sets were to be built, as the daily timetable would require 119 units. At Stage Two, the train mileage would total 10,014,000 miles and the three-car sets mileage 13,312,000 miles. An additional 57 three-car sets would have to be built, with the daily timetable requiring 169 sets in total to be available. The new trains would allow for a 28 per cent increase in passenger capacity during the first stage and a 39 per cent capacity increase on completion of the second stage.

With all the components of the electric railway described, Raworth could now provide an assessment of the total cost of the project. He estimated that the generating station at Angerstein's Wharf would cost £1,492,500, the 33,000V lines £774,000, the substations £636,000, provision of conductor rails £2,690,000 and the electric trains £5,000,000. The Stage Three passenger train electric locomotives would cost £630,000, and the goods train locomotives £610,000. Repair and running sheds would be required for the new sets and locomotives and these would cost an estimated £205,000. This all totalled up, after a 10 per cent contingency allowance had been applied, to a grand total of £13,241,800. This sum, paltry as it seems in the twenty-first century, would have caused a sharp intake of breath by the SECR directors, but Raworth noted that there had been high inflation during the war years and that in 1914 the total cost would have been only £5,480,483.

It is always very tempting to try and indicate by a simple calculation based on inflation as to what 'today's' equivalent of £13,241,800 would be. Given the usual inflation uprating this would be about £670 million in 2019. This would however be misleading as the world in 1919 was quite different to today. Such a project would have been labour intensive but following the end of the war unemployment was high in 1919. Sadly, there was an abundant reserve of unused skilled and unskilled labour willing to take on any work for low wages. Furthermore, workers' expectations were lower: most would settle for a wage packet that gave them and their families a comfortable existence, not a salary allowing the acquisition of material things and the funding of ever more expensive leisure activities.

Essential materials for the scheme, such as copper used for the cabling, could be imported cheaply from countries that were in 1919 outposts of the British Empire, or, if this were not possible could be bought from non-empire suppliers using sterling which then still had a relatively high exchange rate value compared to other world currencies. For example, the pound sterling was worth $3.80 in 1919, even though this did compare rather unfavourably with just under $5 at the start of the war.[18] Also, due to the lack of orders, plant manufacturers were likely to offer lower prices for their products.

Furthermore, there have been significant and expensive changes to the way that projects are governed since 1919. For example, today's health and safety legislation, together with the Construction, Design and Maintenance Regulations that sit in the legislative hierarchy below the Health and Safety Act. This legislation, while being very welcome from the safety aspect, does come with a cost burden. Today the regulations and safety rules make it more difficult to work on the 'live' railway, as each task requires full and detailed method statements and the operatives provided with daily risk assessment documentation as they are 'set to work'. Unlike railway projects today, the SECR would have carried out all the work in-house without resorting to having to employ outside consultants, lawyers or significant numbers of specialist contractors. Each of these would expect to make a profit at the SECR's expense and there would have been a culture of 'project creep' and 'extras' to boost earnings and profits. Unlike today, the SECR owned both the track and the trains, avoiding messy and expensive contractual arrangements between the train operating company and the owners of the railway infrastructure.

Despite the huge capital outlay required, if the SECR could finance the work by borrowing, and if the return on the investment exceeded repayment of capital and the interest on the loans, the scheme would be viable. And, of course, the work would be carried out in stages. Raworth in his report claimed that the annual cost of operating the electric railway would be £1,589,957. The operating cost using steam locomotives on the same routes was £2,011,900. The saving was therefore £421,943, which was not in itself enough to warrant the expenditure. Additional revenue however could be anticipated from the improved services. Raworth estimated that this would be £830,000 (the 'sparks effect' again), giving a much more agreeable saving of £1,251,943 per year. Such an increase could only be subjective, but Raworth cited genuine traffic increase on the SECR's London neighbours who had taken the plunge and pressed forward with electrification schemes. He reminded the Joint Management Committee that his former employer, the LSWR, had reported a nearly 26 per cent increase in traffic in 1916 compared with 1915; a 41 per cent increase over 1915 in 1917; and a 73 per cent increase over 1915 in 1918. This percentage increase was anticipated to be a massive 112 per cent by the close of 1919. He had gleaned from the secretive LBSCR that its electric services had seen an increase in passenger numbers of 87 per cent and the value of its receipts had risen by 105 per cent and the number of tickets sold had increased by 219 per cent.

Recalling what he had witnessed at Grand Central Station in New York, to further press the case for electrification, Raworth stressed the added benefits possible when all steam locomotives were removed from the electrified lines. These he listed as the removal of turntables and locomotive watering facilities and a decrease in the wear and tear on paintwork. Steam and smoke could be eliminated from the railway completely allowing the possibility of an income stream arising from being able to build above the railway as he had witnessed in New York and exactly what was to be done half a century later above London's Cannon Street station.

In his final conclusions Raworth referred to Sir Francis Dent's original concerns about track occupancy. He declared that due to the more rapid acceleration of the electric trains it would be possible to increase the number of trains working through London Bridge station by 50 per cent and this would avoid the need to carry out a track widening programme.

It took two months after Raworth had completed his report before Tempest, having perhaps taken his time carefully to mull over the contents and have informal consultations with his colleagues, submitted it to the SECR for consideration at a regular meeting held on 12 November 1919. No immediate decision was made on the day, but a copy of the report was distributed to each member of the committee for their consideration.[19]

As Raworth's report was being considered by the SECR directors, Percy Tempest continued his work with the

Channel Tunnel Company. This involved the construction of trial holes on either side of the Channel to confirm that the preferred chalk marl continued inland beyond both the English and the French coastlines to ensure that the tunnel could be built along the whole route without the risk of the water ingress that had defeated previous attempts at tunnel construction.[20] Tempest envisaged electrification of the tunnel using Raworth's design and it is a tantalising thought that in 1930 it might have been possible to travel out from Victoria by a train composed of solid and comfortable Southern Railway coaches hauled by a 'Raworth' 1,500V + 1,500V electric locomotive – the whole ensemble in Southern Railway olive green livery lined out in yellow. There would be the pairs of conductor rails, fitting neatly into the single-bore tunnels under the sea, these being used as far as Calais, where, with a few clunks from internal contactors and the raising of the pantograph on the locomotive the train would set off for Paris Gare du Nord under the wires of the new French 1,500V direct current system. Sadly, this could never be, for between 1919 and 1922 the proponents of a Channel tunnel made 19 attempts to get Parliament to consider the question and all proved to be fruitless.[21]

During the gestation of Raworth's report, and perhaps when he had time to spare on his Atlantic crossings, he managed to find time to revisit his father's regenerative control territory by designing and patenting a novel system for 'Improvements in the regenerative control of electrically propelled vehicles or trains'.[22] His proposal was to make a connection between the controlling field resistance and the field coil within a direct current motor to connect via a pickup shoe with a conductor rail of 'appreciable resistance'. When two vehicles come close together the paralleling of the field resistance would result in an increase of current to the fields and the motors would exert a regenerative braking effect. Given that it was trams that frequently came close together this design was more applicable to these, not railway trains.

CHAPTER 9
MEETING THE MEN FROM THE MINISTRY

Following the completion of Raworth's report another unfortunate hiatus occurred. The lack of any immediate progress on the SECR electrification proposals has been in part blamed on the post-war economic situation,[1] but in the post-war world when Alfred Raworth produced his report there was political uncertainty as to the future ownership of the railways. There was a strong possibility that the railways might be nationalised or compulsorily amalgamated into larger companies. This made it impossible for them to finance large capital schemes as shareholders could not be asked for funds in the form of share issues and the banks would not lend money to a company that might soon go out of existence. Nonetheless, the state of the national economy was an added complication. With the cessation of hostilities there had been an, albeit short, economic boom period, accompanied by chronic inflation preceding the stagnation, unemployment and recession that was to blight the next decade. The economic chill in the 1920s has faded from the public memory, being overshadowed by the global Great Depression of the 1930s. The cost of running the railways – including the wage increases that the government had instigated during the war years – had risen, however the shareholders did continue to receive payments from the government at the agreed 1913 dividend rate. In addition to the economic troubles, the Russian Revolution had sent out worldwide waves of discontent which had the 'establishment' – which included the SECR directors – fearing imminent industrial unrest or even violent revolution.

Four days after the signing of the Armistice, Lloyd George called a general election which was held on 14 December 1918. The Liberal leader campaigned with the slogan: 'a country fit for heroes to live in'. The Conservative and Liberal parties split into coalition and non-coalition factions, with the coalition candidates promising to continue the reforms of the wartime administration. The coalition Conservatives won more seats than any other party, acquiring 332 and Lloyd George's coalition Liberal supporters came second, securing 127. Despite the Liberals winning fewer seats Lloyd George continued as prime minister in preference to the Conservative leader, Andrew Bonar Law.[2]

Lloyd George appointed Sir Eric Geddes, the former First Lord of the Admiralty, as minister without portfolio with the remit to reconstruct British transport. The seconded railwayman was not comfortable in the political arena, having never been subjected to what has been described as the 'hardening process', due to his meteoric rise to high office.[3] At the 1918 general election when he defeated the Labour candidate, he had exhibited a lack of political acumen, earning for himself notoriety due to a hustings speech when he appeared to advocate the popular demand for excessive draconian reprisals to punish the defeated Germans. Consequently, the jingoistic comment that the Germans should be squeezed like a lemon 'until the pips squeak' has been unfairly attributed to Geddes.[4] A more careful study of his words however shows that in fact he had a genuine concern for the German workers, and he later insisted that he had been misquoted.[5]

Lloyd George had been impressed with the successful combined operation of the nation's private railway companies by the REC during the war, and he believed that because of this the majority opinion held by the professional railway managers was that nationalisation was inevitable.[6] The politicians who supported nationalisation were divided into two groups.[7] The *left* perspective ideologically craved state control to protect workers' pay and conditions of service. Alternatively the *national interest* viewpoint insisted that the railways should be publicly owned because they had become a utility like gas, water and electricity, and in order to make profits for the shareholders the companies had to charge excessive 'rates' to carry the nation's essential goods traffic, thereby creating an unnecessary on-cost for British business. In this second group was Winston Churchill, at the time a Liberal Party member, who during the 1918 election campaign had spoken out in favour of railway nationalisation. He even implied that the government *was* to nationalise the railways. This revelation at Churchill's Dundee hustings brought horror to the railway industry which, despite Lloyd George's belief, was generally of the opinion that while some form of coordinated cooperation between the railways was required, full nationalisation was not necessary.[8]

Geddes' and Lloyd-George's standpoint came close to the 'national interest' view, believing that state intervention was necessary for post-war reconstruction, although Geddes' preference was for private, not state monopolies. The political imperative at the time was *reconstruction*, but it was evident that both Geddes and Lloyd-George saw this process as a major reorganisation to improve the nation's infrastructure – not a programme to reconstruct the organisational structures which had failed in the past. Geddes introduced to Parliament his all-embracing Ways and Communication Bill – a measure intended to combine the ownership and control of all means of transport, including roads and even electricity supply. The new government's view was that in the light of wartime experience the railways should not return to the pre-war situation with its inherent wasteful competition and duplication of facilities.[9]

Preparation had begun on Geddes' all-inclusive Bill before the 1918 general election, but afterwards coalition politics intervened with a shift in the political climate. While Lloyd George was in France at the Versailles Peace Conference, Geddes – naively assuming that his plans were uncontroversial – was left alone to explain the benefits of using the power of the state to achieve economic reorganisation to sceptical MPs, many of whom for self-interest motives objected to his proposals.[10] Geddes was confronted by many backbench Conservative MPs (including some that had supported the coalition) who were averse to any form of state intervention in industry, particularly such an all-embracing proposal. They desired to return to the pre-war status quo of competing private companies. Objections to the Bill were twofold. First, even Geddes later had to admit that the powers within the Bill permitting the minister to act using 'Orders in Council' were autocratic, for he had misjudged the mood of Parliament when asserting that this arrangement had worked well in wartime. The second objection was that the scope of the Bill was too great – and this resulted in coastal shipping being removed at an early stage in the proceedings.[11] The result was the Ministry of Transport Act of 1919 which was intended to combine the myriad of transport duties scattered around Whitehall into a single ministry and Geddes duly became the first Minister of Transport. Within the scaled-down Act there remained the controversial Section Three entitled 'Power to Control Railways' which effectively continued the government's wartime control of the railways for a further two years to allow time for the new ministry to formulate a policy on railway ownership.[12]

Given the Conservative Party's opposition, it is unsurprising that any possibility of full railway nationalisation had disappeared by March 1919. A resolution from the Federation of British Industries linked to pressure from backbench Conservative MPs resulted in the removal of powers enabling the minister to nationalise any form of transport from Geddes' next Bill in 1921. Consequently the 1921 Transport Act had been systematically stripped of most of Geddes' intentions that would have resulted in a comprehensive transport regime. The new Minister of Transport's ideas – soundly based on his business and wartime experience and approved of by the Civil Service – were dispensed with after he had failed to understand the political landscape at Westminster.[13]

Geddes as the new minister had however acquired for himself powers to direct the level of both passenger fares and goods rates and to also set wage and salary levels. He could instruct the railways as to which lines they should provide services on and conversely on which lines they must withdraw services. He could order improvements in efficiency (and in this area Geddes had 'form', having a reputation for a no-nonsense approach in his dealings with the army and navy during the war) and could to a degree dictate railway traffic by conferring running powers for one company over another's lines. His only real constraint was that he had no power to enforce the railways to carry out any capital expenditure – such as that needed by the SECR for its proposed electrification.[14]

What was most dear to the hearts of the railway company directors however was the effect that this was to have on their dividends. They had the 1913 dividend being paid by the government, but this was a poor reward in 1919. Due to wartime inflation, the pound sterling was worth only 47 per cent of its 1913 value. If the railways had been allowed to increase their rates in line with inflation during the war the 1918 dividends would have been 60 per cent higher than in 1913. The most conservative, and probably the most reliable, estimation at the time was that the true profit from the railways under wartime government control was £30 million, and this amount should have been distributed to the railways. In what was a legal test case, a claim for compensation based on the terms of the 1914 agreement amounting to £616,000 was made by the North British Railway. This, after a prolonged process, was reduced to £430,000. The Ministry of Transport however refused to pay, believing that it was entitled to suspend payments until the railways were de-controlled.[15]

Geddes' new Ministry of Transport intended to be more than just a temporary owner of the railways, it proposed to determine policy as well. Continuing the approach made by the wartime Reconstruction Committee which proposed that any future railway electrification should be carried out as part of an overall national plan,[16] in November 1919 the Ministry of Transport requested that each railway company in London provide details of any proposals they had for electrification schemes. The SECR Joint Management Committee agreed that a letter be sent to the Ministry of Transport accompanied by a copy of Alfred Raworth's report. In the letter the SECR pointed out that it was impossible to raise funds for the scheme due to the current political uncertainties.[17] Nonetheless, the directors were clearly satisfied with Alfred Raworth's report, permitting it to be sent to the government department, and to underline their appreciation of Raworth's efforts, at their meeting held on 10 December 1919, they awarded him a pay increase of £100, raising his annual salary to £700. He did not have to wait long for his next increase in what was a period of rampant inflation, for with the general rise in salaries taking place all around, by September 1920 he was earning £1,200 a year.

Early in 1920, Francis Dent resigned as the SECR general manager following a quarrel with Cosmo Bonsor.[18] Despite the atmosphere of bad feeling Sir Francis was awarded a pension of £2,000 a year[19] and he remained a director on the LCDR board. Later, he was to become an active member of the Southern Railway board. His duties were taken over by Percy Tempest, who, for the remainder of the SECR's independent existence, was to carry out the duties of both chief engineer and general manager.

The change in general manager did not change electrification policy. In May 1920 Percy Tempest wrote to the Ministry of Transport pointing out the serious and ever-increasing difficulties that the SECR was experiencing due to inadequate line and platform capacity. The London railways may have lost a considerable amount of local traffic to the LCC tramways, but the SECR was experiencing a massive increase in traffic from the outer suburban areas. Since the war, the traffic through London Bridge towards Cannon Street and Charing Cross had reached unmanageable proportions, particularly during the incoming morning commute between 9am and 10am. The obvious solution to these woes was to implement the electrification proposals and Tempest requested that the government agree to pay the interest on any loan taken out to finance the scheme while the railway remained under government control, or subsequently following nationalisation.[20] Not surprisingly perhaps, Tempest's request for such financial assistance was ignored, for the official government policy, detailed in a 1920 White Paper, was that the railways should finance the electrification schemes and not expect government assistance.[21]

Later in 1920 Tempest, realising that there was no imminent prospect of electrification, established a committee to advise on the operation of the suburban services and appointed Alfred Raworth to be its chairman.[22] This was Alfred Raworth's opportunity to further expand his knowledge of railway operation, which was to be an asset when, in the late 1920s and 1930s, he was to assist Sir Herbert Walker with the preparation of the business cases for further electrification. While no specific details of the recommendations of the 'General Manager's Committee on Suburban Traffic Operations' have come to light, it is certain that proposals for parallel running came from this committee – the scheme does have the 'Raworth' lateral thinking flavour.

Parallel running was introduced between London Bridge and Cannon Street or Charing Cross.[23] For many years, hapless commuters had not had the luxury of joining trains at, say Dartford, and by choosing which half of the train to sit in, could be transported to either Cannon Street or Charing Cross without change, since the dividing of trains at London Bridge had long been abandoned as it required two paths for the separate portions between London Bridge and Borough Market Junction. With parallel running in operation pairs of trains ran on adjacent lines simultaneously to avoid conflicting movements. For example, on the 'fast' and 'slow' lines between London Bridge and Cannon Street the 6.55am ex-Hastings train would run alongside the 8.22am ex-Beckenham train. The two trains would depart London Bridge at the same time. Where the routes diverged above Borough Market, the route was set towards Cannon Street, so it would conflict with any trains leaving Charing Cross. The two trains would be timed to pass two more parallel running trains – the 8.48am empty train from Cannon Street to Bromley North and the 8.48am Cannon Street to Erith – hence high capacity without conflict. The next pair of parallel workings would most likely involve four trains to and from Charing Cross. By this method, the SECR could run 36 trains within the hour into its two termini. The fact that all trains had to stop at London Bridge – because that station was used by so many season-ticket holders – helped to ensure that the system ran smoothly. This was an ingenious system, but it was a case of allowing the 'tail to wag the dog'. Trains had to be scheduled to fit in with a very rigid and difficult operating system (particularly with steam traction) rather than the operators designing a traffic pattern to meet the needs of the commuter.

Another outcome from the committee's deliberations was the decision to provide new carriages urgently needed for the suburban lines. While the arguments as to when the SECR suburban lines were to be electrified raged on, the railway took delivery from contractors, or built themselves at Ashford works, 66 10-compartment coaches with the view that they might later be converted for use in electric trains. These 10-compartment third class only coaches, all turned out in the drab SECR post-war brown livery, were to be known as the 'long tens' and had seats for 100 passengers. Conversion to electric operation never occurred and despite being cramped and uncomfortable many of these vehicles survived until the late 1950s.[24]

Meanwhile, Sir Eric Geddes, having gathered in all the information on proposed electrification schemes from the SECR and other railways, set up the Electrification of Railways Advisory Committee with the former LSWR consultant Sir Alexander Kennedy in the chair.[25] One important member of the new committee was the Ministry of Transport's own consultant Sir John Aspinall. He was a long-time advocate for railway electrification having introduced electric traction to the LYR when he had been its general manager. Speaking privately to Sir John Aspinall, Sir Eric made it clear that standardisation and interchangeability were the aim of the committee such that it would be possible – to quote the minister's example – 'to travel by the same electric train from Inverness to Dover'.[26]

Other committee members included Sir Phillip Nash, the Ministry of Transport's director general of traffic, and Sir Philip Dawson, who had designed and built the LBSCR 6,700V system. Officially representing the nation's railway companies was Sir Henry Thornton, general manager and consultant to the GER, a company that was also active in the preparation of electrification schemes for its London suburban lines. Perhaps the man who was to prove the most influential member of this learned ensemble was Charles Merz who had prepared the 1913 report for the SECR and had recently been involved with the NER 1,500V direct current electrification. Ominously for the SECR's 'own generating station' plans, also on the committee was Sir John Snell who was the chairman of the recently created Electricity Commissioners, a department of the Ministry of Transport set up to regulate the electricity supply industry.[27]

Evidence was taken from not only British railways, but also from mainland Europe as representatives from Swedish and Swiss railways travelled to London to be questioned by Sir Alexander and his committee. Manufacturers were also consulted including British companies such as British Thompson Houston and English Electric, and some from other European countries, for example the Swiss Oerlikon company which from its extensive works in Zurich had recently provided equipment to some British companies including the LNWR. It is notable that, unlike Alfred Raworth and all the others before him, the eminent committee appear to have found no need whatsoever to make any enquiries to the USA apart from speaking to the British subsidiaries of American companies. Percy Tempest and Alfred Raworth attended the committee sessions on two occasions to give evidence and to promote the SECR electrification proposals.[28]

One important person called to give evidence was the astronomer royal, Sir Frank Dyson. He was custodian of the Greenwich Royal Observatory which had on site its 'magnetic room' monitoring the Earth's magnetic field. Before radio and satellite navigational aids, a full knowledge of the slow movement of true magnetic north was essential for a nation relying heavily on maritime trade and having a navy with a global military presence. The possibility of any stray traction direct current disrupting his readings was of serious concern to Sir Frank. There were electric tramways within a mile or so of the observatory, but these had all been equipped with an additional negative overhead wire clear of earth[29] at the insistence of the observatory which, at the time due to its strategic military significance, was owned and managed by the Admiralty.

In July 1920, the committee produced an interim report proposing that the national standard traction supply should be 1,500V direct current.[30] The mildly mannered but persuasive Charles Merz had convinced the committee – which included engineers who had built direct current conductor rail and a higher voltage alternating current overhead line system – to favour 'his' 1,500V direct current overhead line design. In 1916, Merz and McLellan had successfully completed the Shildon to Newport mineral line electrification for the NER and were preparing the designs for a York to Newcastle East Coast Main Line electrification. During the previous year, the first public electric trains began running between Sandringham and Essendon on the Melbourne suburban system[31] using the 1,500V system as recommended in Merz's reports to the Victoria government.

Just as Merz had recommended to the SECR, a ground-mounted conductor rail could be used where necessary – for example where clearances to bridges and tunnels were restricted. Also, it was advised, 'submultiples' (for example, 750V) or 'multiples' (for example, 3,000V) of 1,500V could be used where this was shown to be economically advantageous. Precise multiples or submultiples would enable trains to run with both voltages by switching individual motors or groups of motors between parallel and series operation. Sir Eric's hypothetical Inverness to Dover train might be equipped to run on any of the voltages and have overhead current collection equipment – which was preferred – but might also be equipped with pickup shoes for use with conductor rails.

The committee recommended the planned LSWR 600V system could be installed and the LBSCR 6,700V alternating current suburban network should be completed – but major extensions to both networks should use 750V direct current. Another important recommendation – again probably influenced by Charles Merz, for he was also an advocate for a 'national grid' integrating public and railway supplies – was that electricity for the railway electrification schemes should be generated by three-phase alternating current generators at a frequency of 50 cycles. Raworth had insisted in his report that 25 cycles generation must be used, but Sir Alexander's report concluded that 50 cycles could be used for railway purposes without any detriment. But there was no actual insistence as to any frequency being used since the committee advised that the railway should use any frequency used by the electricity suppliers in the vicinity of the railway line.

On 28 September 1920, the SECR duly received its copy of the interim report. From Raworth's point of view the advisory committee's conclusions seemed to approve the use of his three-wire system. It ticked several boxes, being a 3,000V scheme – a multiple of 1,500V – and it used the permitted 1,500V conductor rails. What it did not do was constitute a system that other railways' trains could use without infringing the unique three-wire, no earth return feature. This was despite Raworth's rather vague (and Walker-like) comment in his 1919 report that 'in a few cases in which it may be desirable for electrical suburban trains of other railways to work over our lines and vice versa, arrangements can be made to enable them to do so'. But his proposals would, arguably, satisfy the requirements of the astronomer royal. The SECR response was for Percy Tempest to write to the Ministry of Transport on 1 November 1920 stating that the SECR should be allowed to build its system based on the advisory committee's recommendation allowing 3,000V systems as this was shown to be economically justified.[32] As the interim report had agreed the higher voltage could be used with the minister's consent and conductor rails could be used, Tempest formally requested permission to use Raworth's three-wire system. Prompted by Raworth, Tempest also queried the statement in the interim report that a frequency of 50 cycles was practical for railway operation.

The Ministry of Transport however was not to be pressurised by the railway, so before the final report was published Tempest and Raworth were called before a special subcommittee set up specifically to consider the SECR proposals. The chairman was Sir John Aspinall and the members included Sir John Snell.[33] Aspinall seemed a little unclear as to exactly what this subcommittee should investigate. He understood that the terms of reference were only to determine if there was enough economic advantage in the SECR using what was effectively a 3,000V system – a multiple of the 'standard' 1,500V. They decided that there was some advantage of using the three-wire system, but they did not believe that the savings were as large as Raworth and Tempest had stated. The recommendation was that the three-wire system should only be approved when the minister was satisfied that there were enough savings to justify the non-standard system. In response to their doubts, Raworth commissioned for the benefit of the Ministry of Transport a report from the consultants Kennedy and Donkin to verify the cost of his scheme. Apart from cost, the subcommittee did however raise some concerns about extra track maintenance due to the presence of an additional live conductor rail and the inevitable interruptions to the negative centre rail which might not be critical for multiple unit stock but problematic when, in future, locomotives were to be used. They were also concerned about the issue of interchangeability with other electrification schemes.

The final report from the advisory committee was dated 12 July 1921[34] and contained much more detail in respect to the standardisation of components to facilitate inter-railway electric train working. These standards included such details as to the positioning and height of the conductor rails and the parameters for the overhead lines – clearances to structures, height of the contact wire, etc. Sir Alexander's report carefully considered the important issue of inter-railway operation of electric trains. If the SECR's neighbours were to operate the standard 1,500V direct current conductor rail system as allowed for by the committee, then trains could still operate over the SECR using only the positive rail – but there would be problems with voltage drop and the return current would flow back through the running rails. Overall, the SECR proposal did not look too good for interchange of traffic between adjacent electric railways. The GWR's electrical engineer, Roger T. Smith, in his address to the Institution of Electrical Engineers in 1922[35], spoke forcibly about the need to achieve standardisation between railway companies fearing a 'break of gauge' type situation, referring to the problems that the GWR had encountered when Brunel's arguably technically superior seven-foot-wide broad gauge had to interconnect with the British standard gauge. Sir Alexander's committee's recommendations were just that – the Ministry of Transport had no obligation to implement the committee's findings, but Raworth's scheme would be examined against these proposals by its detractors. In truth, his scheme was technically in line with the advisory committee's published conclusions but was some distance away from what Charles Merz, together with Sir Eric Geddes – the main drivers behind the committee's proposals – had in mind.

Meanwhile in 1920, the Ministry of Transport published its White Paper entitled *The Future of Railways*. This resulted in the passing of the Railways Act in August 1921. This momentous Act required the *grouping* of the railways into four large private companies – forever to be referred to as the 'Big Four'. The SECR was to join the LSWR and the LBSCR in the new Southern Group – the Southern Railway – notionally from 1 January 1923. The other three were a much-enlarged GWR and what were to become the London Midland and Scottish Railway (LMSR) and the London and North Eastern Railway (LNER).

The new Act did contain some good news for the railways in that £60 million was to be distributed among them, half of which was intended to compensate for the lack of dividends during the period of government control, and the remainder to cover the costs that had been incurred to catch up with the arrears of maintenance suffered during the Great War. The arrival of some money from the government, although welcome, was not particularly useful when spread across the four new companies and was certainly insufficient to finance any large electrification schemes on any of the British railways, once essential maintenance had been completed. The spectre – as the directors would have seen it – of 'worker director' disappeared to be replaced by provisions for a National Wages Council to protect the livelihoods and wages of the railway workers.

But there was bad news too, for the Railway Act set up the Railway Rates Tribunal that effectively handcuffed the railways commercially, setting rigid rates for the conveyance of goods and not permitting effective competition with the burgeoning road haulier business. For British railways, it was freight that produced most of the income, and this was now in jeopardy. Part two of the Act also made provision for the Minister of Transport to issue an order to instruct the railway companies to standardise on equipment – this being a step beyond the *recommendations* previously emanating from *advisory committees*, and this provision was likely to affect railway electrification schemes such as Raworth's proposals for the SECR.

By October 1921, the political imperative was focusing on the wide scale unemployment in the land. Lloyd George's vision of a Britain which was 'fit for heroes' was a hollow one now that so many of them were unemployed and living in poverty. At the time, due to the depression in world trade and the return from the war of so many men of working age, the unemployment rate in Great Britain was a staggering 17 per cent.[36] The Ministry of Transport wrote to the SECR enquiring as to whether it had any large engineering schemes that would enable it to take on labour to alleviate the then current chronic unemployment caused by the national reduction in trade.[37] This must have caused a few jaws to drop and eyebrows to rise in the boardroom at London Bridge, given that the Ministry itself had refused to assist the SECR with electrification work in May 1920. The SECR management committee promptly replied, reminding the Ministry of Transport that the railway had a fully prepared electrification scheme that would produce immediate employment, as a large element of the overall cost was for wages. It was further stated that the first stage of the proposed suburban electrification scheme was expected to cost around £5 million of which £3.5 million would be spent on wages for 6,500 men over three years.

Given the national importance of reducing unemployment, the SECR directors sat back in the expectation that financial assistance for their bargain-offer project would be forthcoming from the government. They were to be partially disappointed. Just over a week later a letter arrived from the Ministry of Transport clearly stating that there were no government funds available for electrification, but the directors were informed of the Trades Facilities Bill then passing through Parliament, the aim of which was to provide government guarantees for loans on approved schemes. This correspondence went on to say that the SECR would be able to make application to the Trades Facilities Committee as soon as the Bill became law.

The SECR response was for Percy Tempest to write to the newly set up Trades Facilities Committee on 3 December 1921.[38] He stated that the SECR scheme would over the next three years provide employment for skilled men in the mining, iron and steel and engineering industries in the North, Midlands and north-east England. Skilled work would also be provided locally at Ashford, Kent, and the scheme would also generate over 90,000 man-weeks of unskilled local labour. Furthermore, he made assurances that the electrification scheme was not a new project hastily put together to take advantage of the new financial offerings.

This was immediately accepted by the civil servants for, after he had met officials of the committee, the chairman of the SECR finance subcommittee, Sir Robert Kindersley, reported to the main board on 14 December that the draft proposal had been sympathetically received by the new Trades Facilities Committee, which seemed disposed to guarantee a loan for the SECR. Sir Robert did however take this opportunity to make it quite clear that neither the SER nor the LCDR boards were able to take any financial risk, and because of the impending amalgamation of the SECR into the Southern Railway it was not possible to ask the shareholders for the loan capital that the government would guarantee.[39]

On 6 January 1922[40] – just one year before the SECR would cease to exist – the deputy chairman of the SER, Sir Alfred Smithers, supported by senior SECR staff (it was not recorded who they were – but the delegation must surely have included Alfred Raworth) met Sir William Pender, the Trade Facilities Committee chairman. The SECR desired a loan guarantee to the value of £3 million. The proposal was well received, but some members of the Trade Facilities Committee needed some convincing of the viability of the SECR's electrification proposals. One of the members, Sir John Harwood, did not believe the SECR's traffic increase predictions, for he opined that the rise in traffic numbers on the LSWR and LBSCR was not due to electrification at all, but expressed the rather strange view that it was all due to improved passenger facilities at Waterloo and Victoria respectively. Strangely, Sir Alfred Smithers stated that, despite the proposal in Alfred Raworth's report, the railway was *not* going to build its own power station but was to take supply from an outside power company. Perhaps this was a minor ruse to avoid appearing to go against the general government policy opposed to railways building their own power stations. Sir Alfred's plan was probably to get the loan agreed, and then argue about the power supply. The Trade Facilities Committee was therefore left with the understanding that while the SECR had not agreed to any power supply contracts, it was confident that it would have no difficulty in making such an agreement when necessary.[41]

Some of the SECR attendees at the meeting (and probably Sir Alfred himself) would have known that if the SECR wished to build its own power station it now had new legislative hoops to jump through which could seriously delay the whole project. Foremost of these was the need to make an application to the new Electricity Commissioners, now charged with assessing the need for every new power station from a national electricity supply standpoint – not just to suit a single user.

It was soon to become obvious however that despite the need to apply to the Electricity Commissioners, an *outside* electricity supply was not what the committee and its chairman, Cosmo Bonsor, wished for. The SECR had not made any approach to the LESC – the obvious source for an outside supply for a limited Stage One scheme – reinforcing the premise that Bonsor wanted nothing to do with them.

The only recorded discussions that the SECR had with any outside electricity supplier were with the West Kent Electricity Supply Company. This was a small concern founded in 1902 and by 1909 it supplied parts of Kent, namely the western end of Northfleet Urban District Council, the rural districts of Strood, Malling and most of Sevenoaks. The meeting between West Kent and the SECR was not instigated by the railway but was the result of a friend of Percy

Tempest, a senior employee with the West Kent, Oliver Bury, making an informal suggestion that it might be advantageous to both parties if they met to discuss the SECR electrification. Consequently, when obtaining finance for the project seemed to be far in the distance, Percy Tempest and Alfred Raworth duly met representatives of the West Kent at Raworth's office at 19 Railway Approach on 4 March 1920 to discuss an 'outside' supply.[42] The West Kent did not yet have a power station, but it traded in electricity by buying it from a generating company. It did, however, wish to construct its own power station at Belvedere, conveniently located between the River Thames and the former SER line between Woolwich and Slade Green which was one of the lines included in the electrification scheme. The West Kent proposed to build what was then described as a 'super-power station' which would also be available for the proposed LBSCR's electrification extension.

At the meeting it was tentatively proposed that the SECR might pay to the West Kent a proportion of the Belvedere generating station's construction capital and interest costs through a regular 'standing charge' based on the number of kilowatts of agreed maximum demand. Additionally, there would be a further charge per unit of electricity consumed by the railway, based on the current price of coal and wages. The crucial point about this proposal was that the SECR's staged payments towards the cost of the power station, including the interest paid, had to stand comparison with the cost of the SECR building its own power station. This was a complex comparison, since while the larger power station would cost more, what was a fair proportion for the SECR to pay? What electrical load would the West Kent be able to sell to the public? How would the costs be shared with the LBSCR should it wish to take supply from the West Kent? There would also be some additional capital cost to the SECR for installing high voltage cables from Belvedere to Lewisham, which despite the route all being on railway land, was a greater distance than from Angerstein's Wharf to Lewisham. No immediate agreement could be reached because while Raworth could provide the West Kent with some solid load information, the other side of the equation – how much was the Angerstein's Wharf generating station going to cost – was still not known. This was because so far no decision had yet been made to proceed with the railway electrification due to the political climate, and if a decision were made, what would inflation and interest rates be at the time of this decision? Despite there being no resolution on the day of the meeting, the SECR said that at the appropriate time any offer given to them by the West Kent would be given careful consideration. By 1922 further negotiations between the West Kent and Raworth and Tempest had progressed to a more definite offer for an electricity supply. These somewhat informal discussions would later be forensically examined by the Electricity Commissioners.

This public pretence that the railway was to use an outside electricity supply did not last long for the SECR formally applied to the Electricity Commissioners for permission to build its power station on 23 February 1922 – less than two months after it had told the Trade Facilities Committee that it was confident that it would find a power company to meet its needs.

Following the necessary formal application to the Electricity Commissioners to seek approval to build the Angerstein's Wharf Station, the disquiet felt by Cosmo Bonsor was revealed in a private letter[43] of 19 April 1922 to Arthur Neal MP, the parliamentary secretary to the minister of transport. In his letter Bonsor confided to Neal details of the difficulties – as he saw them – that the SECR was confronting in respect to obtaining a suitable contract for the supply of electricity. Bonsor maintained that the SECR directors could not accept the recently received West Kent offer because he believed that the company was only recently established and inadequately funded. He stated that the SECR could not take the risk of entering into an agreement with this company, making it clear in his letter that while he was not opposed to the use of an external generator in principle, he asserted that the only viable scheme that he could see was for the SECR to proceed with Raworth's original Angerstein's Wharf proposal. Bonsor, correctly as events were to show, explained to Arthur Neal that in his opinion the unsatisfactory offer made by the West Kent was a major obstacle to an early approval by the Electricity Commissioners for the SECR's Angerstein's Wharf generating station.

Bonsor and the SECR seem to have established a good relationship with the Ministry of Transport through Arthur Neal MP. Sir Eric Geddes had moved on, having been appointed by the Chancellor of the Exchequer to be chairman of a committee to give advice on all questions relating to national expenditure. Geddes had been replaced at the Ministry of Transport by Lord Peel, but continuity was achieved through Arthur Neal who was coalition Liberal MP for Sheffield Hillsborough, and during his term of office was to prove to be a good friend of the SECR.

The possibility that the SECR was to build its own generating station did not concern the Trade Facilities Committee, for far from being put off by any such suggestion, it had no qualms about increasing the loan guarantee by £1.5 million to finance the Angerstein's Wharf station – subject, of course, to Ministry of Transport approval.

At the suggestion of the Trade Facilities Committee it was agreed that the way forward was to create a new, separate company with limited liability to finance and to construct the electrification scheme and then to lease all plant and electric rolling stock to the SECR – thereby avoiding any charge on the SER and LCDR (and subsequently Southern Railway) shareholders, thereby alleviating Sir Robert Kindersley's concerns. The 'rent' that the SECR was to pay to this new company was to cover the interest on the capital raised, a sinking fund of £80,000 per year and any taxation due. As the proposed new construction company would have a government guarantee, it would be able to borrow the capital at a low interest rate from the Bank of England – and this was confirmed by Cosmo Bonsor, a member of the board. The rent paid by the SECR would repay the capital over 25 years, with the first payment commencing on 1 January 1925. The SECR reaffirmed to the Trade Facilities Committee its commitment to the Raworth 3,000V three-wire electrification proposals, despite hearing criticism about the proposals from its future partners at the LSWR and LBSCR. The SECR declared at the meeting with the Trades Facilities Committee that it intended to immediately apply to the Ministry of Transport for approval to use the Raworth system, since it believed that this was the best system for its railway. Also present at the meeting was a representative from the Ministry of Transport who promised that the SECR application would be promptly considered.[44]

CHAPTER 10
MAKING THREE FIT INTO ONE

With the imminent creation of the Southern Railway which would own and operate the newly electrified railway, the Ministry of Transport hoped for the agreement of the two other railways before giving approval for the SECR scheme which was required by the Trades Facilities Committee before the loan guarantee could be given. Unsurprisingly, the LSWR led by Sir Herbert Walker desired a single electrification system across the new railway's territory and given that it had a working and successful system could argue that the standard Southern Railway electrification should be the 600V direct current scheme. Alternatively, the LBSCR wanted to develop its 6,700V alternating current scheme by extending it – as it had always intended – to Brighton and other south coast towns. Even before the Trades Facilities Committee had agreed in principle to guarantee the loan the scene had been set for an acrimonious debate. (See plate 6 (top))

To add to this electrification controversy, ancient tribal rivalry meant that the SECR and the LBSCR bitterly opposed having to work with the LSWR. It is now difficult to comprehend why it was that the company directors, who were all rich and well-educated men, often aristocrats, could appear to be so obsessive and insular to the detriment of good business sense. The important question as to what, if any, was to be the Southern Railway standard for railway electrification was naturally focused in the short term upon which system was to be used for the imminent SECR suburban electrification scheme. The Ministry of Transport represented by the parliamentary secretary Arthur Neal MP thought, initially at least, that it was best for him to stand aside and allow the three railways themselves to make this crucial decision on their own, and only step in if agreement could not be reached. The electrification policy was only one of many issues for which intercompany committees were to be established.

Initially there had been a half-hearted attempt to tackle the issue with the formation of a committee of consultant engineers but the LSWR, which could only gain if no agreement was reached before the grouping, had vetoed this proposal following a lack of agreement as to who was to be appointed as the independent chairman.[1] To restart talks civil servants at the Ministry had advised Percy Tempest that he should take the lead. Tempest arranged for the establishment of a new Group Electrification Committee following consultation with the LSWR and LBSCR general managers Sir Herbert Walker and Sir William Forbes. Members appointed to this committee were operating officers and engineers – including Alfred Raworth and Percy Tempest – from the SECR, the LSWR and the LBSCR and were specifically charged with establishing a policy in respect of the very contentious electrification issue. To progress matters Percy Tempest suggested that the committee chairman should be the SECR's chief operating officer, Edwin Cox.[2] Given the difficulties that the SECR was having in London at the time, Edwin Cox would probably have settled for *any* electrification proposal!

It was soon evident that each company was insisting that only its own system was suitable for use on its existing lines and, by inference, suitable to become a potential Southern Railway standard should, indeed, any such standard be even deemed necessary. At the very first meeting on 31 October 1921 it was proposed that the general managers should not incur any expenditure on electrification – even if approved by the Ministry of Transport – until the committee had completed its deliberations and announced recommendations. It was then the LBSCR's turn to be obstructive and not agree, and at only the second meeting of the committee on 2 November their representative announced that Sir William Forbes had insisted that not only were they forbidden to discuss the relative merits of the rival electrification proposals, but also that notwithstanding what the committee was ultimately to decide, the LBSCR was to extend its single phase alternating current system out towards Coulsdon and Cheam. Following this dissent there was little else to be done other than adjourn the meeting.

Following correspondence between Percy Tempest and Sir Herbert Walker the committee reconvened on 14 November with only the SECR and the LSWR represented. The revised remit was to consider the relative merits of the SECR proposals compared with the LSWR scheme and to report on how, or how not, these two electrification schemes could be unified. Raworth informed the Group Electrification Committee about the report he had commissioned from Kennedy and Donkin to verify the cost of his scheme for the benefit of the Electrification of Railways Advisory Committee and a request was made at the meeting that the scope of this report should be expanded to review the comparative costs of using the 600V direct current scheme and the 6,700V single-phase alternating current scheme.[3]

In December 1921, with the committee having been in existence for 12 weeks, Sir Herbert Walker had informal discussions with Sir Phillip Nash,[4] when he made reassurances, not exactly accurate, that each of the constituent companies was making a serious attempt to reach agreement on the thorny electrification issue. Sir Herbert characteristically stated that the comparative costs of each scheme had to be known before a decision could be made, but the LSWR was keen to avoid yet another system of electrification being introduced onto the new group's network.

The LSWR attempted to take control of the situation by writing to both the SECR and LBSCR on 10 December 1921 requesting financial cooperation between the three companies in respect of dividend payments and long overdue monies from the government, and again insisted that no electrification work involving large expenditure should be undertaken without consulting and gaining the approval of all three railways. The SECR reply to the LSWR was curt and to the point, flatly refusing to enter into any such formal agreement.[5]

On 15 December 1921 Kennedy and Donkin sent their expanded report to Alfred Raworth[6] which compared the 'first costs' of four possible schemes to electrify the SECR lines.[7] Only the actual electrification elements were included for comparison, that is cabling, substations and conductor rails (or in the LBSCR scheme, overhead wires). Not included were any rolling stock or power station costs which might reasonably be similar. The four options were: Raworth's 3,000/1,500V system, the Ministry of Transport's proposed 'standard' 1,500V system, the LSWR's 600V system and the LBSCR's 6,700V alternating current system. For each system *complete scheme* and *Stage One only* costs were provided. 'Complete scheme' was defined as all lines within a 30-mile radius of London, 241 route miles; and 'Stage One' was defined as lines within a 15-mile radius of London, amounting to 94 route miles.

For the complete scheme Raworth's 3,000/1,500V proposal would cost, according to Kennedy and Donkin, £3,771,000; the 1,500V scheme £5,187,000; the LSWR 600V scheme £7,058,000 and the Brighton overhead scheme £5,759,000. For Stage One only, the respective costs were estimated to be £1,384,000, £2,123,000, £2,874,000 and £2,261,000. In both cases the SECR scheme was cheaper and the LSWR the most expensive. For the running costs, the SECR costs were calculated also to be less. The savings hinged upon the relative costs of conductor rail compared to the cost of substations – and Raworth had specified only one substation, at Lewisham, for the first part of the electrification of the SECR lines. Kennedy and Donkin stated that their cost comparisons would only be changed if there was a significant increase in the cost of conductor rail accompanied by a simultaneous drop in substation equipment costs – and this they considered unlikely.

The report recommended that to avoid interference with the Greenwich Observatory, both the standard 1,500V and the LSWR 600V conductor rail schemes would require insulated running rails near Greenwich to comply with the conditions of the SECR's original 1903 Act, but this would not be necessary for the 3,000/1,500V system as only a negligible current would flow back through the running rails and earth.

Kennedy and Donkin however were not totally uncritical about Raworth's scheme, stating three causes for concern: the presence of two live conductors on each track; the risk of disruption to services due to the length of the conductor rails from the single substation; and the difficulty in designing locomotives and multiple unit trains to operate on other types of electrified lines. They did however assert that these disadvantages were not insurmountable. It was agreed that the presence of two live rails might be a hindrance during track maintenance and hence extra cost would be incurred, but this extra cost was an issue for the SECR permanent way engineers, and it was confirmed that the live rails would be protected against accidental contact by timber shields. As for the length of the conductor rails it was recommended that they be split into sections to allow any faulted section to be isolated for repair. Kennedy and Donkin had allowed for the extra switches and links that would be necessary for this in their cost estimates. In respect to interchangeability with other electrified lines, Kennedy and Donkin concluded that there would be no difficulties with interchange with the standard 1,500V system but it would be more difficult to interchange with the 600V system. Nevertheless, in December 1921 Percy Tempest, with more hope than solid reasoning, informed the Ministry of Transport that it could easily be arranged for SECR trains to work over the LSWR, LNWR and MDR lines.[8]

The report from Kennedy and Donkin was duly placed before the Group Electrification Committee on 16 December 1921. It was anticipated that a final decision would be made at the next meeting on 29 December. Despite the overwhelming recommendation for the use of the SECR system, Percy Tempest predicted that there would be a failure to agree when he wrote to the Ministry of Transport hinting that he had a proposal that would allow the scheme to proceed.[9] In the event the meeting scheduled for the end of December had to be postponed until 5 January because Alfred Raworth, obviously a key participant, had unfortunately contracted influenza.[10] He may not have been laid low by the same strain of Spanish flu that had resulted in millions of deaths worldwide between 1918 and 1919, but his absence due to this reason was not a trivial matter.

Following Raworth's recovery, the meeting finally convened on 5 January 1922 at the Charing Cross Hotel with Edwin Cox in the chair. The three SECR representatives did not include Percy Tempest, but Alfred Raworth was accompanied by one of the railway's senior civil engineers, George Ellson, who would advise on any track maintenance issues. At the committee's request, a representative from Kennedy and Donkin also joined the SECR delegation. A three-man representation from the LBSCR deigned to attend – they had boycotted the two previous meetings, but the three general managers had just agreed that all three railways must be represented at such an important meeting. Leading the LBSCR contingent was its chief electrical engineer, Robert Houghton. The LSWR sent its chief engineer, A.W. Szlumper, and Herbert Jones.

The areas where opinions diverged soon became obvious. There was existing electric traction near the observatory which consisted of the LCC trams – some with 'insulated' return – and the relatively nearby East London Railway which used the fourth-rail system. The SECR insisted that the 600V earth return system with an uninsulated return path could not be used as it would contravene the 1903 Act of Parliament, whereas the three-wire system could be deemed a non-earth return and would therefore be compliant. This may not have been an unreasonable stance, but the LSWR wished that it should be placed on record that it did not agree. Neither would the LSWR accept the Kennedy and Donkin cost estimates. The LSWR opined that 600V direct current should be used for all suburban services – including the SECR lines – and that a single-phase overhead alternating current system might be used for the longer distance main lines. The voltage for this overhead system would probably be at 11,000V rather than 6,700V – the higher voltage was then being proposed by the LBSCR for its longer distance extensions. This solution was derided by the SECR and supported by Sir Alexander Kennedy, who insisted that the 600V system was an earth return system and that the suggested dual system solution would involve duplicate equipment – overhead lines and conductor rails – throughout much of the suburban network and furthermore this arrangement would be expensive to install, maintain and operate. The SECR concluded by restating its view that the Raworth three-wire system was more economical than any other system and particularly suited to its needs.[11]

So, as Percy Tempest had predicted, there was a massive failure to agree and a complete deadlock reached. On 6 January 1922, Tempest wrote to the Ministry of Transport reporting that he had duly followed its suggestion and arranged meetings with the other railways. Unfortunately, he reported, agreement had not been reached as the LBSCR had refused to discuss details about the costs of its system and the LSWR had refused to accept the conclusions of the Kennedy and Donkin report. Furthermore, the suggestion that a mix of direct current conductor rail and high voltage alternating current overhead line for the longer distance routes was in total contradiction to the conclusions reached by the Ministry's Advisory Committee on Railway Electrification. The SECR general manager formally requested that the Ministry of Transport approve the SECR electrification scheme as it was a multiple of the standard 1,500V and it had been conclusively proved by the Kennedy and Donkin report that there was a financial advantage to the railway of using Raworth's 3,000V proposal. He reasserted that the 3,000/1,500V three-wire system was the only one that the SECR was justified in adopting.[12] Tempest was now confident that the Ministry of Transport would give its technical approval to allow the Trades Facilities Committee to formally agree the necessary loan guarantee.

The nervous Ministry of Transport labelled the Kennedy and Donkin cost analysis as controversial, which is of course ministerial code for not being accepted by everybody, and deemed that agreement between the three railways was essential. On 18 January 1922, following another indecisive meeting on 12 January in an attempt to end this stalemate, Arthur Neal MP together with four senior Ministry of Transport officials including Sir Phillip Nash had a meeting with the chairman and general managers from the three railways at the Ministry of Transport. The SECR was represented by Percy Tempest but, to add to the confusion, the SER faction was represented by Sir Alfred Smithers (deputy chairman of the SER) and the LCDR directors were represented by Lord Chilston. The LSWR was represented by the chairman, General Sir Hugh Drummond, and Sir Herbert Walker. The LBSCR was represented by its new chairman, Charles Macrae, and Sir William Forbes.

Arthur Neal in his capacity as private secretary to the Minister of Transport took a somewhat consolatory and understated position, declaring that his motive in calling the meeting was to try to remove any difficulties that there might be preventing the three companies progressing with their electrification proposals. Sir Hugh Drummond said that he and the LSWR welcomed the grouping but reiterated they were concerned that the new Southern Railway might have three separate electrification systems.

Percy Tempest seemed to have been prepared to compromise and proposed that the SECR might be persuaded to downgrade its three-wire system to a 1,500/750V specification. This proposal was viewed by the civil servants at the Ministry as a sound engineering compromise[13] offering many of the advantages of the three-wire system but allowing through trains from other railways to use the SECR's 750V positive rail. If the LSWR was to uprate its system to 750V the SECR trains could run on the LSWR lines after some switching on the negative power bogies to allow connection to the single conductor rail. Raworth may have reluctantly originated the proposal himself, or at least acquiesced to it, attempting to move events forward. This proposal which would, ultimately, allow a degree of integration, was accepted by Sir Herbert Walker and Sir Hugh Drummond as an important concession. But perversely Sir Alfred Smithers – speaking only for the SER directors – did not agree, insisting that the original proposal had been fully endorsed by the Kennedy and Donkin report and should be implemented. He maintained that there was hardly any interchange of traffic between the three railways, therefore the SER directors could see no reason why the railways could not each implement their own electrification proposals. The Ministry of Transport was now having to deal with four, not three, railways.

The LBSCR also seemed to be equally as adamant to act independently as it seemed to be thinking in terms of a grouping in name only. It asserted that, grouping or no grouping, it would press ahead with its ambitious overhead electrification scheme and proposed to expend, with or without Trades Facilities Committee assistance, £1.5 million immediately to complete its suburban electrification to Cheam and Coulsdon and furthermore proposed – heads-in-sand for this would be after the formation of the Southern Railway – to spend a further £10 million to complete the long awaited main line electrification to Brighton and Eastbourne. Charles Macrae also said that as there was little interchange of traffic between the LBSCR and the LSWR, the main issue as he saw it was the line to Redhill which was also used by the SECR. His solution was a simple one – all the trains between Victoria or London Bridge to Redhill should be run by the LBSCR using the overhead system.

Sir Hugh Drummond was becoming exasperated. Again he complained that the new Southern Railway was to have two systems of electrification and if the SECR had its way there would be *three*, which would not result in future economies or improved efficiency as intended by the grouping. But it was Charles Macrae, presumably confident about the suitability of the Brighton system, who suggested that the whole matter be placed before another new committee. Arthur Neal was not impressed; he did not want yet another committee of 'experts' each promoting their own preferred scheme. He made a more nuanced suggestion by proposing a new committee that would look closely for any areas of possible agreement and for any *errors* in the existing arguments. At this point, Arthur Neal and his Ministry of Transport entourage left the meeting in the hope that at the very least the railway companies could agree on terms of reference for yet another committee.

To Neal's surprise, the railways agreed to form a committee comprising of a representative from each railway with a remit to consider in the light of the forthcoming grouping the existing systems of electrification, how these systems should be extended, and what was the best way of proceeding to secure economic installation with the maximum of interchangeability. The committee should be chaired by an independent chairman nominated by Arthur Neal.[14]

Neal agreed with this offer and proposed that the chairman should be Sir Phillip Nash. The LBSCR nominated its long-time consultant Sir Philip Dawson; the LSWR proposed Mr Theodore Stevens – a London consulting engineer who had electrification experience both in Britain and abroad.[15] The SECR was to be represented by Sir Alexander Kennedy, the engineer who had written the original Ministry of Transport advisory report, had been a consultant to the LSWR and had so recently approved the Raworth three-wire electrification scheme. It is significant that apart from Theodore Stevens, all the committee members – including the chairman – had served on the 1920 Ministry Advisory Committee.[16] Theodore Stevens was the outsider. He was a member of the three main engineering institutions – civil, mechanical and electrical, and had his office at London House, London Street, in the City of London, next door to Fenchurch Street station.[17]

This committee may have been established to examine the practicalities of interchangeability – not the necessity for interchangeability. To make sensible recommendations however, some idea as to the order of magnitude of any future interchangeability was crucial. Enquiries were duly made to the three general managers. Sir Philip Dawson approached the LBSCR general manager, Sir William Forbes, who replied with a short letter with a long report attached. Sir William gave his personal view, ruling out widespread interchangeability because two of the companies had already electrified so much of their local services. He recommended that steam traction could provide any future interchangeability. His attached report listed all the LBSCR routes and recommended that on the joint lines dual operation with the alternating current system installed above the proposed SECR 3,000/1,500V conductor rail system was the solution. Not official LBSCR policy, but Sir William was advocating that his railway should continue what was best for his railway, and if others wanted to place conductor rails below their overhead equipment – then that was fine![18] Once again it would seem that the railway management at the time – with the notable exception of Sir Herbert Walker – seem not to believe that the imminent grouping was ever going to happen in any meaningful way. Life would go on exactly as before, with each company working independently as a part of a federal Southern Railway.

The committee looked at the cost of equipping the trains for dual operation over both the LSWR 600V lines and the SECR 3,000/1,500V lines. Sir Alexander Kennedy estimated that it would cost £800 to modify an SECR three-car set to run on the 600V system and £1,600 to modify an LSWR three-car set to run on the 3,000/1,500V lines. The possibility of electrifying the SECR suburban lines using the 6,700V system was discussed, but Sir Philip Dawson commented that 'his' system was unsuitable for suburban electrification. From the introduction of the 'elevated electric' – as the LBSCR marketed its suburban lines – it had always been the intention to extend the wires along the main line to Brighton, and had he been asked to design a suburban-only electrification scheme he would have designed a direct current alternative.[19]

Walker told the committee that he believed that each company should have access to all the 'Southern' London termini. His rationale was that operational flexibility would be achieved, avoiding the need to invest money in expanding these stations, if traffic could be shuffled around among the existing termini. But Sir Herbert did concede that adequate connections between the three railways did not currently exist and that it might take 10 years before significant track alterations were completed to achieve the level of interchangeability that he envisaged. One example cited by Sir Herbert was a variation on his earlier Waterloo tube ambition, as he proposed the construction of a new line tunnelled under the River Thames from a new junction just outside Waterloo to the City termini – Holborn Viaduct and Cannon Street. This, he hoped, would relieve Waterloo station which, despite the recent investment creating a virtually new station, was already suffering from excessive traffic.[20]

Sir Herbert did not accept the additional cost of converting existing or the building of new rolling stock to be able to operate on both the LSWR and SECR lines. He opined that no railway management would tolerate the idea of running two types of direct current system on the same railway. This was typical Walker bluster – he would have known that the NER was operating third-rail suburban trains in Tyneside and that a 1,500V overhead system was being proposed by the same railway for its main line, which would have resulted in two types of direct current system at Newcastle Central station. Sir Herbert's words did not go down too well with Sir Alexander Kennedy, who could not reconcile Sir Herbert's aversion to two direct current systems, the trains from which might by the simple operation of change-over contactors be able to operate on either system, while accepting the notion that long distance trains could be operated by a high voltage alternating current overhead line electrification installed above the LSWR third-rail system. It would be impossible to economically equip trains to work on both direct and alternating current systems. Sir Herbert further amazed the committee by claiming to be completely unaware of the possibility of using high voltage direct current for his trains – that is 1,500V from conductor rails or overhead lines as recommended by the Advisory Committee on Railway Electrification and the system clearly dismissed by Jones in his 1912 report. Sir Alexander did however point out to Sir Herbert that in his view 1,500V was suitable for suburban and main line electrification.[21]

The committee completed its work in early March 1922. It produced two reports, the main majority report and a minority report written on behalf of the LSWR by Theodore Stevens. The main report summarised the situation. The LBSCR had 111 track miles electrified using the 6,700V overhead system and it was considering adding another 662 track miles as part of its main line electrification to the south coast (to Brighton and Eastbourne, an ambitious main line scheme), and the LSWR had 178 track miles electrified using the 600V direct current and wished to add another 252 track miles (to Guildford via Cobham, as well as to Farnham and Windsor – an extension to its suburban lines). The SECR, of course, had no lines electrified, but Raworth's three-stage proposal would result in 648 track miles of electrification.[22]

The main report provided the cost comparison between the SECR and the LSWR systems for use on the SECR suburban lines. Costs for the LBSCR system were not calculated as Sir Philip Dawson had said that only a direct current system should be used for suburban traffic. The committee's calculations were for the total of the electrification proposals and not on the same basis as the earlier Kennedy and Donkin findings. These first costs were for electricity distribution infrastructure, alterations to permanent way, conductor rail, bonding and protection, multiple unit trains and electric locomotives. The estimated total cost using the SECR system was £7 million and for the LSWR system £10 million. It was agreed that the operating costs of the two systems would be about the same.

Approval again then for the Raworth system as far as cost was concerned – but what about the interchangeability issue? Three possible options were proposed.

The first – to use the Brighton system – was dismissed out of hand thereby avoiding converting the existing 600V lines to 6,700V overhead electrification. The second option was to adopt the LSWR third rail across the complete system; this would involve expenditure of £1 million to convert the already built LBSCR electrified lines and to impose upon the SECR lines a system which had been shown to be more expensive. The third option was to use a higher voltage direct current scheme (and in the terminology of the day this meant 1,500V direct current) but this too would mean scrapping all the existing LBSCR and LSWR equipment.

The report reiterated that with the two systems already in use, any adoption of a single electrification for the Southern Railway would involve the expense of dismantling existing plant. Contrary to Sir Herbert Walker's vision, the report did hint at the doubtful necessity for standardisation. Given the traffic patterns appertaining at the time, how necessary was it for trains to run across old company boundaries? The operators may see advantage in interconnection but changes to route patterns would be at a cost since new track and signalling arrangements would have to be installed. The committee, which had been set up to examine the practicalities of interchangeability – not the necessity for interchangeability – admitted that it was not able to comment on the economics of such extensive interchangeability.

The majority report recommended that for any future electrification the new Southern Railway should abide by the conclusions of the Electrification of Railways Advisory Committee and adopt 1,500V as standard for main line electrification. Higher direct current voltages had been successfully used elsewhere in the country (for example on the NER Shildon line), and the Southern Railway should follow this lead. Just as the LBSCR was eventually to be extended to the south coast, so too would the LSWR and SECR lines. The committee was adamant that when such extensions were required the 600V third-rail system would be unsuitable. New 1,500V rolling stock could be built, at a reasonable additional cost, for dual voltage operation. To facilitate this, it was recommended that the existing 600V system be over time uprated to 750V, since this would later facilitate 'dual voltage' operation when 750V-rated motors could be operated in parallel at 1,500V or in series at 750V.

If a conventional 1,500V 'two-wire' system – that is a single 1,500V conductor rail or overhead line with the return through the running rails – was to be adopted as the Southern Railway standard, this would not preclude the use of Raworth's proposed system on the SECR suburban lines. It was also concluded that the LBSCR 6,700V system was suitable for main line operation and the committee confirmed, as per the Electrification of Railways Advisory Committee's recommendation, that the then current proposal for a modest extension should go ahead.

Despite in general concurring with the earlier Ministry of Transport Advisory Committee's views, the latest committee's majority report nonetheless recommended that the SECR scheme should go ahead with Raworth's proposed scheme as it was believed that the adoption of this system was justified on the grounds of economy, since it could be installed at considerably less cost than the alternative systems and the three-wire system would meet the requirements of the SECR's 1902 Act of Parliament by avoiding excessive stray currents.[23] Perversely though, it was recommended that the SECR contribute £30,000 to the Royal Observatory for the relocation of its magnetic equipment to Abinger in far-away Surrey.[24]

This main report therefore completely vindicated Raworth's proposals and had given both the SECR and the LBSCR exactly what they wanted – but the mighty LSWR had lost out. Unsurprisingly, Theodore Stevens did not agree with the recommendations and wished to submit a minority report. On 4 March 1922 Sir Alexander Kennedy wrote to Sir Phillip Nash complaining that they were still waiting for Stevens' minority report which was delaying the submission of the majority report. While Sir Alexander had at first been of the opinion that they should await Stevens' submission, he finally conceded that the report should be published forthwith, as it was unacceptable that the minority report was not yet ready and furthermore it was not going to be possible to reach any mutual agreement with Theodore Stevens. An angry Sir Alexander believed that instead of simply stating his alternative opinion on matters in the main report, Stevens was adding in all sorts of other issues that the committee had not discussed. After the protracted final meeting lasting late into the previous evening, a tired and weary Sir Alexander had returned to his office in Victoria Street, Westminster, to find awaiting his return Alfred Raworth in a state of great anxiety – 'tearing his hair out' –desperate to proceed with his electrification scheme and bemoaning yet another week of delay.[25]

But despite Sir Alexander's pleading, Sir Phillip did await the arrival of Stevens' minority report and duly sent a copy of both reports to Arthur Neal on 10 March 1922.[26] In his covering letter Sir Phillip stated that Theodore Stevens' dissent went much further than he had led the rest of the committee to believe, and that other committee members had not been able to comment on the views expressed due to them not yet seeing Stevens' report. Furthermore, Stevens had not fully presented his objections at the appropriate time during committee meetings.

In his minority report Stevens first severely criticised Raworth's high voltage three-wire design. He stated that the highest voltage yet used on a railway conductor rail in Britain was 1,200V on the LYR Manchester to Bury line. His view was that two 1,500V conductor rails creating 3,000V between them would be hazardous because any contact with either rail by railway workers or others would certainly result in fatalities. Furthermore, it would be impractical to safely install insulated and protected rails with a potential difference of 3,000V. Most of us on first hearing of Raworth's '3,000V' system using ground mounted conductor rails instinctively tend towards this opinion. But while flying the safety flag, he in part contradicts himself by asserting that due to the lower voltage on the LSWR conductor rails, protection was not necessary. This viewpoint, even allowing for the pre-health and safety ethos of 1922, is quite surprising. It is relatively easy to get 'across' the 600V direct current supply when the single, unprotected charged rail is only centimetres from the 'return' running rail. It would be much more difficult if the live rail, albeit at a higher voltage, was provided with some form of timber protection, and almost impossible to accidently get 'across' the 3,000V due to the distance between the two protected rails. It could be argued that the Raworth 3,000V system with two protected rails was safer than a 600V system without any rail protection.

Next in his minority report Stevens attacked the principle that with the three-wire system no harmful return currents would flow through the running rails. He stated, correctly, that the negative centre rail would be broken frequently at pointwork when the train's negative bogies would momentarily have to coast across the gaps; and that in sidings only one overhead wire for freight locomotives, either positive or negative, would be used. In both instances, due to the infrequency of substations, current might flow up to 10 miles through the running rails to return. This is a very good point, but on the three-wire system, compared to the conventional earth return system, there would be a massive reduction in the return current with a proportional reduction in the small amount of stray currents that might cause damage. There would however be excessively high-voltage 'spikes' on the rails due to the length of the return path. He then denigrated the SECR claims to special circumstances on its lines, stating that the track conditions on the SECR were no different from those on the LSWR, citing observations from the LSWR chief engineer, who had once been employed by the SECR (Alfred Szlumper's first appointment had been an engineering assistant for two years with the SECR). Stevens also stated that, based on the LSWR's practical experience, the danger to third parties' equipment had been exaggerated – no doubt the National Physics Laboratory at Teddington would have a view on that assertion!

Stevens then cast doubt on the practicality of using multiple unit control systems at a voltage as high as 3,000V.[27] A strange and spurious objection since Raworth's design would have used 1,500V control equipment. The massive 3,000V freight locomotives employed on the Chicago, Milwaukee and St Paul Railroad using 1,500V motors in series across the 3,000V supply made use of the Sprauge-General Electric Type M control system which facilitated the multiple unit operation of these massive beasts in pairs.[28]

Theodore Stevens had presented to the committee his estimation of the likely costs of electrifying the SECR suburban lines. He based his calculations on the costs to electrify part of the New York Long Island Railroad (LIRR). Stevens maintained that to build Raworth's 3,000V system it would cost £13,895,000 and to build the equivalent 600V scheme it would cost £13,095,000. In the *Electric Railway Journal* published later in the year, Stevens is credited with costing calculations for a 'British suburban electrification' (and this can only be the SECR scheme) using data based on a route length of 90 miles. He showed that the comparative costs based on 'dollars per single track mile' for the small 600V conductor rail scheme worked out at $143,000 and for the equivalent 3,000V system $160,000.

Theodore Stevens had presented a table to the committee based on the LIRR to compare power consumption and train mileages with the costs that Raworth had provided for an alternative 600V scheme for the SECR.[29] He pointed out that the LIRR electrification was for a compact network of lines, none with a route length longer than 30 miles. The LIRR, like the LSWR, used 11,000V three-phase alternating current to feed the substations equipped with transformers and rotary convertors from which direct current at about 600V energised the conductor rails.

Matters seem to be getting personal because in the report presented to the committee, the SECR proposal table is headed 'Raworth', not 'SECR'. Stevens hoped to convince the committee that for a similar track mileage (227 miles for the LLIR, 210 for the combined Raworth SECR Stages One and Two) the American system used only 14 substations while Raworth's latest estimate for a 600V scheme was 46. This is obviously where the massive cost discrepancy comes in and Stevens showed that the total capacity of the LLIR substations, spaced out along the line every four and a half miles, was 55,000kW while Raworth was insisting on SECR substations spaced at just over two and a half miles apart with an aggregated capacity of 144,000kW.

This may seem all very damning for Raworth's calculations but was Stevens really comparing like-for-like? There is scope for some subjectivity surrounding both the traffic density (number of trains making frequent stops and starts) and the sizing of rotary convertors. It was usual to provide machines with enough capacity to accommodate the substantial short duration overload required to provide short bursts of high demand as trains started away from stations. Also, given the high amount of maintenance that the rotary convertors required (removing powdered copper dust from around the brushes, cleaning up the arcing damage on the commutators, etc.) how much plant do you install to be held in reserve for use when the plant is out of service for maintenance?

Stevens also claimed that the proposed British trains were lighter than the American trains; the SECR trains were expected to be 325 tons, while on the LLIR the maximum train weight was 672 tons. Speeds too were higher along Long Island – an average of 32mph compared to the projected 25mph on the SECR.[30] But here again questions must be asked – were there a comparable number of stations along the route? Frequent stops and restarts would also impact on the power output required from the substations – and reduce the average speed.

It might be argued that because Stevens based his calculations on 'as built' costs, not estimates, they were likely to be more accurate. It is questionable however whether as-built costs of a similar project are safe to use when related to a scheme in another country where there may be any number of local factors to distort the calculations.

Stevens does not mention batteries. Significantly, at least one substation on the LIRR, number five at Hammel, had a huge battery installation built in 1905 to keep trains running in the event of a power failure.[31] Many railway and tram companies used batteries to supplement the power supply at times of peak load or as standby for outages due to faults or planned maintenance, removing the need for spare rectifier units. Other, smaller, lines such as the Manx Electric Railway (MER) from 1893 onwards used batteries to provide a power boost when required, recharging taking place during periods when the trams were not running.[32] Furthermore, the LIRR had built four portable substations – rail wagons containing substation equipment – which could be strategically positioned on the network at times of heavy seasonal traffic, such as race meetings or in the summer when crowds flocked to the beaches.[33] Raworth, of course, never intended to use batteries or portable substations, so these were probably the reasons why Stevens was not comparing like-for-like when calculating the amount of substation plant required.

The main difficulty with accepting Stevens' figures from the LIRR data is why he referred to a 17-year-old electrification scheme in a foreign country when he could have used data from the railway that was employing him as a consultant – the LSWR – which had relatively recently completed a similar electrification scheme. Ultimately Theodore Stevens failed to convince Sir Phillip Nash and the rest of the committee that his calculations formed a valid argument.

Another interesting conclusion in Stevens' minority report is that while he states that 600V is ideal for suburban electrification on a network like the LSWR's (at the time most engineers would concur with that point of view), he also maintained that 1,500V, as recommended by his committee colleagues who had previously sat on the Ministry's Advisory Committee on Electrification, was totally inadequate for railway electrification. This view had been formally stated by Herbert Jones and was also later to be reinforced by Alfred Raworth. Stevens made it quite clear that it would be pointless to use a new standard of 1,500V as a compromise voltage as it was neither suitable for suburban or for main line operation.

Stevens concluded that the existing LBSCR 6,700V alternating current system using overhead conductors should be extended. But then Stevens even questioned the very need to electrify *any* of the longer distance railways. He asserted that in Britain, with adequate cheap coal and no available water power for hydroelectric schemes, there was little prospect of general electrification and he saw no advantage that would justify the capital outlay.[34] This was consistent with his own publicly expressed views, for later in 1924 he stated at an Institution of Electrical Engineers meeting that in 1912, when acting as a consultant for a foreign electrification scheme which intended to take advantage of a new hydroelectric plant, he had advised the promoters that they would get a better return on their investment in the power plant if the electricity was sold to the nearby towns and the railway remained steam operated. He concluded that electrification should only be undertaken when there are substantial reasons to warrant the capital expenditure.[35] Taking forward his logic, he seemed to be saying that unless there is a good *other* reason to electrify – such as the SECR had needing to overcome track occupancy problems around London Bridge – stick with steam.

Stevens' wide-ranging and assertive minority report had many 'Walker' features – superficially convincing but lacking reliability. Therefore it may be suspected that Stevens' rather unprofessional behaviour in delaying his report allowed time for him to consult with Sir Herbert Walker and Herbert Jones, allowing himself to be coerced into putting forward new objections in the time between the committee's last meeting and the publication of the reports. To be fair to Stevens, his report showed consistency with a lecture he gave later that year at the LCC's School of Engineering and Navigation,[36] when he compared unfavourably the LYR's 'high voltage' 1,200V conductor rail system with the LSWR 600V system.[37]

But even at the SECR there was some concern about the perceived dangers associated with the Raworth 3,000V system. The Kennedy and Donkin report had raised concerns as to the safety of track maintenance staff, and the SECR permanent way engineers were duly consulted. It is probable that the pure 'Raworth' design would have used only single boards either side of the live rail (as the LSWR used in safety critical locations such as at stations). This would have allowed electrical contact at the top of the rail from a single central shoe on the middle, negative rail and made interchangeability with other railways' systems much easier. The assumption has been however that the rail would be side contact and shrouded in a similar fashion to the arrangement used by the LYR on its 1,200V scheme. George Ellson, one of the SECR resident engineers who had previously attended the Group Electrification Committee meetings with Alfred Raworth, was consulted. (See plate 5 (bottom)) Ellson, a man who was later to become chief engineer for the Southern Railway,[38] then designed his 'live rail protector'. It is somewhat quaint that the description of the arrangement implies that it is the rail that is to be protected – not the workforce from high voltage electricity. In February 1922, while awaiting formal approval for the scheme, Percy Tempest had requested permission from the SECR directors to proceed with a patent application for the rail protector in the joint names of George Ellson and the SECR. This side contact arrangement may not have been to Alfred Raworth's design (or even liking) but it was his cousin and former colleague at Raworth's Traction Patents, John Raworth, who was appointed as agent for the patent application for the SECR.[39] The 'Ellson' side contact rail does bear a close resemblance to the LYR design,[40] the major difference being the way that the side boards were held in place – this might have saved the SECR if the LYR had demanded royalties.[41]

With the Trades Facilities Committee still patiently awaiting confirmation of the Ministry of Transport's technical approval prior to instigating any financial arrangements with the Treasury, on receipt of his copy of the main report Percy Tempest wrote immediately to the Ministry of Transport pointing out that the SECR electrification scheme had been approved and that it was time for a decision on the financial arrangements to be finalised.[42] The Ministry knew it had to move fast, for it was under severe political pressure to remove any obstacles to the electrification of the railways south of London; specifically from the House of Commons Industrial Group of MPs (which included the recently elected Conservative MP Sir Philip Dawson) who claimed that this would alleviate the then current stagnation in trade.[43] The depressed economy was continuing to cause mass unemployment in the land which, while it had been falling slightly, remained at over 14 per cent.[44]

Arthur Neal wished to get everybody together to settle the matter once and for all, to avoid excessive correspondence and endless meetings. So, only six days after he had received the report, he called a meeting at the House of Commons where, with his four Ministry of Transport colleagues, he faced a delegation formed of the chairmen and general managers of the SECR, LSWR and LBSCR.

He opened the meeting by confirming that the committee set up to consider the Southern Railway electrification options had completed its work and the majority report had recommended the adoption of the SECR system and the extension of the LBSCR system. He reported that he had informed the Trades Facilities Committee that the proposed financial arrangements could now be set in place. Charles Macrae, the LBSCR chairman, stated that the LBSCR had been awaiting the outcome of the report, and subject to any change of mind resulting from this meeting, would proceed immediately with an application to the Trades Facilities Committee for finance to commence the modest suburban extensions to its electrified railway. Cosmo Bonsor commented that he was pleased that the SECR scheme had at last been given the go-ahead and that he understood that in the immediate pre-grouping times consideration had to be given to the views of the future partners in the Southern Railway. General Sir Hugh Drummond on behalf of the LSWR concurred with Bonsor but did comment that there remained certain safety concerns associated with the SECR scheme and it was unfortunate that the new Southern Railway, as such a small group, would have to have as many as three different electrification systems on its network. The meeting closed in an apparent, but deceptive, air of warm self-congratulation.

A few days later at the SECR board meeting held on 22 March, it was recorded that the railway had finally received from the Ministry of Transport formal permission for its electrification scheme and that the draft legal agreements were being prepared to initiate the Trades Facilities Committee loan guarantee arrangements with the Treasury.[45]

Despite the apparent earlier outbreak of goodwill, a letter from the LSWR was received at London Bridge clearly indicating that the SECR may have won the battle but it had not yet won the war. At first the letter, signed by General Drummond, congratulated the SECR on receiving approval for its electrification scheme and reaffirmed that the LSWR had no objections to the SECR proceeding with this proposal. He agreed that the SECR had every right to apply for a loan on the terms agreed with the Trades Advisory Committee. But then the LSWR chairman outlined in the strongest terms the LSWR's negative opinion of the Raworth electrification proposal. Its first objection was that the design did not conform to any existing electrification system. Furthermore, if the SECR went ahead and electrified its suburban lines as proposed, the new 'Southern' group would have three separate and incompatible electrification systems to manage. This would, he claimed, be detrimental to traffic and would incur additional cost. It was also claimed that the Raworth scheme did not comply with the suggestions made by Sir Alexander Kennedy's original report and that such a system with two live, high-voltage conductor rails would constitute a danger to employees and the public. Having Damned the proposals from the technical viewpoint, Drummond went on in his letter to point out what was probably the LSWR director's primary concern, which was that while the LSWR had consistently refused to incur heavy capital expenditure at the current inflated prices, the SECR was now to carry on regardless with expenditure that would ultimately fall upon the new Southern Railway when established.[46] This extravagant broadside may have been fired by the financier Drummond, but the gun was probably loaded by Sir Herbert Walker, for it has all the hallmarks of his combative style, relying on dubious assertions.

The committee was not to be easily dissuaded. A special meeting was held with the SECR shareholders who gave their approval of the terms offered by the Trades Facilities Committee. Consequently, formal Treasury approval for the guaranteed loan was received in April 1922: £5 million for the Stage One electrification and £1.5 million for the generating station at Angerstein's Wharf.[47] As per the terms of the offer a separate company, the South Eastern and Chatham Power and Construction Co. Ltd, was established with a limited liability of £10,000. Cosmo Bonsor was the chairman and there were six other directors including Percy Tempest.[48] The first formal meeting of this new board was held on 2 May 1922. Later in the month the *Railway Gazette* reported that the SECR intended to electrify its suburban lines and that the three-wire proposals had met with the approval of the committee that had been set up under Sir Phillip Nash by the Ministry of Transport.[49]

Now the SECR electrification project could commence. First, additional staff had to be recruited and more office space found for Alfred Raworth's electrical department. His office accommodation at 19 Station Approach was duly enlarged by the renting of the adjoining rooms at number 20 Railway Approach where the present occupier, Mr Boyle, was paid £150 compensation to relocate himself to number 31.[50] With his new managerial responsibilities, it was agreed that Alfred Raworth was to have his salary increased to £2,400 to take effect from 1 March 1923, which was two months after the SECR was due to cease to exist. Yet another example of how the three railways at this time seemed to believe that the grouping was in name only and that it would have little practical effect on the real railway business.

In June 1922, to assist Alfred Raworth with his electrification project, an additional assistant engineer was required and consequently Mr Frank Farly was engaged with a salary of £700 a year. A new chief draughtsman was also needed, and this was to be Mr W.C. Moore whose salary was to be £450 per year.[51] Moore was destined to continue to work very closely with Raworth throughout the forthcoming Southern Electric era,[52] and to assist him another draughtsman, Mr C.R. Wringly, was appointed on £425 per annum. Every project office needs good clerical backup and a new chief clerk, Mr W.G. Young, was duly appointed on a salary of £400 per annum. He was to be another long-standing member of Raworth's staff. The Joint Management Committee decreed that 'other draughtsman, typists and a messenger' were to be 'employed at weekly rates'. Meanwhile Kennedy and Donkin were contracted to provide a consultation service to check calculations, prepare specifications and provide other general assistance for the following three years for a very meagre fee – compared to what they had been paid to assist the LSWR – of £2,000.[53] This lower fee suggests that on this occasion Raworth's staff – which now included a small drawing office – were to do much more of the work in-house.

CHAPTER 11

THE BATTLE OF ANGERSTEIN'S WHARF

The site for the proposed SECR generating station was at Angerstein's Wharf, situated on the River Thames at Charlton. The wharf had been named after John Julius Angerstein, the local landowner. A short connecting railway, one mile in length, had been built by the wharf owners to a junction with the SER between Blackheath and Charlton. This line was leased to the SER from 1852 until 1898, thereafter the railway company took over full ownership of the branch line and the wharf. The 26-acre site offered 20 acres of space on which to build the generating station, with a wharf frontage of 675 feet (206 metres) along the River Thames. The site was suitable because it had an abundant water supply from the river which, at this location, is three quarters of a mile wide with more than 20 feet (6 metres) tidal rise and fall. Coal could either be delivered from coastal vessels delivering up to 2,000 tons at a time or alternatively by rail along the Angerstein's Wharf branch.[1] Raworth had written in his report that there was not only sufficient space for the proposed power station but also room for any future expansion and, from the electricity distribution perspective, the location was only three and a half miles from the proposed Lewisham substation where two-thirds of the station's output was required. Raworth believed that the site was perfect for his proposed scheme. The branch, and the wharf, remain in use today – now simply Angerstein Wharf – as an aggregate terminal and in the past has been used as an oil terminal and by the United Glass Company.[2]

Despite Raworth's proposal for the SECR to build its own power station, it was clear by 1920 that there would be difficulties with the newly established Electricity Commissioners, whose view was likely to be that the railway should take its electrical energy from the public electricity supply. Despite his report proposing a railway-owned generating station, Raworth acknowledged that a Bill then before Parliament might prevent one being built. He did however believe that if proposed amendments to the Bill were enacted the SECR would be allowed to build its own generating station.[3]

The SECR was not inclined to deal with any of the local electricity supply companies and part of this prejudice may have originated during the many disputes with the electric power companies concerning lighting supplies to stations. As in 1912, the only practical 'outside' supplier was the LESC, the owner of the pioneering power station at Stowage Wharf in Deptford, yet once again the SECR seemed to look away from this solution.[4] The very cosy relationship between the LESC and the LBSCR and the continuing animosity between the SECR and the LBSCR might explain why the SECR appears never to have made any enquiries about taking an electricity supply from the Deptford generating station.

The SECR now had to make a formal application to the Electricity Commissioners before commencing work on its hoped-for generating station. For the background to this new legislation, it is necessary to wind back three years to when Sir Eric Geddes was attempting to get his Ways and Communications Bill passed by

Parliament. Sir Eric had made it clear at the second reading debate in the House of Commons that he believed in the concept of unified control applied to both railways and the electricity supply industry.[5] He cited the situation that existed on the railway of which he had recently been the deputy general manager – the NER. In Newcastle, where the NER operated its third-rail direct current Tyneside suburban lines, he maintained the electric power company had worked hand in hand with the railway, leading to a growth in trade to the benefit of both companies. Geddes argued along the same lines as Raworth, that the larger and more diverse the load on a generating station the cheaper the power. But Geddes was arguing that the railways should not own the power stations, but that they should instead obtain their electricity from the 'public' electricity companies. Geddes asserted that if the main line railways were electrified, 20 per cent of all the electricity generated could be used for traction. This reasoning was used to justify his original intention to combine transport and electricity supply into a single Act of Parliament and thus a single government ministry. But Parliament was suspicious of such an all-powerful organisation. Consequently, the coalition government backed off and decided to make its changes using smaller steps. It published two separate bills, the Ministry of Transport Bill and the Electricity Supply Bill. When the second Bill was passed by Parliament the resulting Electricity Supply Act established the Electricity Commissioners, who acquired powers previously held by the Board of Trade. The aim was to integrate and regulate the electricity supply industry and to set up local joint electricity authorities. By this roundabout route Sir Eric did get his way, as it was agreed that the Electricity Commissioners were to be part of his new Ministry of Transport.

Charles Merz, with his association with the NER and the local electricity company that had been founded by his father, was a confidant to Sir Eric Geddes. In 1917 he had played a large part in the preparation of a report for the Coal Conservation Committee in which he had been critical of the way the railways had built power stations at locations often badly sited for water supply. Merz's view was that the building of the railway power stations in London, including the LSWR's station at Wimbledon, had been carried out in a most irrational and haphazard fashion.[6] He had, of course, recommended Angerstein's Wharf to the SECR, but his explanation for this proposal was that it was not possible to obtain a suitable local supply *at that time*. Despite the apparent split in the legislation procedure, the government's view on the relationship between railways and power companies remained the same and was quite clear – power companies were to generate the electricity for the railways wherever possible. The new Electricity Supply Act did not prescribe that the railways could not generate their own electrical power, but it gave the responsibility to the new Electricity Commissioners to adjudicate on any applications in line with the new government policy. Sir Eric Geddes had asked Charles Merz to be chairman of the Electricity Commissioners, but he had declined, preferring to continue his work as a private consultant, but he did give advice as to who should be the new commissioners. He was a quiet and unassuming man who hated public speaking, so it was completely in character for him to decline such a high-profile government appointment.[7] The first chairman of the Electricity Commissioners was to be Sir John Snell.[8]

Given that the whole impetus for the legislation was to encourage the spread of the public electricity supply with standard voltages and frequency, and that large public schemes could be accomplished on the back of the needs of large users such as the railways, the outlook seemed bleak for the SECR project. Fearing possible refusal of the SECR application and acutely aware that time was running out – the loan guarantee offer from the Trades

Facilities Committee was due to expire in November – Cosmo Bonsor made a request for assistance to Arthur Neal asking him to make representation to Sir John Snell on the SECR's behalf. In his reply to Neal, Sir John did state that the SECR request to build its own power station could only be refused if there was an electricity supply company that was willing and able to provide a supply to the railway which was of adequate quantity, was reliable and was at a cost no greater than would have been incurred by the railway itself. Therefore, any commercial and technical offer being made to the SECR had to be better than Raworth's Angerstein's Wharf proposal.[9] To get exactly what he wanted, all Bonsor had to do was instruct his electrical engineer and general manager to produce a financial case that cost less than any 'outside' scheme. This should not be difficult as the SECR would not have to purchase any land; Bonsor was prepared not to book any profit to the Angerstein's Wharf generation as any private company would have to do – the SECR would make profits from increased passenger revenue alone; and the SECR had a better credit rating than any of the power companies, which would ensure a lower interest rate when capital had to be raised.

Furthermore, the SECR believed that its wide-ranging 1902 Act already gave it all the approval it needed to build a generating station at Angerstein's Wharf. Unfortunately, there was an unexpected technicality as Angerstein's Wharf had not been mentioned as a possible site in the original 1902 Bill. This was very irritating because the 1902 Act had been quickly passed through Parliament with little scrutiny of the 'named' sites. But technically the six sites had been approved by Parliament. The resulting Act may have allowed the SECR to build on any other site if required or necessary, but it was argued that due to the lack of scrutiny the Angerstein's Wharf site required an Order as per the later 1903 Railways (Electrical Power) Act. This long-forgotten and completely ignored piece of legislation, intended to scrutinise what we would now understand as 'planning permission' issues without necessitating an Act of Parliament, should have been used by the LSWR in 1912 and its failure to do so at the time had been negligently tolerated by the Board of Trade. Given the animosity that the SECR's future partners had towards the electrification scheme and that it would have been easy for it to be innocently overlooked by the government, it is possible to suspect that somebody somewhere had maliciously called 'foul'. But it appears that it was not the LSWR or the LBSCR that had noticed this anomaly – the prime suspect is the West Kent Electricity Supply Company. The SECR justifiably felt upset by having yet another unwanted hurdle to cross and the news came as a shock. Cosmo Bonsor duly took legal advice only to be informed that an Order as per the 1903 Act was necessary.[10]

Bonsor wrote a formal letter to Sir Philip Nash at the Ministry of Transport in respect of an Order under the 1903 Act.[11] He clearly stated that the SECR had no objections to making the application for the Order but was concerned about the time that the procedure required by the Act would take. A long wait was unacceptable because the guaranteed loan arrangement from the Treasury was to provide immediate work for the unemployed. This was a constant mantra by the SECR, but of course the directors and officers of the railway primarily wished to get their scheme underway before they were thwarted by the forthcoming grouping. Nonetheless, Sir Phillip seemed very keen for the SECR scheme to commence as soon as practical. The implications of the 1903 Act were discussed with Bonsor who requested that the rules be relaxed so that an Order could be raised as soon as possible. The rules as they stood required that formal notices must be displayed in the *London Gazette* and local newspapers for two consecutive weeks, copies of plans must be deposited three weeks after the first press notice and a further three months wait was decreed to allow objections after the first press notice.

Resulting from these discussions Alfred Raworth had a meeting with Sir Phillip Nash at the Ministry of Transport on 2 May 1922.[12] The 1903 Act placed responsibility on the Ministry to follow a consultation procedure and Raworth was advised by Sir Phillip to speak directly with the authorities who would need to be consulted, principally the chief mechanical and electrical engineer for the Post Office, Sir William Noble, and to go to the Ministry of Works to meet Hubert Bains to see if the process could be 'fast tracked'.

Meanwhile, Arthur Neal spoke informally to his officials whose initial view was that it would take longer to change the rules than follow the prescribed process, but they did cooperate by agreeing to speak to any of the bodies that might have legitimate grounds for objection to see if they would accept less time to respond. In the spirit that time was of the essence because of the urgent need to find work for skilled workers in the electricity manufacturing industry, the canny civil servants contacted the appropriate organisations. They obtained an answer from the GPO that stated that as it was by now used to dealing with railway electrification schemes and had in place standard procedures to deal with issues such as possible interference due to induction, it was able to turn around a standard reply straight away. The Admiralty, as custodians of the Greenwich Observatory, seemingly discounting the 'magnetic room' issue, merely wanted the same power station chimney conditions as had been applied to the LCC tramway power station in Greenwich. This covered issues such as chimney height, type of fuel and flue gas temperature in order that their view of the stars would not be obscured. One other possible objector was the Port of London Authority which it transpired merely required reassurance as to how the work affected the river. It appears that the back-door methodology was equally effective as Alfred Raworth directly appealing to the titular heads of the Post Office and the Ministry of Works. The only sour note came from the Ministry of Works saying that three months was not enough to establish the likely effect of stray currents on the royal palaces and other important buildings, citing the bad experience with the LSWR due to its electrification. It took the opportunity to remind the Ministry of Transport of the railway's lack of cooperation at the start of that project.[13]

The SECR, having formally applied to the Electricity Commissioners for permission to build its power station on 23 February 1922, did not have long to wait, for, as decreed by Section 11 of the Electricity (Supply) Act 1919, a public inquiry was arranged, and this was opened at the Middlesex Guild Hall in Westminster at 10.30am on Tuesday, 30 May 1922.[14] The Guild Hall stands just off Parliament Square, and in 1922, despite being in the City of Westminster, was the property of Middlesex County Council and was relatively new, having been rebuilt in what was described as an 'art nouveau Gothic style' between 1912 and 1913. The inquiry was chaired by Sir John Snell. The updated SECR application was for a 60,000kW power station which would provide for the first two stages of Raworth's electrification proposals, but the site had space for future expansion. At the same inquiry, the Electricity Commissioners were to examine the related application received from the West Kent Electricity Supply Company that wished to build its rival 150,000kW power station at Belvedere, Kent. Given that the location of the Belvedere site was close to the proposed SECR electrified line it therefore suited the government's intentions precisely, in that West Kent was an independent electricity supplier that wanted to develop its network and saw the SECR as the perfect large customer to justify the construction of a new generating station. Also present were representatives from both the LESC and the County of London Electricity Supply Company (CLESC), as following the publicity that the inquiry procedure had created, they both wished to bid for the SECR railway electrification business.

The CLESC wished to supply the SECR from its proposed Barking power station situated on the north bank of the Thames which was on the other side of the river from the SECR railway lines.[15] Not to be discussed was Alfred Raworth's preferred alternative proposal which was that the Electricity Commissioners sanction a 'public' 'super-station' at Angerstein's Wharf which would be three times the size of Lots Road, generating at 25 cycles per second to supply local industries and the SECR – and ultimately the LBSCR and the GER.[16] (See plate 7 (top))

The SECR went to the inquiry in a combative and confident mood. The railway's lawyers were Mr W.B. Clode KC and his two learned assistants. The SECR technical experts were to be Alfred Raworth and, from the appropriately named firm of consulting engineers Charles P. Sparks and Partners, their senior partner Mr Charles Pratt Sparks MICE MIEE. One of the many counsels present asked Sir John Snell just how long he thought the proceedings would take as the following week being Whitsuntide, holiday arrangements had been made by many attendees. The reply from Sir John was that he hoped it would take two days. However, so intense and convoluted were the arguments that it was to be 10 working days before the completion of the proceedings.[17]

The SECR was full of confidence even if its evidence was at times irrational. When W.B. Clode made his opening statement on behalf of the SECR, he maintained that over the last 20 years the SECR's suburban traffic had been effectively 'killed off' by competition from bus and tram operators and the SECR needed electrification to recover the situation. This surprising remark, later affirmed by Cosmo Bonsor from the witness stand, did not quite tally with the SECR's dire track occupancy difficulties caused by a massive *increase* in outer suburban traffic. Clode maintained that the SECR had always been keen to electrify its lines but circumstances such as the Great War and other calls on the SECR's time between 1903 and 1920 had caused the railway to somehow, quite mysteriously, 'forget' about its electrification scheme. It seemed that the SECR's memory had in part been jogged by the Metropolitan Borough of Woolwich which had recently passed a resolution requesting that the SECR get on with its parliamentary approved electrification scheme and electrify the North Kent loop.

The SECR, Clode claimed, believing that it had powers to build a generating station on its own land, did not, until only weeks previously, understand that it needed to apply for an Order as prescribed by the Railways (Electrical Power) Act 1903. This oversight was quite reasonable because no railway to date had ever been compelled to make use of this piece of obscure legislation. Given these unfortunate circumstances it was put to the Electricity Commissioners that as an application had now duly been made, and that the 1903 Act allowed for an inquiry if deemed necessary, would not the present inquiry cover all the likely issues? Sir John Snell, without completely concurring with this, did agree that the proposal for a dual-purpose inquiry did seem to make sense.

Clode told the inquiry that the loan guarantee arrangement with the Trades Facilities Committee was only valid until November – less than five months hence – and if not accepted by then it would be withdrawn. Furthermore, he asserted the SECR was confident that it could not obtain its electricity supply from any outside source at a price that was less than the anticipated cost from its own installation at Angerstein's Wharf, nor would any outside source be as reliable. Another advantage he cited was that the SECR, and the Southern Railway to be, would in 25 years' time *own* the generating station that it had built. Future development at the site would provide electricity to other parts of the Southern Railway, for example the LBSCR lines where the 17-year contract with the LESC was due to run out in five years' time. Clode explained the SECR's reluctance to accept any offer from the West Kent Company was because it was a small, indebted company. This was Bonsor's main concern since the West Kent was effectively owned by the Metropolitan Carriage Wagon and Finance Company.[18]

This reason for not agreeing a contract with the West Kent was just not good enough in the eyes of the Commissioners. The counsel for the West Kent Company, Mr Tyldesley Jones KC, maintained that the SECR was being unreasonable and just making excuses. Jones claimed that his client had not been given enough information to be able to make a reasonable offer for an electricity supply. When Clode was questioned by Jones and Sir John Snell himself, they perceived that the crux of the issue was that the SECR had deliberately withheld information and failed to fully specify its requirements so that the power companies could not make a realistic offer. Naturally Clode would not accept this – but the gloves were off!

Bonsor was the first witness to speak for the SECR. He took the stand full of pompous confidence, which did not impress Sir John Snell, who opined later that the inquiry had been made to *suffer* the testimony of H. Cosmo O. Bonsor. The SECR's chairman's main argument, which to him seemed perfectly reasonable, was that the SECR had received a government backed guarantee of a loan to electrify suburban lines and that this offer included the generating station; and since the aim of this guarantee was to provide jobs for the unemployed, it was the duty of the Commissioners to effectively rubber stamp the SECR's request and to approve the scheme forthwith. Sir John took great exception to this standpoint for he had been charged with regulating and integrating the nation's electricity supply provision and felt his inquiry would be reduced to a farce if he was not to look at the SECR's proposals in the context of the nation's overall electricity supply provision. Jones, representing the West Kent Company, argued that his client had made a serious offer based on a fixed charge to cover the capital equal to what the SECR would have to pay for its own power station; but in his evidence Bonsor insisted that no formal offer had been received. So, what did go on at the negotiations between the SECR and the West Kent? Bonsor limply said that he himself had taken no part in the discussions as this was all done by his general manger, Percy Tempest. This prompted Sir John to ask if Tempest was to give evidence, and Clode had to admit that it had not been his intention to call him. This annoyed Sir John and he promptly insisted that Tempest must be called later as a witness.

Bonsor was questioned about the loan arrangements with the Trade Facilities Committee. He told the inquiry that it had deemed the SECR's request as one of the best schemes that had ever been put before it. The first negotiations, Bonsor affirmed, had been for only a £3 million guarantee but later, when it became apparent that the SECR would not be able to obtain an electricity supply from an outside source, the Trade Facilities Committee was happy to add an extra £1.5 million to cover the generating station costs. After further questioning, Bonsor explained that he suspected that the £1.5 million for the power station might be excessive, but he would only borrow the money as and when required and that it might not be necessary to take the full amount. Asked about relations with the other partners in the forthcoming grouping, Bonsor, with a degree of mendacity, claimed that the LSWR and LBSCR were now in favour of the scheme and that their only objection had been that the SECR was proposing yet another electrification system, but he claimed that it had been *proven* that there was no interchange of traffic on the suburban lines.

Sir John Snell and the other legal councils queried just how serious the SECR had been in obtaining an 'outside' electricity source. Bonsor was asked to comment on quotes from first Sir Francis Dent, who had long since moved away from the

SECR decision-making process but had apparently stated that it was perfectly acceptable for the SECR to take an outside supply, while Sir Alfred Smithers had apparently expressed the opposite view by asserting that it was as ridiculous to source electricity from anybody else as it would be to suggest that others own and run the SECR steam locomotives. The SECR's opponents in the Middlesex Guild Hall were obviously minded that the latter opinion was held by the SECR. Bonsor could only confirm that both these gentlemen were representatives of the SECR and in principle the railway would accept an outside supply, but only if the terms were right.

Later in the week, when proceedings had already exceeded Sir John's estimate as to how long the inquiry would take, Alfred Raworth took the stand. The intention was that he was to enlighten the inquiry on engineering matters, but as he had been with Percy Tempest during discussions with the West Kent and the general manager was yet to take the stand, Raworth had to soak up many awkward questions relating to the discussions about a possible supply from the proposed Belvedere generating station. Raworth appeared to be brusque and tetchy, in contrast to the confident young Alfred Raworth who spoke in 1906 to the Leeds branch of the Institution of Electrical Engineers giving full and lucid responses to the technical questions concerning regenerative control.[19] But, Bonsor had already given too much away, and Percy Tempest was yet to speak on the contentious issue of the negotiations with the West Kent, so it made tactical sense for him to say as little as possible and avoid being tripped up by the clever lawyers. Consequently, he provided only a very minimal response to long and convoluted questions. Raworth was showing that he could 'stonewall' when asked difficult questions at public inquiries just as well as his past and future boss Sir Herbert Walker. Sir Herbert was a past master at managing such inquiries, and while there is no record that Sir Herbert attended the Angerstein's Wharf inquiry, it is inconceivable that he would not have had his 'spies' in the hall so as to receive a full account of the proceedings. So, if Alfred Raworth was stonewalling to protect the interests of the SECR and to avoid further embarrassment for his chairman, his conduct under interrogation would not have done his reputation any harm in Sir Herbert's eyes.

Raworth got off to a bad start. Unlike Charles Sparks and Herbert Jones (whose evidence earlier to the inquiry largely supported the technical aspects of the SECR scheme), he had never found the need to progress to full membership of any of the engineering institutions. He was not asked by the barrister to confirm his 'professional qualifications' which would have endowed his evidence with a veneer of unassailable competence. For this reason, outwardly at least, Alfred Raworth appears to have been given a tougher time during questioning and gave his answers in brisk, short sentences.

The main aim of the Electricity Commissioners and the West Kent's barrister, Mr Tyldesley Jones KC, was to probe into the details of the 'negotiations' between the SECR and the West Kent. Alfred Raworth had been with his general manager as advisor during the discussions. Jones asked Raworth if it was only when it was found that an outside supply was not available that it was decided to build the SECR power station. Raworth confirmed this was the case, a truthful answer, since he had come to this conclusion back in 1919 when he wrote his report. Next, he was asked if it was on his advice that the decision was made to build the generating station. Raworth confirmed that it was – truthful, but this was not a recent decision, it had been part of his 1919 report. Raworth was subjected to questions designed to ascertain his competence, for having admitted recommending the building of the SECR generating station, he was then forced to admit that he did not have any managerial experience of operating a power station.

Jones continued to try and tease out the timing of the decision to build the Angerstein's Wharf generating station. He suspected that the SECR had never really wanted an agreement with any outside electricity supplier. He alleged dishonesty by the SECR in respect of its dealings with the West Kent company, since it had been deliberately vague about its requirements. He strongly implied that both Raworth and Tempest wanted the freedom to get on with their electrification scheme, untrammelled by having to rely on a third party for such a critical element as the electricity supply, and in consequence had made it difficult for the West Kent to honestly make a viable offer once the Trade Facilities Committee had made it possible for the railway to build the whole scheme itself.

Raworth was asked if there had been any negotiations with any of the other London power companies, and he had to admit that the SECR had not had any serious negotiations with anybody other than the West Kent. Letters from Percy Tempest to Mr G. Frederick Fox, the general manager of the West Kent– for which Alfred Raworth would have advised on the wording – were read out. They seemed to indicate that way back in 1920 at the meetings held at Raworth's office the SECR was agreeable to the West Kent supplying the SECR. The questioning went on for some time as Tyldesley Jones picked away at the edges, trying to prove that it was only the offer of a loan guarantee from the government that had swung the SECR back to building its own generating station, thereby forcing out the West Kent– a position that the Electricity Commissioners, having a wider strategic brief, would not accept. Raworth was quizzed as to the exact date that the final decision had been made, and despite it only being a few months after the event, he stonewalled with the usual abrupt, short answer maintaining that he could not remember the exact date. At one point, he did provide a much longer answer – concerning the negotiations as to exactly how the West Kent's capital costs to make up the standing charge to the SECR were to be calculated – but it appeared that he was reading his answer. Sir John Snell broke in with a stern comment that he hoped that Raworth was reading from notes that he took at the time. Raworth was asked if an adequate supply could be taken from the LESC's power station at Deptford. The usual long and verbose question was answered with an abrupt one-liner saying that this was unsuitable as the power station was not by the railway. He did not elaborate that if Deptford was used, there would be extra cable installation costs and a risk of delays obtaining permission to use private land.

When Percy Tempest did, belatedly, come to the stand, a farcical situation ensued when the same questions that had been put to Raworth were repeated to the SECR general manager, and often the previous answers were read back for Tempest to confirm. Embarrassingly, Tempest had to agree that in February 1922 when the formal application to the Electricity Commissioners to build a railway generating station had been made, there would appear to be no record in the board minutes that the directors had been made aware of the alternative proposal from the West Kent. Tempest rigorously affirmed that the directors had duly received reports of the negotiations and of the subsequent offer – it was just that no record had been written into the minutes. Percy Tempest maintained that he and Alfred Raworth regularly saw Cosmo Bonsor at the London Bridge offices and had been able to discuss aspects of this and other matters of the railway's business daily. Tempest resolutely maintained that all the SECR directors and Cosmo Bonsor believed that the West Kent Company was too small and was undercapitalised. Obviously, Bonsor knew all about the negotiations – he had given his private views about the West Kent Company in his confidential letter to Arthur Neal – but Percy Tempest was unable to openly make any reference to this at the inquiry either because he respected confidentiality or because he was unaware of the letter's existence.

Sir John Snell was getting exasperated. The inquiry was dragging on and any thoughts of a Whitsuntide holiday break had evaporated. Tired of what seemed like obfuscation from both Raworth and Tempest and the fact that the barristers seemed to be going round and round the same points, he put it to Percy Tempest that he and Raworth were averse to getting a supply from anybody else and that they had resolved to proceed with their own station, and had not even deemed it necessary to take the opinion of their directors into account. This Tempest vigorously denied and again insisted that the SECR directors, or at very least the chairman, had always been kept fully informed.

Tempest was asked about negotiations with other power companies, and he replied that only the West Kent had come to the SECR with serious proposals. This raised the obvious question that if it had not been determined to build its own generating station, why was no effort ever made to enter into negotiations with any of the other London power companies? Percy Tempest was vague and maintained that he and some of the SECR directors had met representatives of other power companies socially from time to time but from these informal meetings none of them had thought it worthwhile, apart from the West Kent, to positively come forward to offer their services to the railway.

Cross-examined by his own council, W.B. Clode KC, Tempest partly excused the SECR by stating that he would agree with the sentiment that the electrification of the suburban lines was much too important to his railway to risk the whole project collapsing should any outside electricity supplier fail to deliver as promised, after the railway had expended so much capital on the scheme.

During the inquiry but outside the walls of the Middlesex Guild Hall, a revised offer to the SECR was received from the West Kent. This resulted in Raworth and Sparks having further negotiations with the electricity company and this revised offer was duly presented to the SECR Joint Management Committee on 14 June 1922, one day before the public inquiry closed. Again, the offer was rejected but this time the decision was properly minuted. Also at the inquiry, proposals were tabled by the LESC and these too were discussed at the same SECR committee meeting, but it was agreed that any decision would have to be deferred until Raworth and Sparks had carried out a detailed examination of the offer.[20] Once this offer had been examined however it was also deemed unacceptable to the SECR on financial grounds – and again the committee's decision was properly minuted.

Compared to the outside supply issues the technical issues were less contentious. The SECR electrical consultant, C.P. Sparks, was questioned about the technical aspects of the proposed Angerstein's Wharf generating station. He confirmed the suitability of the site and that the power station would use large capacity generating sets with low steam consumption and have, at the first stage, a load factor of 35 per cent. The load factor is usually calculated as the percentage of the average actual load generated compared to the maximum possible. The average load factor for municipal power stations at the time was about 30 per cent. Railway generating stations were performing better – the LSWR's Durnsford Road for example was achieving 41 per cent, and this was confirmed by Herbert Jones when giving his evidence later in the inquiry. Sparks maintained that, during the second phase, if supply were to be offered to the ex-LBSCR lines, the load factor at Angerstein's Wharf could rise to 44 per cent. Raworth had in his 1919 report predicted a load factor as high as 46 per cent at the third stage of the SECR electrification.[21]

The representative of the LBSCR was its electrical engineer Mr Robert Howard who confirmed that the LBSCR had always been satisfied with the reliable service that it had received from the LESC. The LBSCR had a huge interest in the outcome of the tribunal because if, as planned, it was to extend its electrification to the south coast it would not be possible to obtain sufficient electricity from Stowage Wharf, so it was looking towards possibly obtaining an additional supply either from the SECR or from the West Kent's proposed super-station at Belvedere.[22]

Alfred Raworth had an easier time with the technical questioning. It started easy enough with the hoary issue of the alternating current frequency. Raworth maintained that the main reason to use 25 cycles alternating current was because a 50-cycle system would either increase cost or reduce efficiency, and that 25 cycles was the frequency then being used by the SECR's future partners in the southern grouping. Probed further he maintained that 50-cycle rotary converters were less reliable and that motor generators were more expensive and less efficient.

Data sheets had been prepared for presentation to the inquiry detailing the SECR proposals. Raworth was questioned as to the transmission voltage to be used and stated that it had not yet been decided whether to operate at 11,000V or 33,000V; 33,000V had been an important feature of Raworth's 1919 proposal, but times had moved on since then and with the grouping nigh, any electrification beyond the currently proposed first stage would be under the auspices of the new Southern Railway, and the LSWR was using 11,000V as the transmission voltage. If that voltage were to become the standard, it made no sense to use expensive 33,000V equipment for the 3.5-mile connection from Angerstein's Wharf to Lewisham. One version of revised data indicated that seven 11,000V three-phase cables would be installed between Angerstein's Wharf and Lewisham substation.

The inquiry then entered its second phase – examination of the West Kent's proposal for Belvedere power station. A surprise witness for the West Kent was the LBSCR's consulting engineer Sir Philip Dawson. He provided technical details of the proposed Belvedere generating station which would generate 25,000kW at 50 cycles for industrial users, and a further 5,000kW at 25 cycles for local industry and for the railway companies' use, amounting to 45,000kW for the LBSCR and 30,000kW for the SECR.

The third contender for the SECR's business was the CLESC and its protagonist was none other than the 'Father of the Grid', Charles Merz. CLESC's proposal, as might be expected from the forward thinking Merz, was more holistic. Despite his dislike of public speaking, Merz forcibly castigated the West Kent for stating earlier in the inquiry that it would not proceed with the Belvedere power station without the SECR business, saying that it was guilty of 'small-mindedness'. Questioned about the SECR's Angerstein's Wharf proposal, and in particular about a small hint that Bonsor had previously dropped that perhaps the SECR generating station could be hived off to another trader company in due course, Merz denounced this proposal saying that in terms of integration into any future national grid – which such a trader would ultimately be required to do – the Angerstein's Wharf generating station, a site that he himself had once recommended in other circumstances, was in the wrong place and would operate at the wrong frequency.

The CLESC proposal presented by Charles Merz was for a 'mini grid' involving the installation of nine 33,000V cables to be laid through a tunnel beneath the River Thames from a proposed new power station at Barking, operating at Merz's preferred 50 cycles per second frequency. Five of these cables would run to a new substation at New Cross, primarily to supply the LBSCR, and four would run to Lewisham to supply the SECR. An interconnecting link of two cables would be laid entirely in public highway, not touching railway land, to connect New Cross and Lewisham substations to form a 'ring'. Additionally, two further cables would run from New Cross substation to the LESC's existing generating station at Stowage Wharf, Deptford. Two more 33,000V cables would serve the local community in Barking. Operating at 50 cycles per second Barking would – and did for the generating station was soon to be built – become part of the proposed National Grid. When questioned about the rather ambitious tunnel below the river, Charles Merz said that it would cost £190,000 and that the CLESC would not proceed with this part of the scheme if the SECR did not accept the 'county' offer. He also promised the SECR that it would be indemnified for the additional costs incurred from having to take supply at 50 cycles per second.

Mertz's grand proposal and its offer were also examined by Raworth and Sparks away from the public inquiry. The SECR view was that it was by no means certain that this company could provide the supply in time and that it was unhappy that so much of the cable route would be over third parties' land and not on railway-owned land. Bonsor pointed out to the Joint Management Committee that the SECR – and following amalgamation the Southern Railway – would be contracted to make high leasing payments to the South Eastern and Chatham Power and Construction Co. Ltd from 30 June 1925 and any delays would not be tolerated, as it was essential that income for the new electric services was flowing in to meet this financial commitment. So, the CLESC offer was also rejected, and an explanation was provided by letter, dated 19 July 1922, from Cosmo Bonsor to Arthur Neal. Bonsor, on the advice of Raworth and Sparks, listed the shortcomings of the offer, of which first and foremost was that the cost of the units supplied would be greater than from the SECR generating station. He also made the point that in view of the forthcoming amalgamation of the SECR, LBSCR and LSWR into the Southern Railway in less than six months, there would be much greater potential savings to the whole group from using electricity from Angerstein's Wharf, and that this was in line with one of the important aims of the Railways Act 1921 which intended that operating costs would be shared. The proposal was discussed with the chairman of the SECR's future partners in the Southern Railway, and both agreed that the offer from CLESC should not be accepted.[23]

In his closing statements to the inquiry, Mr W.B. Clode KC, speaking for the SECR, emphasised the need for his client to have certainty that an electricity supply would be available ready for the commencement of its electric train services at the end of 1925, when it would have to start loan repayments, and that only by building its own power station could it have this necessary certainty. Mr Tyldesley Jones KC, winding up on behalf of the West Kent, seemed to be amazed that the SECR should wish to take a course contrary to the views of some of the most experienced railway electrical engineers such as Charles Merz who were convinced of the advisability of power companies supplying the railways rather than generating their own traction supply.

The public inquiry finally closed on 15 June 1922 after 10 days of deliberation. Now it was the time for the SECR to nervously await the Electricity Commissioner's report. The Commissioners did not provide a full ruling immediately but wrote to the Ministry of Transport on 21 August 1922. They concluded that the Angerstein's Wharf site was suitable for Stages One and Two of the SECR electrification proposals but cast doubt as to there being sufficient free space to be able to extend the plant on the site to meet the needs of any third stage, but they affirmed that this had been conceded by SECR technical witnesses (Raworth and Sparks) and they had duly agreed to modifications to the site layout. The West Kent Electricity Supply Company's scheme did not fare well under the Commissioners gaze. Agreeing with H. Cosmo O. Bonsor's sentiments, it was stated that the company would have to start from nothing to build the power station and that it had hitherto operated on only a very small scale. The CLESC had consent to build its 150,000kW station at Barking but would have to construct a cable tunnel, six feet in diameter, below the Thames. The SECR, the synopsis said, had given evidence stating that it did not believe that enough supply could be afforded in time or at a supply cost less than the railway's own generation station cost. Trying to be helpful perhaps, the Electricity Commissioners sat on the fence and while appreciating the urgency of the matter wished to give the companies a chance to agree on terms before a final decision. The conclusion of this interim statement was that the Commissioners hoped that in a spirit of goodwill the SECR could reach an amicable agreement with one of the power companies.[24]

This was an unsettling and unsatisfactory situation for the SECR. Knowing that time was running out, it was impatient for a decision to be reached. It had been six wasted months since it had made the application to the Electricity Commissioners and thus far it had only received a rather feeble non-committed response from them. Arthur Neal arranged a meeting between Percy Tempest and Sir John Snell to resolve the matter, which did result in revised offers from the power companies. These were however also rejected by the SECR Joint Management Committee on the basis that the prices offered by the power companies were higher than were estimated for the railway's own scheme.

On the morning of 6 September 1922 – less than four months before the grouping was due to take effect – the SECR directors met to discuss the impasse. It was agreed that a letter be sent to Arthur Neal explaining that they were not able to accept that any outside supplier could provide them with electricity at a lower price and, more importantly, with the certainty that a full connection could be ready in time for the electric services to commence. If another supplier was used and defaulted by not meeting the deadline, it would not be possible beforehand to take out any form of insurance to protect the railway from the financial consequences of having to pay for the electrification without any new income. In their letter Neal was reminded that the Angerstein's Wharf proposal had been approved at a special meeting of the SECR shareholders the previous February. Any change would have to be approved by the same shareholders, but in the circumstances the directors did not feel that they could recommend such a change. Furthermore, even if such a proposal were to be put to the shareholders, feedback that had been received from them indicated that the proposal would be rejected. The letter claimed that the lack of approval to build the Angerstein's Wharf generating station would jeopardise the company's aim to provide large-scale employment during the coming winter and operate improved services for the travelling public.[25]

Every string had been pulled, but this last plea was to fall on deaf ears. On 28 September, the SECR received the final judgement from the Electricity Commissioners, signed not by Sir John Snell but by R.T.G. French, the Commissioners' secretary. In his letter, French reported that the latest offers of supply from the West Kent Electricity Supply Company and the CLESC had been examined by the Commissioners and that after a full review they regretted that they were unable to give their consent to the establishment of the proposed generating station at Angerstein's Wharf. French continued with a patronising paragraph explaining to the SECR directors the government's policy in respect of the benefits to the whole nation of public supply generating stations in comparison to railway-owned generating stations. The Damning formal refusal ended with the offer of assistance to the railway with its negotiations with the power companies.[26] The directors were not impressed, believing the SECR had won the argument but had been refused permission for purely political reasons. Even after this formal rejection of the Angerstein's Wharf proposal, the SECR directors continued the fight. Bonsor and Tempest unsuccessfully pressed for a revised decision in their favour by lobbying anybody who would listen to them including Sir John Snell who they pursued to his weekend retreat in Derbyshire.

On 12 October 1922, six months after the application date, the South Eastern and Chatham (Electric Power) Order 1922 was issued as per the 1903 Act. The Order contained several clauses aimed at protecting the interests of the various third parties: the Port of London Authority, the GPO and the Admiralty. But the final condition was that the generating station could not be built without the consent of the Electricity Commissioners in accordance with the Electricity (Supply) Act 1919.[27] This was the first, and only, Order that was ever made in respect of the 1903 Act.[28]

The Treasury loan guarantee had been provisionally agreed in January 1921, before the Trades Facilities Bill had become law. Following the outcome of the Electricity Commissioners inquiry the formal legal documents were drawn up and duly received the SECR company seal in November 1922. All references to the SECR generating station proposal were expunged from the provisional legal documents. It was a complex legal and financial situation due to the number of separate legal entities involved. There had to be an agreement between the Treasury and a new company, which was wholly owned by an existing company that was in fact two separate companies contracted to work together. A formal agreement was drawn up between the Treasury and the new SECR 'Power and Construction Company', agreeing that the Treasury would guarantee a loan not exceeding £5.5 million and that the interest on this loan must not exceed 6 per cent. This agreement was to cover the first stage electrification works as detailed in Raworth's 1919 report. The loan guarantee was to be for a period of 22 years, commencing on 3 October 1923 and, on 31 December 1925, or sooner if work had been completed, the SECR Power and Construction Company would lease all plant and rolling stock back to the SECR until completion of the loan term. The SECR had the option to buy outright the leased equipment in 1927, 1932, 1937 or 1942. The SECR obligations would be transferred to a successor organisation, by now certain to be the Southern Railway.[29]

Despite the Electricity Commissioners seeing the West Kent and County of London electricity undertakings being potential suppliers to the railway, it was to the LESC that the SECR turned for an economical and reliable supply of electrical energy. Perhaps that was where they should have gone in the first instance. The LESC was quick off the mark by applying to the Electricity Commissioners for permission to extend its generating station at Stowage Wharf, Deptford, and this was granted with astonishing speed on 4 October 1922[30] – clearly indicating that a request had been made much earlier and the decision only awaited the outcome of the Angerstein's Wharf decision, but it was to be many months before the new Southern Railway was to agree a contract price for this supply of electricity.

Many commentators have viewed the Angerstein's Wharf saga as an example of government incompetence, when one set of civil servants countermanded another group of civil servants. Each of the government departments involved had laudable aims and in life compromise and pragmatism must be employed if disparate intentions are to be met as fully as possible. The Trade Facilities Committee was charged with obtaining finance for capital schemes that would provide work for the unemployed masses. The committee correctly judged that the SECR scheme would provide additional employment, but what it was not competent or authorised to do was make a technical assessment of the scheme. The Ministry of Transport traffic department, after first hoping that the railways could agree among themselves, accepted the responsibility and approved Raworth's electrification proposal.

The Electricity Commissioners were in a more difficult position being seen at the time as ineffective as they were having difficulties in establishing the local joint electricity authorities and needed to display their own authority. Conversely, there was a growing belief that the all-embracing Ministry of Transport was working against the spirt of the grouping, by imposing conditions on the railways preventing them from making economies of scale such as combining their efforts to build large power stations. The interim report issued by the Electricity Commissioners came close to approving the Angerstein's Wharf generating station and perhaps it was hoped that following this statement those at the pinnacle of the Ministry of Transport hierarchy would step in and approve the scheme. The civil servants who issued the South Eastern and Chatham (Electric Power) Order 1922 carried out their duties faithfully, but this was an irrelevant sideshow, probably mischievously instigated by the West Kent Company.

CHAPTER 12

THE BIRTH PANGS OF THE SOUTHERN RAILWAY

Despite Alfred Raworth's SECR electrification proposal being approved by the Ministry of Transport, the LSWR continued its objections, claiming it had been misled about the SECR's financial agreement with the government and announcing that it now intended to block the scheme.[1] The LSWR had consistently argued that no capital expenditure commitments should be made by any of the constituent companies on the very eve of the grouping, and furthermore the Treasury was only to guarantee the loan of the capital which would be a charge on the new Southern Railway. Collusion must be suspected, for at the same time the LBSCR also withdrew its support for the SECR when the Brighton directors also queried the financial arrangements. Bonsor, when replying to the LBSCR chairman, Charles Macrae, on 2 August 1922 argued that there was no financial risk to the Southern Railway. He argued that the Treasury would guarantee both the principal and the interest on the loan and with this level of security he claimed that the railway would be able to raise the money in the short loan market at less than the current bank rate.[2] With the fateful date of 1 January 1923 fast approaching, a new committee of directors was established which was for all practical purposes the Southern Railway board in waiting. To placate the LSWR, this committee put pressure on Cosmo Bonsor to delay commencement of work on electrification until after a meeting scheduled for October 1922. The SECR Joint Management Committee responded by reiterating that the electrification scheme had been approved by the Ministry of Transport, and since at that time it was anticipated that the Electricity Commissioners would approve the proposal to build the Angerstein's Wharf generating station, by October the work would have been started.[3] The SECR would not be moved from this position and despite the increasing objections from its future partners, it appeared to accept no criticism or contrary advice, claiming that it was to carry out its scheme in the 'national interest' to get unemployed men back to work. The committee had already set Raworth and his staff to work and decreed that they would all be paid by the SECR (or rather what they anticipated to be the nominally independent SECR division of the Southern Railway) with appropriate cost transfers from the construction company to the SECR. The promise was made to Alfred Raworth that on completion of the construction work he would continue in office as the SECR electrical engineer.

Following the refusal of the Electricity Commissioners to grant permission to build the SECR generating station, a further letter was sent by the SECR to the LSWR in October making it quite clear that the electrification would continue, as to go back on the agreements made with the Treasury would bring the railway into disrepute.[4] This issue of 'honour' would have been important to H. Cosmo O. Bonsor. The LSWR, perhaps as a stalling manoeuvre, replied in October, claiming that it had been agreed the previous June that the SECR would provide details of the proposed financial deal with the Treasury. The LSWR asserted that to date it had not received any such information.[5]

An ad hoc committee of directors established a huge subcommittee to examine the electrification issue, which was given the imaginative title of the 'Committee of Twenty-One' due to the number of its members: eight directors from the LSWR, eight from the SECR and five from the LBSCR.[6] Through this committee the LSWR, with the help of the LBSCR, continued to call 'foul', concentrating the attack on the funding for the SECR scheme which would commit the new group to substantial capital expenditure. To proceed with this argument, they decided that now was the time to involve the man charged with adjudicating upon such financial intricacies – the president of the Railway Amalgamation Tribunal: Sir Henry Babington Smith.

The 1921 Act creating the railway grouping required the establishment of the Railway Amalgamation Tribunal to referee the complex financial and legal issues that the grouping process would generate. On 11 April 1921 Sir Eric Geddes wrote to Sir Henry Babington Smith requesting that he join this tribunal.[7] Sir Henry was a distinguished civil servant who had loyally represented his country in financial matters. He had been educated at Eton and Trinity College Cambridge and served abroad in South Africa and Turkey and at home in the Treasury and at the GPO. He had been knighted in 1908. His most important claim to fame was that he was the private secretary to the Viceroy of India, the 9th Earl of Elgin whose daughter, Lady Elizabeth Bruce, he married.[8] Sir Eric proposed in his letter that the tribunal was to be made up of three members; these should be Sir Henry for his financial expertise, assisted by an accountant and a lawyer. This tribunal was not to include any member that had any railway operational or engineering expertise. Sir Henry did not accept at once, being concerned that the tribunal would be expected to propose company amalgamations, a task for which he did not have the necessary experience.[9]

Appreciating Sir Henry's concerns, Sir Eric explained that the operational details of the amalgamations were to be agreed between the railways and the Ministry of Transport, and the tribunal members were to employ their expertise to settle the terms and embody them into financial and legal schemes for each of the new companies.[10] After this reassurance Sir Henry accepted Sir Eric's offer,[11] and he became the president of the Railway Amalgamation Tribunal assisted by Sir William Pender and Mr G.J. Talbot KC.[12] Despite these reassurances from Sir Eric, Sir Henry was reluctantly soon to become indirectly embroiled in at least one operational dispute – the SECR electrification proposals.

Before the Great War many of the larger railway companies had wished to amalgamate with the intention of avoiding unnecessary competition, achieving increases in efficiency and reducing costs. This proved problematic because to achieve it they would have to follow the private member's Bill procedure. Such Bills would inevitably be challenged by members of both Houses of Parliament who were ideologically opposed to any reduction in competition and by those who saw amalgamations as a route to less directorships and managerial posts. But Sir Henry Babington-Smith was empowered to carry out mini-amalgamations to pave the way and simplify the ultimate grouping proposal. Some advantage was taken of this arrangement, for example, the LNWR and the LYR were amalgamated prior to becoming part of the LMS.

The financial health of the SECR had changed since the formation of the Joint Management Committee and many shareholders believed the relative value of their investments in both the SER and the LCDR – plus many legacy shares held in respect of railways making up the two main companies – did not reflect the true situation after two decades of progress. The powers granted to the Railway Amalgamation Tribunal seemed to create the perfect opportunity for many anomalies to be resolved by the method of an acceptable formal amalgamation of the SER with the LCDR. In July 1922, the SER and LCDR stockholders had a meeting to discuss the shares arrangement for such an amalgamation.[13] This meeting broke up without any consensus being reached, but a reconvened meeting in August 1922 did finally produce an agreement[14] which was approved by the SECR Joint Management Committee on 9 September 1922.[15] So, Sir Henry Babington Smith now had the task of approving a mini-amalgamation before producing a revised version of the Southern Railway amalgamation scheme – which was contentious enough without this subsidiary task. His first draft proposal for the 'Southern' scheme had allocated the new shares based on previous SER and LCDR share ownership.[16]

Meanwhile, the SECR continued to promote Raworth's electrification scheme, and despite it now being obvious that it was contrary to government policy, continued to argue the need for the Angerstein's Wharf generating station. There were those at the Ministry of Transport who were losing patience with the SECR. According to C.F. Klapper in his book *Sir Herbert Walker's Southern Railway*,[17] H. Cosmo O. Bonsor was summoned to meet Sir Henry Babington Smith to endure what Klapper describes as a 'stern reprimand'. It is highly unlikely that Sir Henry, given his initial reluctance to get involved in any decisions as to the functional conditions of the amalgamation, would have wished or felt competent to be able to become deeply involved in any engineering or operational controversy given the overall magnitude of his main task. But the attack on the SECR scheme had turned towards financial issues and furthermore the SECR was requesting Sir Henry's assistance in approving the complex mini-amalgamation of the SER and the LCDR.

The SECR shareholders were more concerned about an equitable resolution of the share distribution than they were about electrification schemes. Leverage appears to have been applied such that any late changes to the financial arrangements between the LCDR and SER investors would only be accepted if the SECR, and its shareholders, agreed to abandon their plan to build a generating station at Angerstein's Wharf and thereby fall into line with the government's overall energy policy. When offered the choice between their own power station and a better financial settlement it is not surprising that the shareholders easily gave way to Ministry of Transport pressure, despite the insistence from the SECR that its hands were tied because the *shareholders* had approved the Angerstein's Wharf scheme. Furthermore, Sir Henry's influence was to be further used by the SECR electrification scheme detractors. This was because his draft amalgamation scheme had included a limit on the amount of capital that the new Southern Railway could borrow[18] and while technically the capital was not to be borrowed by the railway itself but by the construction company, this was merely a device to circumvent any liability upon the shareholders. So, at his meeting with Bonsor, Sir Henry seems to have put some pressure on the SECR chairman to fully comply with the wishes of the new Committee of Twenty-One. Perhaps it was the diplomat in Sir Henry who proposed to Bonsor and the other railways' managements that an outside adjudicator should be consulted to determine the merits of the SECR scheme in the context of the unified Southern Railway. Ideally this should be undertaken by a consulting engineer who was completely independent and had not been tainted by the conclusions of any of the previous committees. When the October meeting of the Committee of Twenty-One chaired by Edwin Cox was convened, it was proposed and agreed that to finally resolve the electrification issue, yet another consultant be called in to adjudicate on the Southern Railway's electrification policy.[19] Reluctantly, the SECR communicated to the other two railways that it was prepared to accept any of the proposals in the new electrification report.[20]

The chosen consultant was to be George Gibbs, an American with excellent credentials who had worked as an employee and consultant for several railroads in the USA and had taken charge of the Westinghouse Company's interests in Europe. It was with Westinghouse that Gibbs had become involved with projects such as the Paris Metro and the Mersey Railway in Liverpool. Joining forces with another electrical engineer L. Rowland Hill, they set up the consultancy firm of Gibbs and Hill. Thereafter George Gibbs was associated with the Pennsylvania Railroad's main line electrification and the New York Subway.[21] If the LSWR had not suggested George Gibbs – and the suspicion must be that the company had some part to play in his appointment – it would certainly have approved of the choice as he was an associate of the LSWR's consultant, and critic of Raworth's proposals, Theodore Stevens. Stevens was to provide information concerning the costs of suburban electrification to Gibbs for his prestigious presentation to the International Railway Congress to be held at Rome the following April, and both men were completely averse to any voltage significantly above 600V being used for conductor rail schemes.[22]

George Gibbs' report recommended that the Southern Railway adopt the Ministry of Transport approved 750V direct current third rail for all its suburban services, including the SECR lines, and that the 6,700V (with a possible uprating to 11,000V) alternating current overhead system should be used for the longer, main line routes. This would ultimately involve a dual electrification for the LSWR, LBSCR and SECR suburban routes which also carried long-distance traffic. This was a proposal that Sir Herbert Walker had previously suggested would be acceptable to him and the LSWR when he gave his evidence to Sir Phillip Nash's Ministry of Transport Committee. Gibbs recommended that the LBSCR suburban electrification should be extended as planned, but any future extensions to the LSWR suburban lines should be at 750V direct current.[23]

With the Gibbs report on the table, it was the SECR man, Edwin Cox, as chair of a preliminary departmental meeting, who performed the final act to kill off Raworth's electrification scheme by formally recommending to the emerging Southern Railway board that they should fully accept George Gibbs' proposals.[24] This decision was formally agreed by the Group Committee of Directors on 21 November 1922. The SECR directors duly received their copy of the Gibbs report[25] and were therefore made fully aware of the contents. The arrival of the report was recorded in a somewhat muted tone in the formal minutes of the now powerless Joint Management Committee. They agreed, after consultation with Alfred Raworth and Sir Alexander Kennedy, that Kennedy and Donkin should commence the massive task of preparing the drawings for what was now a completely new design. Raworth and Kennedy were asked to comment on the Gibbs report, but while no record exists of their observations, it is easy to guess what they concluded. Two years of design work were to be thrown away; the new drawings would involve not just the simple removal of one of the conductor rails but a full set of proposals for the electricity distribution to several new substations. The management committee voted to transfer £30,000 to South Eastern and Chatham Power and Construction to pay for the redesign. Ironically, it fell to Alfred Raworth to formally provide notification of this important change of design in a letter he wrote to the Ministry of Transport on 6 December 1922.[26]

Meanwhile, now that the SECR had accepted the recommendations of the Gibbs report, the amalgamation of the SER with the LCDR into the full SECR could be attended to. Sir Henry Babington Smith and the SECR agreed that it made little sense to merge the two companies at such a late stage, so the terms agreed with the shareholders in August would be incorporated into the final version of the Southern Group Amalgamation Scheme.[27] The amalgamation proposals were duly placed before Sir Henry Babington Smith on 11 December 1922. There were to be tough discussions over the next two days before a substantive agreement with Sir Henry was reached, and this was subject to some very minor issues that were temporarily placed to one side for future consideration.[28]

It was, however, not quite the end of the electrification decision-making process. On 1 February 1923, the temporary 'Court of Directors', the new Southern board in all but name, decreed that, contrary to what had been previously agreed on 21 November 1922, that the SECR lines would be electrified using 750V direct current. It was now intended that both the SECR and the LSWR lines would operate at a maximum pressure of 660V at the substation busbars.[29] In all future references to the voltage of the Southern Railway electrification this busbar voltage would always be quoted. This small rise in the nominal voltage on the LSWR network was technically possible by selecting a different 'tap position' on the transformers' high voltage windings. Now that the Southern Railway was established and freed of any influence from the Ministry of Transport and Sir Henry Babington Smith, a unilateral decision could be made for a standard direct current system. It is probable that Sir Herbert Walker, Alfred Raworth and Herbert Jones met together to sensibly agree that the use of two nearly compatible systems on one railway was unacceptable from everybody's point of view. Raworth then had to make an appointment to visit the Ministry of Transport to explain this modification to a proposed scheme that it had so recently given amended approval for.[30]

History, it has often been claimed, is written by the victors. The view that has percolated down through the decades is that the Southern Railway was *lucky* because the SECR scheme had not commenced as the new railway would have been cursed with three electrification systems. The perceived wisdom seems to indicate that the SECR was stubborn when insisting that the Raworth system must be used. Charles Klapper refers to the insistence of the SECR to electrify its lines with another electrification system as a folly. But imagine for one moment, a significant proportion of the suburban lines south of the River Thames today being equipped with two conductor rails, properly encased in bright yellow insulating material with, for extra safety, a shutter arrangement actuated by the pickup apparatus. Operating at 3,000/1,500V such a system would be capable of delivering over four times the power to the trains as the Southern Railway's 660V system. More substations than Raworth envisaged would of course now be required, but such an arrangement would have significant advantages today when the requirement is not only traction, as in Southern Railway days, but also to provide the modern necessities such as air conditioning, power-operated doors and computerised messaging systems.

The Raworth scheme had been clearly shown to be more economical than the other two alternatives. Sir Hugh Drummond, edged on by Sir Herbert Walker, castigated the SECR for proposing the Raworth system, as his scheme did not conform to any existing electrification system. Drummond, and his other directors, were not scientists or engineers and probably both did not understand and feared this new-fangled electricity technology. What they did understand was that the 600V third-rail network was working and literally paying dividends. The LSWR based its decision on the degree of commercial risk rather than on any difficult to understand engineering issues. Raworth had acquired a reputation as a competent engineer but so too had his late father, and the nervous directors might have feared that this latest Raworth invention might well end in disaster just as the regenerative control project had – failure not due to lack of brilliant innovation but due to unfortunate matters of detailed design and manufacture. The directors could not risk such a failure due to any unexpected issues that they could not foresee nor even understand should they unfortunately arise.

Would three different electrification schemes have been detrimental to the operation of the Southern Railway? Much was made of the need to have 'standard' electric trains. The changeover arrangements to allow a Raworth train to run over LSWR lines would have involved a reconfiguration of the motor connections. Without complicated switching, it is doubtful if this would have been easily achieved by simply operating a lever in the driver's cab. But the altering of solid link connections might have been achieved in the depot overnight and the train then towed by a locomotive to the 'other' territory when required. This may not seem a very practical solution in railway operating terms, but with the advantage of decades of hindsight it can be noted that services did *not* cross the old boundaries and that any migration of electric stock from one area to another was rare. For example, the 2-HAL units built for the Maidstone electrification scheme in 1939 (an ex-SECR route) were all transferred to former LSWR and LBSCR lines from 1959, making way for new British Railways stock being built for the Kent coast electrification scheme. Furthermore, those who claimed just before the grouping that there would be no interchange of suburban traffic have been vindicated, for today in the twenty-first century we witness very few changes to the traditional LSWR, LBSCR and SECR routes, with all services running to the traditional London termini, and these traffic patterns have now been enshrined in franchise arrangements imposed by the government after privatisation. The major change has of course been the cross-London Thameslink services using the former LCDR route through Blackfriars.

But what can be said about Alfred Raworth? How did he feel about being plucked from relative obscurity to design a major engineering project, only to find every aspect of his proposals picked away at and then discarded? He had done what every competent design engineer should do, that is, he proposed a scheme that would economically meet his general manager's specification. It was not an unrealistic flight of fancy by an overenthusiastic young engineer. Sir Francis' requirement, fully met by Raworth, was for an electrification system suitable for both suburban and main line service and, possibly, later to be suitable for use in a Channel tunnel, and furthermore that would comply with the SECR's 1902 Act. Another new electrification system design was not what his former patron, Sir Herbert Walker, wanted, but Alfred Raworth at the time was working for Sir Francis Dent. For Sir Herbert Walker, Jones, with the assistance of Raworth, had in 1912 provided a scheme that suited their general manager's specification – a sound economic proposal to electrify only the LSWR suburban lines. Was Alfred Raworth bitter about the outcome? Most probably he was, just a bit, but hopefully not for too long. All good engineers love a challenge, and in the following years he was to be responsible for electrifying what was to become the world's greatest suburban electric railway.

CHAPTER 13
COMPLETING THE LEGACY SCHEMES

The bitter arguments between the three railway companies were to continue, and it was several months after the official grouping that the Southern Railway management structure was finally resolved, and the first formal board meeting convened. As might be expected in such a major reorganisation, the decision-makers from each of the former railway companies were jockeying to retain their own position, or to take over control of the new Southern Railway. General Sir Hugh Drummond was the LSWR mouthpiece but the real drive behind the company was his general manager, Sir Herbert Walker. Walker may have been the youngest of the general managers from the three constituent companies but, as the wartime leader of the REC he was the obvious first choice to be the general manager of the new Southern Railway. Despite what with hindsight must seem to have been an inevitable and sensible outcome, it was not a pushover for Walker and Drummond who were to become the new general manager and chairman. It took Sir Herbert's threat to resign and the strong will of Sir Hugh to force the issue to create a single Southern Railway board of directors, which did not meet until July 1923 – seven months after the formation of the company. Percy Tempest stayed on until the end of the year as joint general manager with Sir Herbert, but the LBSCR general manager, Sir William Forbes, was forced out of office.[1] Several important positions were filled by ex-SECR men however – the chief mechanical engineer was Richard Maunsell, and the chief operating superintendent was Edwin Cox. Herbert Jones was appointed as chief electrical engineer. Sir Francis Dent returned to the scene as a Southern Railway director and was soon to take up duties as chairman of a subcommittee and to remain an influential figure for over a decade, becoming a 'crotchety relic' from the pre-grouping era.[2]

After confirming all the most senior Southern Railway appointments, the minutes of the first Southern Railway board meeting state that 'Mr Raworth has been appointed by the South Eastern and Chatham Management Committee as electrical engineer in charge of electrification (South Eastern and Chatham section).' An interesting turn of phrase, for was the minute writer hinting that Raworth's appointment was slightly disapproved of as it was now a *fait accompli* and thus had to be tolerated; or was it that the appointment was such a prominent position, ranking with all the other principal officers, that it had to be recorded? Certainly, among the board members there were former LSWR directors who probably saw Raworth as being slightly suspect – tainted with the controversial SECR electrification proposals they had recently so vehemently rejected.

Raworth however, despite possible misgivings from some directors, had been appointed to manage the SECR electrification and between 1923 and 1925 he slipped into relative obscurity to undertake this task.[3] Despite Raworth's appointment to the South Eastern and Chatham Power and Construction Co. to electrify the SECR lines, Sir Herbert left it in no doubt that *he* was to oversee the scheme and therefore needed a man at the helm whom he could trust. Raworth had at first to concentrate on the SECR electrification as this was a new design, whereas the work on the second phase of the LSWR suburban electrification had been planned years before by Herbert Jones and himself. As the months passed, and certainly with Sir Herbert's approval, Raworth was incrementally to take on the responsibility for all new schemes, leaving Jones to manage the day-to-day activities. Details of the proceedings of the South Eastern and Chatham Power and Construction Co. have been lost, but within the minutes of the engineering subcommittee the 'Engineer' is often mentioned in relation to the second phase of the LSWR electrification. Most likely the minute writer refers to Herbert Jones, but as the project progressed there are many references to purchases – for cables and plant such as rotary convertors – being ordered as extensions to the contracts appertaining to Raworth's Eastern Section electrification.

Another important decision at the first meeting of the Southern Railway board was to authorise the Southern Railway to loan the South Eastern and Chatham Power and Construction Co. £10,000 to enable Raworth's work to commence. This first loan was to cover the provision of conductor rails, bonding of rail joints on the running lines, the purchase of cables and the commencement of substation construction. The Southern Railway was to make further loans at regular intervals: £100,000 in February 1924, £200,000 in July 1924, £200,000 in November 1924 – all at 4 per cent interest. This rather silly situation needed to be changed, for this was not the scenario that the Trade Facilities Committee had envisaged when it negotiated the arrangement with the SECR. The decision was made by the Southern Railway board in November 1924 to wind up the Construction Co. and the solicitor was duly given his instructions. This proved to be a complicated business, dealing with the Treasury, directors who had left the scene and owners – that is the former railway companies – which no longer existed. This formal winding-up was not finally completed until April 1925.[4]

Of course, now that the SECR electrification scheme was effectively 'earth return' via the running rails, the issue of protecting the Royal Observatory's equipment at Greenwich to comply with the original 1902 Act of Parliament had to be addressed. The conclusion was that any measures that the new Southern Railway would need to carry out to fully insulate the return paths would be prohibitively expensive and that the only practical solution was for the railway to fund the relocation of the magnetic room. Previously, Percy Tempest had some preliminary negotiations with both the Board of Trade and the Ministry of Transport but, it took a high-level meeting with the Admiralty to reach a final agreement. The Admiralty played host to a delegation comprising of the Southern Railway's Chairman, Deputy Chairman, and Alfred Raworth.[5] The outcome of their discussion was that the Southern agreed to provide the Observatory with £30,000 to finance the removal of the magnetic room from Greenwich to Abinger in Surrey.[6]

With the fully established Southern Railway resolutely set upon the task of electrifying both the SECR and the LSWR lines using the 660V direct current, top contact third-rail system, the railway's directors, managers and engineers might have believed that they could carry on without any interference or attention from the Ministry of Transport. Unfortunately, this was not to be so, for as early as December 1923 the Minister of Transport set up what he called his 'technical conference' to continue the effort to realise Sir Eric Geddes' vision of a national standard for railway electrification. The chairman of the conference was the Ministry of Transport's chief inspector of railways, Colonel Sir John Wallace Pringle. Since 1840 the railway inspectors had all been either serving or retired officers from the Royal Engineers because these men were competent engineers whose judgment would not be compromised, since they were not, nor ever had been, employed by any of the railway companies.[7] Sir John had been the first chief inspecting officer for railways to be appointed by Geddes in 1919, when the important role of inspecting new railway schemes and investigating accidents had been transferred from the Board of Trade to his new Ministry of Transport.

Attending Colonel Pringle's electrification conference were representatives from the four new railway companies. The LNER was represented by Mr O. Bulleid,[8] a future Southern Railway chief mechanical engineer. He was at the time assistant to H.N. Gresley – later Sir Nigel Gresley – the first chief mechanical engineer of the LNER. The LNER at this time was using 1,500V overhead electrification, but after the initial success of the Shildon mineral line, the NER's 1,500V main line scheme espoused by Charles Merz and Sir Eric Geddes was summarily dropped by the 'grouped' LNER. A beautiful 53-foot (16 metre) long 4-6-4 (2-B-2) 1,300 horsepower (970kW) electric locomotive had been built by the NER as the prototype for a fleet to haul heavy passenger trains of 14 bogie coaches at 65mph from York to Newcastle and then later, it was anticipated, to Edinburgh.[9] The Southern Railway was represented at the conference by Sir Hugh Drummond, Sir Herbert Walker and Herbert Jones. The LMS and GWR sent representatives and another important attendee was Sir John Snell, the chairman of the Electricity Commissioners.

Colonel Pringle soon made it clear that the Ministry of Transport's objective was the standardisation of railway electrification using the 1,500V direct current – not stated explicitly but implied – using the overhead system as had been recommended by the Ministry's advisory committee. When questioned about their future intentions, both Sir Hugh Drummond and Sir Herbert Walker must have sat with their hands behind their backs, fingers tightly crossed, when they stated that the Southern Railway would seriously consider using the 1,500V system to electrify its lines beyond the suburban area. Sir Herbert also insisted that the Southern was considering extending the LBSCR overhead scheme and he promised to keep the minister informed of any proposed extensions to Southern electrified lines. Subsequent events were to show that the Southern Railway had little intention of following either course. Herbert Jones confirmed the directors' intentions by stating that the Southern Railway was to only electrify its lines at the nominal 660V direct current within a 30-mile radius of London, and he asserted that it would use the higher voltage beyond the suburban area and thus use both direct current schemes in the London area.[10] Another vague proposal never to be implemented.

The recommendations emanating from Sir John Pringle's conference were that extensions to the 600V network should be limited to an area within 20 miles of Charing Cross and that any use of the 600V system, except for 'tube' lines, must be deemed temporary pending conversion to 1,500V as soon as possible. He reaffirmed main lines were expected to be electrified at 1,500V direct current using overhead lines.[11] Sir John proposed restrictions on future 600V electrification to minimise the cost of later conversion to what he deemed to be the standard 1,500V system.

Meanwhile, Sir Herbert Walker was busy giving details to the SR directors as to how he intended to fund the SECR area 660V third-rail electrification scheme. This was to be by a mixture of new capital and dipping into the various renewal budgets. He explained that the whole scheme was budgeted to cost £3,939,000, made up of £2,138,000 capital expenditure, £150,000 from the locomotive renewal fund, £786,000 from the carriage renewal fund, £320,000 from rebuilding of premises fund and £545,000 from a suspense account. Sir Herbert felt free to do this because following electrification there was to be a lesser need for new steam locomotives and rolling stock.[12]

The revised SECR scheme was divided into three stages, approximating to Stage One proposed in Alfred Raworth's 1919 report. First to be electrified was the former LCDR main line from Victoria and Holborn Viaduct to Orpington making use of the SECR connection at Bickley. The 'Catford loop' and the Crystal Palace branch were also included at this stage. The second stage electrified the former SER main lines from Charing Cross and Cannon Street to Lewisham, Addiscombe Road and Orpington. Also included was the Bromley North branch, the Hayes branch and the connection onto the former LCDR Catford loop at Beckenham. The third part of the scheme was to electrify all three routes to Dartford via Woolwich, Bexley Heath and Sidcup. The total scheme involved the electrification of 58 route miles – 210 track miles. (See plate 8)

The LESC was to generate the electricity at Deptford and to install seven 11,000V cables to the new Southern Railway substation at Lewisham where the supply was to be metered. The supply agreement stipulated that the LESC was to provide the Southern Railway with electricity for 50 years. Initially, on 1 July 1925 the requirement was to be only 10,000kW, but this would increase to 20,000kW on 1 December 1925 with a further increase to 30,000kW by 1 May 1926. Looking to the future 90,000kW was to be made available to the Southern Railway 'when required'.[13]

The overall electrical distribution design was similar to that designed by Alfred Raworth and Herbert Jones for the LSWR before the Great War but was more 'radial' and without a 'ring' arrangement. The 11,000V switchgear in the huge brick-built Lewisham substation was of a similar design to that at Durnsford Road generating station. The seven circuit breakers for the incoming cables were the property of the LESC and the Southern Railway owned the 20 circuit breakers for the outgoing cables – two of which fed the local pair of rotary convertors. Cables operating at 11,000V connected Lewisham substation to 19 other substations which were usually equipped with two rotary convertors, but some had three. The total installed capacity of the rotary convertors was 90,000kW.[14] The nominal rating of each convertor was 1,250kW, but they were designed to withstand an overload of 50 per cent for two hours, 100 per cent for five minutes and 200 per cent for 20 seconds. Additionally, a 'momentary' overload of 400 per cent was possible and the manufacturers had to guarantee that no flashovers at the commutator would

The massive Lewisham substation built to house the main 11,000 Volt switchgear and the rotary convertors. The building is still in use in 2021 but with modern equipment. (Author)

Typical Eastern Section substation at Nunhead. The building still survives in 2021 but re-equiped with modern plant. (Author)

occur during these severe overloads.[15] Such overloads could be tolerated by the sturdy conductor rails but would be problematic if small-sized overhead wires were to be used. This very stringent 'overload' specification was needed in the railway environment where the plant is subjected to short periods of high loads when the trains start to move away from rest and other periodic instances of high load during certain times of the day – such as during the rush hour.

A total of 125 miles of 11,000V cable was manufactured and supplied by Johnson and Phillips. At each substation busbar arrangements permitted operational flexibility. The high voltage cables reached the outlying substations by seven cable routes. They were installed by running them out from cable drums mounted on railway wagons and fitting them onto hooks on concrete posts along the side of the tracks. At stations or at overbridge locations, the cables were either installed in bitumen-filled timber troughing or through fibre ducts. On all routes at least two cables were installed to ensure continuity of supply in the event of a failure or during maintenance. At some locations extra cables were run, for example three cables were installed from Lewisham to Cannon Street substation, from where two cables looped back to Holborn Viaduct substation. Four cables were laid to Nunhead substation from whence two cables were installed to Victoria substation via Loughborough Junction substation and two to Upper Sydenham substation. These Upper Sydenham cables were routed along the Crystal Palace branch, and at the point where this line crossed above Penge tunnel, a cable shaft was excavated down into the tunnel to allow the 11,000V cables to reach the main line below. This shaft was also used for the direct current feeder cables, which were routed back up the shaft from Upper Sydenham substation to feed the conductor rails on the Crystal Palace branch.

By September 1924, 133 miles of conductor rail had been installed and 123 miles of running rails had been bonded across the rail gaps to provide the return path for the traction current. Cable for the 660V direct current connections had been manufactured, inspected and wound onto drums ready for installation. All the 11,000V cable had been delivered and was being installed at several locations. Work was progressing on substation buildings at 18 sites despite delays due to a strike – an all-too-common feature of industrial relations in the depressed 1920s. Progress was being made with the manufacture of substation plant – the transformers, rotary convertors and switchgear – with many items under test at the factories. By August 1925, Alfred Raworth had spent £1.5 million of the railway's money installing 115 miles of conductor rail, cabling and building the substations at Holborn Viaduct, Victoria, Loughborough Junction, Lewisham, Nunhead, Upper Sydenham, Shortlands, Catford and Chislehurst. After the necessary staff training, runs with the new electric trains had been completed, with the first public services commencing on 12 July 1925 from Victoria and Holborn Viaduct to Orpington and along the Crystal Palace branch.

This speed with which the electrification was carried out is made more remarkable because of the range of other work that was deemed necessary to complete the scheme. Early in 1923 it had been estimated that to accommodate the anticipated electric trains it would be necessary to lengthen and raise platforms at no less than 56 stations on the former SECR network at a total cost of £35,000. A total of £7,100 was spent on improvements at Holborn Viaduct and £17,000 at St Paul's. Improvement works at Victoria station provided direct connections between the former SECR and LBSCR platforms and an enlarged concourse.[16] At Loughborough Junction, the track layout was modified to allow the trains on the Catford loop to pass through the junction at the same time as trains on the main line. A new island platform was built for the main lines to give space for these alterations which cost the Southern Railway £11,900. At Brixton, new covered platforms were built and the Catford loop platforms were demolished. Herne Hill, the station at which the LCDR express trains had once been divided into 'City' and 'West End' portions, was adapted for its new role in the electrified era, with the provision of an interchange platform to enable London-bound passengers to change between Holborn Viaduct and Victoria trains. At Nunhead a whole new station to facilitate realignment of the junction was built at a cost of £22,500. At the Bromley North terminus £32,200 was spent to completely rebuild the original SER station. It was not just the major stations on which work was carried out, for several minor works were also identified, such as a halt to be provided at Green Lane, Orpington, in anticipation of a new housing development.

The use of the dangerous exposed live rail was causing some public concern. The Southern Railway agreed that a pedestrian subway should be built under the line between Well Hall and Shooter's Hill stations to eliminate an existing pedestrian level crossing across the railway tracks, the cost of which was to be shared between the railway and the local council.[17] Between West Dulwich and Sydenham Hill, the former LCDR line passed along the boundary of the playing fields of Alfred Raworth's old school, Dulwich College. The school governors, mindful of the errant behaviour of small boys who might trespass on the electrified railway, requested improvements to the fencing along the boundary. It was agreed that the college could remove the existing fence and install a six-foot-high 'unclimbable' fence to a design approved by the railway, and that the maintenance of this new fence would henceforth become the responsibility of the railway.[18]

Sir Herbert had been unimpressed by the complicated track formation on the approaches to Charing Cross and Cannon Street through London Bridge[19] and he ordered extensive alterations to the approaches to these termini, as well as alterations to the platforms and track layouts before the second and third electrification phases could be completed. The track alterations at Charing Cross and Cannon Street necessitated some disruption to train services. The work at Charing Cross was carried out one section at a time, working around the existing services, but to completely relay the entire track the closure of the whole station was necessary. Electric services planned to commence from Cannon Street and Charing Cross had to be postponed in November 1925 due to the inability of the LESC to supply enough electricity from its power station at Deptford. As a consequence the Southern Railway had to issue a press release apologising for the two-month delay to the introduction of its new electric services into London Bridge and Charing Cross. It was asserted that while the Southern had completed all the construction work, the necessary electrical installation by the electricity supply company had not been completed. The situation was made worse because track alterations to suit the new electric services had been completed on time, but the new track layout was not suitable for the reinstatement of the former steam-hauled services. A restricted service for commuters would be provided in the meantime. This, the Southern Railway's statement insisted, was directly due to the railway not being able to build its own power station.[20]

Electrification is today usually accompanied by a modernisation of the signalling, and prior to the grouping the SECR had plans to re-signal both Cannon Street and Charing Cross stations. The SECR had concerns about the condition of the signalling and points at Charing Cross, where many shunting movements had to be carried out by hand signals, due to insufficient interlocking and consequently risks were daily being run. Following the grouping the SECR re-signalling proposals were put to one side and the first electric trains to run on the former SECR lines were to be controlled by the existing signalling. The reason for the delay to such urgent work was due to Herbert Walker's desire to completely remodel the former SECR termini and also because the whole question of the installation of improved signalling was the subject of a new report due from the Institution of Railway Signal Engineers. The Institution had set up a committee in March 1922 with the task of investigating various issues relating to signalling heavily trafficked routes. Herbert Walker and the Southern Railway were to take note of this eight-member committee which included the Southern Railway's new signal and telegraph superintendent, W.J. Thorrowgood and his assistant, W. Challis.[21]

In 1923, it was believed that the best way to control an intensely worked suburban railway was to install 'three-position' signals. These semaphore signals indicated not only a 'stop or go' function but also an indication of the status of the next signal rather than the use of a separate 'distant' signal. The extra 'position' of the signal arm would allow safer close running of trains using more signalling 'blocks' for them to run through. The SECR had installed three-position signals in its power signalling installation at the LCDR half of Victoria station. This type of signal was common in the USA, but no British manufacturer was prepared to provide the SECR with these signals, so the equipment had been bought from the General Railway Signalling Company, Rochester, New York.[22]

The Institution's 'Three-Position Signalling Committee' presented its report at a general meeting in December 1924, but through Thorrowgood and Challis, the Southern Railway obtained an early sighting of its contents. By this time a few 'daylight colour signals' had been introduced on main line railways, therefore the report advised that the term 'three-aspect signalling' should be used.[23] The report recommended that the colour light signals should have three lights, or aspects, showing a red light to mean 'stop'; a single yellow light to mean 'caution, be prepared to stop at the next signal'; and a green light to mean 'proceed – all right'. If necessary, an additional yellow aspect could be used such that both yellow lamps showing would mean 'attention – run at medium speed'.[24] The single yellow indication replaced the semaphore distant arm and the double-yellow indicated that the next signal was at 'caution', therefore the train's speed could be adjusted to suit. Fundamentally this system of signalling is still in use almost a century later.

Therefore, in line with the Three-Position Signalling Committee's report it was now acceptable to go one better than 'three-position' signals and advance to a 'four-aspect' signalling scheme. In 1924 – too late to be ready in time for the commencement of the first electric train service in 1925 – the new Southern Railway's engineering sub-committee approved a new version of the SECR plan for power signalling by authorising the expenditure of £68,282 to install colour light signals between Holborn Viaduct and Elephant and Castle; plus £59,126 for colour light signals between Charing Cross and Borough Market Junction and a further £52,000 for colour light signals between Borough Market Junction and Cannon Street.[25] The first stage of this, the first ever colour light signalling scheme in Great Britain, was put into service in March 1926 between Holborn Viaduct and Elephant and Castle station on the original LCDR main line into London.[26] Seven manual signal boxes were closed. At Holborn Viaduct a new 86-lever power frame was installed within the existing box and a new 120-lever frame signal box built at Blackfriars Junction. (See plate 7 (bottom))

The colour light signals significantly increased track occupancy and allowed the new Southern Railway electric trains with their much superior acceleration and deceleration to run safely at higher speeds and much closer together. The beam of light from these new signals, was brighter than the pathetic glimmer from the traditional oil lamp desperately trying to shine through the grubby spectacle glass on a semaphore signal and was more visible both at night and during times of poor visibility in the infamous London smog. In June 1927, more four-aspect signalling was brought into use following the platform and track alterations at Cannon Street and Charing Cross. A new 107-lever signal box, spanning all the tracks, was built at Charing Cross and a new 143-lever signal box was built at Cannon Street. Both had power operated frames but the existing 60-lever manual box at Metropolitan Junction was retained.[27] The signals were of two patterns, either with the aspects arranged vertically or arranged in a circular cluster depending on the location. At the junctions, many of the early installations were like the old semaphore signal configuration inasmuch as each dividing route had a separate four-aspect signal array, which was placed side by side on a single post (complete with an elegant, pointed finial);[28] or alternatively a single signal was used with an illuminated 'route indicator'. The Southern Railway with its usual frugal approach mounted many new signals on existing posts and gantries which had previously supported semaphore signals.

The route indicators (displaying with a white light the number of the line the signal referred to) were placed on the top or to the side of the signal, which was arranged with the aspect lights from the top: green, yellow, red, yellow, the two yellows being placed apart to ensure a clearer distinction between the single and double aspect. Shunting signals were two small red and green aspect lights below the main signal, or in isolation at track level as required. Interestingly the modern arrangement for a four-aspect signal is, again from the top, yellow, green, yellow and then red. The rationale for this change was that the all-important red aspect is closer to the driver's eye-line and this light will not be obscured by any build-up of snow over the lens hood of the aspect below. To assist with platform capacity at both Cannon Street and Charing Cross 'intermediate platform signals' were installed on some of the platform roads. These permitted a 300-foot (91 metres) space for a short train to stand between the buffer stops and an intermediate signal allowing two trains to use the platform.

Due to the direct current electrification, the entire track circuiting equipment had to be converted from direct current to alternating current. Traditional track circuiting, used to indicate the presence of a train on the track, uses low voltage direct current applied to the running rails. A train standing on a section of track completes the electrical circuit as the current flows from rail to rail through the wheels and axles. Once the running rails become part of the direct current traction circuit the two electrical systems must be segregated. This was achieved using alternating current circuits with blocking arrangements to allow the passage of only the alternating current to the track circuiting equipment.

A temporary timetable of electric train services commenced from Cannon Street and Charing Cross starting on Sunday, 28 February 1926. Trains were run to Orpington, Bromley North, Hayes and Addington. Over 100 trains were operated successfully, all except three arriving no more than one minute late. The next day, 1 March, a Monday and the start of the working week, the Southern Railway announced that the first rush hour service had run without a hitch – only four trains were more than one minute late.[29] To achieve timekeeping the Southern Railway expected its passengers to promptly exit and board the trains in order that each station stop lasted no more than 20 seconds.

The subsequent closure of Cannon Street to carry out the track alterations, re-signalling and platform improvements, took place between 5 and 28 June 1926. While Cannon Street was closed from 6 June all the Dartford services were diverted into Charing Cross. Between 28 June and 7 July, it was possible to partially reopen Cannon Street and operate a temporary timetable using a mix of steam and electric trains. The electric service final phase was intended to commence on 11 July 1926,[30] but the full service with trains from Cannon Street and Charing Cross to Plumstead, Dartford, Orpington, Bromley North, Beckenham, Hayes and Addiscombe ran from 19 July 1926.[31]

The Southern Railway issued a press statement congratulating its engineering and electrical staff who had worked continuously by day and night. Furthermore, it claimed, the project had been completed early and within budget.[32] The final cost of the SECR electrification was £3,987,000.[33]

Previously, in March 1923, the short-lived LSWR 'territory' board had resurrected its pre-war proposals devised by Jones and Raworth and anticipated restarting the electrification programme.[34] This was to electrify the main line from Hampton Court Junction to Guildford via Woking; the New Line from Claygate to Guildford and from Raynes Park to Guildford via Epsom, Leatherhead and Effingham Junction. The line between Epsom and Leatherhead had previously been jointly owned by the LSWR and the LBSCR. The estimated cost of this work in 1923 was £740,000, and while this expenditure had been approved by the outgoing LSWR board, the scheme stalled in June 1923 due to the railway not being able to find suppliers able to agree to the railway's previously estimated prices following a general rise in all costs.

Sir Herbert was asked by the directors of the new Southern board to look at the figures and he proposed that £508,000 be spent as capital expenditure; £50,000 to be paid for from the locomotive renewal fund; £210,000 from the carriage renewal fund; £10,000 from the building renewal fund; and £54,500 from a suspense account, arriving at a grand total for expenditure of £833,000.[35] To achieve this figure – still above the previous estimate – it was necessary to remove the 'main line' element of the scheme from Hampton Court Junction to Guildford via Woking. But Sir Herbert did propose a significant addition to the scheme which was to include the line from Leatherhead to Dorking. Leatherhead is at the end of the former LSWR and LBSCR joint line section, and the line onwards to Dorking had been owned solely by the LBSCR. This extension was justified to enable a through service of trains to be run from 'LBSCR' Dorking to 'LSWR' Waterloo via Leatherhead and Epsom. This was exactly the 'interchangeability' that Sir Herbert had fought for during the struggle to agree on an electrification policy. Whether the good people of Dorking ever really wanted to travel to Waterloo or not, Sir Herbert had to make his point about integration. (See plate 9)

Durnsford Road generating station was to have additional plant installed to supply the next phase of the LSWR area electrification. Two additional cables were installed trackside from Durnsford Road to Raynes Park substation from where two cables were run on to Epsom substation. From Epsom, a pair of cables were run to Effingham Junction substation; and along this section two pairs of cables were 'teed' off to feed Leatherhead and Dorking substations. Beyond Effingham Junction two cables ran to Guildford substation and two cables were teed off to Clandon substation. A further pair of cables ran from Hampton Court Junction substation to Oxshott substation. At these substations 11,000V switchgear was installed allowing, through the substation's usual busbar arrangement, operational flexibility. Two spare switches were installed at Guildford in anticipation of further electrification.

One of the significant objections raised by detractors of low voltage direct current schemes was that a high number of substations were needed which would be required to be manned at considerable cost to maintain and operate the plant. The new substations built at Guildford, Effingham Junction, Epsom and Dorking were to be conventionally manned but at Oxshott and Leatherhead they were to be 'automatic' and controlled remotely. For Oxshott

On 5 April 2016, a class 455/7 – set number 5724 – arrives at Guildford with the 14.33 Waterloo to Guildford service via Surbiton and the New Line. The train waits at the signal alongside the original substation building which once contained the high voltage switchgear, transformers, rotary converters and direct current switchgear. The class 455/7 multiple units were built by British Rail Engineering at York in 1984/1985. (Author)

substation control wiring was installed from a control panel at Hampton Court Junction substation.[36] The rotary converters were described as 'self-synchronising', allowing them to be unattended. This was a significant advance, even if at this stage it was only a limited degree of automation.

As for the Eastern Section electrification, many platforms were extended to accommodate the eight-coach trains, with £3,000 spent on minor improvements at Claygate, Oxshott, Stoke D'Abernon and Cobham, Horsley, Clandon and at Guildford's 'other' station, London Road.[37] More extensive work was carried out at Dorking and Guildford costing £6,300 and **£9,750,** respectively. As part of the integration of the former LSWR and LBSCR lines at Leatherhead the LSWR station was closed and a short connecting line was constructed to divert the LSWR trains into what had been the LBSCR station. As this new connection required parliamentary approval[38] this was not to be brought into use until 10 July 1927.[39]

To celebrate the completion of the New Lines electrification project, a civic reception was held at Guildford on 9 July 1925 with the full public service commencing on 12 July 1925. The start of these services neatly coincided with the commencement of electric train services on the former SECR lines.[40]

To operate the new train services new electric trains were built. Both the LSWR and SECR had designs for new trains in hand at the time of the grouping and while they were similar, Sir Herbert's wish for a standard design was not yet to be realised.

For the SECR area electrification 29 new three-car sets were built, numbered 1496 to 1524. These three-car sets were like those proposed by Raworth in his 1919 report. They had two motor brake thirds with a small guard's area and a driving compartment, and the centre trailers were composites with seven first and two third class compartments, reflecting the general reduction in first class travel. These new trains were the first to be built with carriages constructed to what was to become the railway's standard length for electric trains of 62 feet (19 metres), with 9-foot (2.7 metre) wide bodies as Raworth had recommended in 1919. The motor coaches were built by the Metropolitan Carriage Wagon and Finance Company at a cost of £2,970 each and the trailers by the Birmingham Railway Carriage and Wagon Company at the cost of £3,050 each.[41] Not included in the price was the electrical equipment; each of the motor coaches were equipped with two Metropolitan Vickers 275 horsepower (205kW) four-pole axle mounted motors, described as being totally enclosed and self-ventilating.[42]

These 29 new three-car sets for the SECR electrification were not enough to run all the proposed services. At the first Southern Railway board meeting in July 1923 the decision was made to convert some existing steam rolling stock into electric trains. Very much to the relief of the SECR area season ticket holders, approval was given for the withdrawal of 515 of the ancient SECR four- and six-wheelers to be rebuilt as electric stock. Typically, one and a half old coach bodies were to be refurbished and placed onto one completely new standard length 62-foot underframe to make 285 bogie coaches. The new underframes for the converted motor coaches were built by the Metropolitan Carriage Wagon and Finance Company at a cost of £1,035 each, and the underframes for the converted trailers by the Birmingham Railway Carriage and Wagon Company at the cost of £628 each.[43] The number of sets would have required 315 bogie coaches, so it follows that a few more old carriages must have been found from somewhere. The balance might possibly have come from the 140 ex-LBSCR nine-compartment bogie thirds. Other 'new' carriages were either formed into two-car trailer units, or occasionally used singly to temporarily strengthen the three-car sets.[44] The withdrawal of the old coaches left the Eastern Section short of rolling stock and in the short term replacement carriages were transferred from the LSW and LBSC sections – a useful early advantage gained from the creation of the enlarged Southern Railway.[45]

Completely new trains were built for the Guildford electrification. These were similar in appearance to the earlier LSWR conversions from the 'block' sets, but they had flush steel panelled sides and therefore lacked the ornate wood panelling. These LSWR designed units had shorter carriages than the Eastern Section stock, resulting in a few less third-class seats. Twenty-six new three-car sets, numbered 1285 to 1310,[46] were built comprised of two motor brake thirds and a first and third class trailer. The motor coaches retained the bulbous front end and were also built by the Metropolitan Carriage Wagon and Finance Company. The trailers were built by the Midland Railway Carriage and Wagon Company. The motor coaches had seven third class compartments and the composite coaches had three third class and six first class compartments. For the Western Section services more two-coach non-driving trailer sets were to be used, marshalled between the three-coach sets to form eight-coach trains at busy periods. These were more conversions from existing 'steam' carriages – each with eight third class compartments.[47] The now standard electrical equipment – motors, control gear, etc. was supplied by Metropolitan Vickers at a cost of £5,444 per three-coach set plus an additional £512 for the necessary auxiliary equipment fitted into the two-coach strengthening sets.

At first, these new trains were painted bright green and were fully lined out in

Publicity material for Eastern Section electric services featuring one of the new three-car sets. This poster is one of the earliest reference to 'Southern Electric' – the wording that was to become the 'trademark' of the Southern Railway's electric services. (Author's Collection)

A former three-car (3-SUB) multiple unit which had originally been built for the Eastern Section electrification is shown here at Bromley South with an Orpington train c. 1960. The set has been strengthened with an all-steel trailer to become a 4-SUB. (Southern Railways Group/John Sutton Collection)

yellow and black, but later economies imposed the more sombre Southern Railway olive green livery without lining. Later in the 1930s the three-car sets were to be designated '3-SUB' in the railway's 'telegraphic' class identification.

To accommodate the new Western Electrification trains at Durnsford Road depot additional inspection pits had to be built.[48] A new carriage shed was built at Effingham Junction at a cost of over £35,000.[49] This six-line shed was built to provide servicing facilities for the electric trains and is 600 feet (183 metres) long. Effingham Junction was originally built as a simple interchange point between the two lines, so with the arrival of the carriage shed major track-work modifications were required, necessitating a new 29-lever signal box being built (of traditional LSWR appearance – the style of signal boxes is always a pointer to which of the old companies owned a station). The new signalling arrangement comprised traditional semaphore signals – the Southern Railway's frugal approach to electrification did not necessarily mean colour lights, particularly in the country areas.[50]

Concurrent with the work on the Eastern and Western projects, work was taking place to extend the LBSCR overhead system. Prior to the grouping in 1923 the LBSCR had announced plans to extend its suburban network by electrifying the lines as far out as Coulsdon North on the main line to Brighton and as far as Cheam on the line from Norwood Junction to the south coast via Horsham and Arundel.[51] This proposal also included the electrification of the main line from London Bridge to Norwood junction and from Balham through Streatham Common to Selhurst. This scheme, unlike the LSWR electrification, was halted by the Great War. This left the LBSCR with several miles of line equipped with overhead structures awaiting the wires. This uncompleted installation was from Balham to West Croydon via Streatham and Selhurst; and from West Croydon as far as Beeches Halt (soon to be renamed Carshalton Beeches) just before Sutton.[52]

The Ministry of Transport's Electrification of Railways Advisory Committee had in 1920 approved in principle the pre-war suburban area extension of the LBSCR overhead electrification and the later committee set up by Arthur Neal concurred with this view. Following the meeting in 1922

A 4-SUB unit comprising of vehicles built for the Western Section 1925 electrification is shown here c. 1960 with an additional all-steel trailer. (Southern Railways Group/John Sutton Collection)

The full electric service on the 'New Line' commenced on 12 July 1925. On that day, a mixed group of proud Southern Railway staff pose in front of one of the new three-car sets. (John Scott-Morgan Collection)

between the three railways and Arthur Neal at Westminster – when the SECR was given the go-ahead for its scheme – the LBSCR felt that there was now no restriction on it resuming work on its electrification scheme. It announced that it too would apply for loan guarantees from the Trade Facilities Committee for not only was it proposing to complete its stalled suburban electrification plans but it would also make full use of the 'main line' capability of the 6,700V overhead electrification to extend the wires to Brighton and Eastbourne.[53] But unfortunately for the railway, when Sir Phillip Nash – acting on its behalf – approached the Trades Facilities Committee to discuss the finance for the whole scheme to electrify to Brighton, he met with a negative response, only gaining acceptance for a loan guarantee to complete the suburban lines extension.[54] Charles Macrae, the penultimate LBSCR chairman, had declared in 1922 that he was convinced that only by electrifying the line to Brighton could the payment of any dividend be achieved.[55] Furthermore,

he announced – obviously assuming his board would retain authority within a 'federal' Southern Railway – that the work would continue with or without a government loan guarantee. This news was greeted with great delight by the good folk in Brighton. But any joyous anticipation soon turned to bitter disappointment when it became clear that the newly formed Southern Railway had absolutely no intention of carrying out any such expensive electrification to the south coast. The aggrieved Brighton Borough Council requested a meeting with the Southern Railway to air its views and, hopefully, seek assurance that the railway did intend to carry out electrification to the town and other neighbouring towns.

A meeting was held at Waterloo on 28 December 1923.[56] The deputation from Brighton was led by the mayor supported by seven others. Sitting at the other end of the table were the eight-strong Southern Railway group led by Sir Hugh Drummond, which included Sir Herbert Walker and Herbert Jones. The Brighton delegation asserted that they had been

COMPLETING THE LEGACY SCHEMES • 159

assured of electrification by the LBSCR and that the Southern Railway had reneged on this promise, leading them to believe that only projects instigated by the SECR and LSWR were being allowed to go ahead. Councillors also expressed concern about the high local unemployment due to the economic situation and hoped that such a major engineering project would alleviate things.

The Southern Railway representatives politely listened to the Brighton delegation. Sir Hugh Drummond responded by saying that the Southern Railway had fully committed its resources to electrifying the suburban lines, which was where the need was the greatest and this did include some former LBSCR routes. Despite the council's assertion that the Brighton area was 'practically suburban', Sir Hugh insisted that there was already a 60-minute journey time service between Victoria and Brighton and that electrification would not improve this. The Brighton delegation agreed that there was a fast service, but electrification would provide more trains. The council representatives were told that electrification to the Brighton area would cost £10 million and that such 'main line' electrification was not Southern Railway policy and furthermore it was also pointed out that the LSWR and SECR main lines terminating in the electrified areas were not to be electrified. As to the question of employment for the near destitute living in Brighton, Sir Hugh declared that all the manufacturing would have taken place away from the South of England and that any unskilled labour would have been performed by the railway's own staff. Sir Hugh did however promise that the former LBSCR workshops at Brighton would not be closed, even if the precise nature of the work carried out might change – indeed this proved to be the case as Brighton works remained and steam locomotive construction continued there into the 1950s. The meeting concluded with both sides having made their case, but nothing was to change.

The new Southern Railway had restarted the LBSCR suburban electrification programme under the direction of Sir Philip Dawson whose services were retained until the end of June 1923. With the pre-war work having reached Beeches Halt, an early approval was given to complete the Beeches Halt to Sutton link, as extensive housing development was about to commence in the locality.[57] But the full LBSCR suburban electrification scheme was never to be completed for only the Balham to Selhurst, Norwood Junction to Coulsdon and Selhurst to Sutton sections ever saw the overhead electric trains – that is the sections that had been partially equipped with overhead equipment before the war. The proposal to electrify the main line from London Bridge to Norwood Junction and the Sutton to Cheam line was abandoned. More 6,700V overhead electrification was precisely what Sir Herbert Walker did not want, but it was an active scheme that he inherited on taking office and, of course, the new Southern Railway now had to find the money for this half-completed venture. Again, once he was fully in control as the general manager, Sir Herbert set to work and allocated £680,000 for the scheme: £295,500 from the capital budget, £40,000 from the locomotive renewal fund, £230,500 from the carriage renewal fund and £114,000 from the suspense account.[58]

There had been many changes to the electricity supply arrangements for the extended LBSCR 6,700V electrification following the initial 'experimental' electrification of the South London Line. At the completion of the project in 1925 the electrical distribution system had grown to the extent that the LESC provided no less than nine incoming single-phase supplies. A large 'switching cabin' at Peckham Rye Junction was now the main intake point. The original Queens Road intake point had been relegated to a standard switching cabin with the two original feeder cables owned by the LESC removed and extended to the Peckham Rye intake point. Radial feeders branched off the ring to Victoria and to London Bridge. Another radial feeder ran from Tulse Hill to Streatham along the non-electrified LBSCR railway line on that route. A further intake point had been established near Norwood Junction. This was achieved in a most ingenious way from two LESC supplies to a new railway substation at New Cross Gate – a location not on the electrified lines, but close to the Deptford generating station at Stowage Wharf. Sir Philip Dawson's scheme had these two single-phase supplies at New Cross Gate connected to a pair of single-phase transformers each rated at 5,000kW. The transformers' secondary windings produced 64,000V, and four single-phase cables were connected to the four secondary winding connections on the two transformers. The route of these four extremely high voltage cables was about four miles long and they were terminated at the new substation near Norwood Junction. At this substation two more transformers reversed the process and a pair of single-phase cables made connections through the various switches to the overhead wires above the railway lines out towards Streatham Junction, Tulse Hill and the new extensions to Sutton and Coulsdon.

The section of high voltage cable provides a clue as to how Sir Philip Dawson might have extended the LBSCR electrification to Brighton: 6,700V single phase was not a high enough voltage to transmit the power from Deptford to the Brighton area, and neither was it possible to have such a lengthy 'return' since there would be an unacceptably high voltage on the rails at the coastal end of the line. By interposing a section of much higher voltage these two problems could have been overcome. Furthermore, if as rumoured, Sir Phillip wanted to use 11,000V on the overhead wires this could be achieved using such an intermediate transformer arrangement.

The new trains for the Sutton and Coulsdon services were usually comprised of five vehicles, four passenger coaches and a powered van. These short vans, nicknamed 'milk vans', ran on two bogies and were fitted with a pair of bow collectors, and despite having carriage-type bodies for parcels or luggage accommodation, they were strictly speaking electric locomotives. They could, if required, be driven as a single vehicle as they had a driver's compartment at both ends. For normal usage however they were placed in the centre of the train and the end 'trailer' carriages had driving compartments. The vans were fitted with four 250 horsepower (186kW) motors. At the time of the opening some of the stock was painted in the LBSCR umber as ordered by that railway, but others did correctly sport the now standard Southern Railway green livery.[59]

Despite Sir Herbert's and the directors' disdain for the method of electrification for their new electric trains, they were to prove to be successful. In anticipation of an increase in passengers the new timetable introduced on 1 April 1925 required the electric trains to Coulsdon North and to Sutton to run a total of 17,370 miles on weekdays (which included Saturdays) and 2,100 miles on Sundays. This compared to the 6,874 miles previously run by steam trains during the week and 701 on Sundays.[60]

In 1926 following the completion of the legacy schemes, the Southern Railway would proudly boast that it now operated over 700 miles of electrified railway and was running three electric trains for every steam train previously run on the suburban lines. The total investment, it proclaimed, had been £8 million.[61]

With the three electrification schemes nearing completion, Alfred Raworth was effectively promoted to 'electrical engineer new works', taking charge of all the new electrification schemes, removing this responsibility from the Southern Railway's chief electrical engineer, Herbert Jones.[62] The January 1926 edition of the *Southern Railway Magazine* printed the annual list of all the principal Southern Railway officers and confirms Alfred Raworth's new appointment. It has often been hinted that this appointment was a 'reward' for his agreement to the scrapping of his

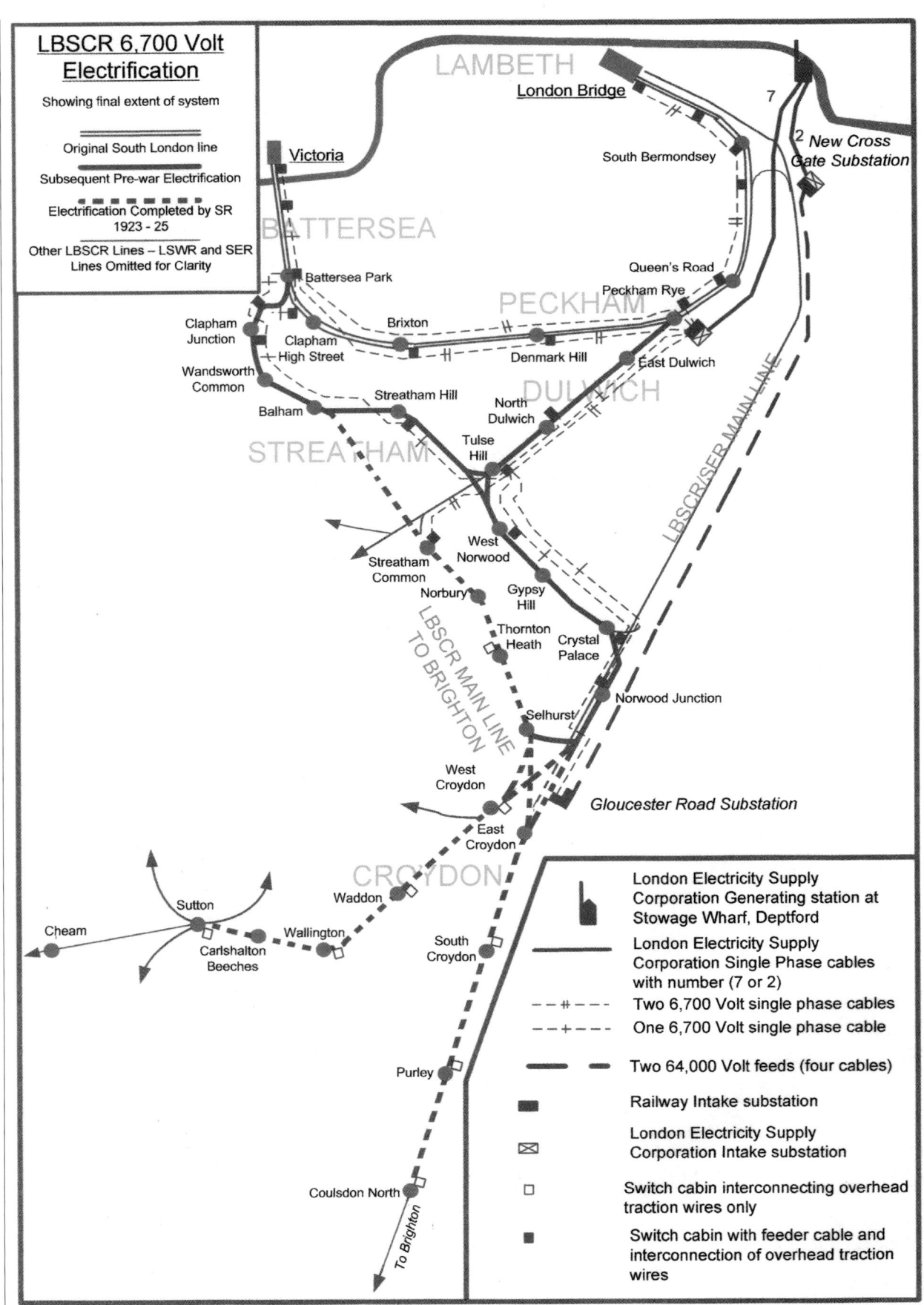

Final extent of LBSCR 6,700V electrification. (Author)

A five-vehicle, 6,700V train comprised of four trailers and a centre motor van on its way to Coulsdon. This service was introduced by the Southern Railway, but nonetheless the train is in full LBSCR livery. (John Scott-Morgan Collection)

electrification proposals and subsequently stepping into line with the LSWR thinking, but more likely this promotion was earned by his loyalty and by his successfully managing the south-eastern suburban electrification.

Following his new appointment, Raworth remained based at his London Bridge office – a significant distance away from the Southern Railway's chief electrical engineer Herbert Jones who resided at Waterloo on the other side of London. It has been alleged that this physical separation was instigated and encouraged by Sir Herbert in an endeavour to avoid conflict between Jones and Raworth.[63] The fact that their respective offices were at opposite ends of the metropolis, when a closer working relationship might have been advantageous, does give some credence to this belief, but Walker may have believed that such a separation ensured that each remained focused on their particular task.

On 1 August 1924 Sir Hugh Drummond died at his home in South Devon. His untimely passing increased the perceived SECR influence on the Southern Railway management structure when the deputy chairman and former SER director Brigadier-General the Hon. Everard Baring was voted in as the new Southern Railway chairman.[64] Baring was a small, thin man who would appear somewhat emaciated when standing beside the towering bulk of Sir Herbert Walker.

An early duty for the new chairman was to officiate at the opening of the extension to the 6,700V suburban electrification. This was of some embarrassment to him because he told the shareholders at the Southern Railway AGM in 1925 that his board was

COMPLETING THE LEGACY SCHEMES • 163

unhappy that the Southern Railway had two systems of electrification – a change of heart perhaps; Baring had been an SECR director when it had been pressing for a third electrification system. Baring announced that the Southern Railway 'experts' were considering what the railway directors and Sir Herbert Walker deemed to be a 'problem'.[65] Consequently at the Southern Railway board meeting held on 29 October 1925, only six months after the extension's opening, Sir Herbert informed the assembled directors that his 'electrical engineers' – that is Herbert Jones and Alfred Raworth – had agreed to the removal of the overhead equipment and that any further electrification should use direct current third-rail and not the higher voltage alternating current overhead wires. It was accepted that on some lines both forms of electrification would be used for a brief period.[66]

Two reports were prepared for the board. The first by Jones and Raworth proposed third-rail electrification on all the former LBSCR suburban routes, and the second by Raworth alone detailed proposals for the removal of the 6,700V equipment. The new scheme would cost £3,767,000, just over £2 million to be charged to the capital expenditure, the remainder to the various renewal funds and a suspense account.[67] Raworth's report, submitted and approved by the directors in April 1928, outlined the cost of the removal of the old LBSCR overhead system including £157,510 for the conversion of existing electric vehicles and £204,050 to be spent on new direct current motors and control gear. Also approved by the board was the provision of additional 11,000V cables to supply Norwood substation (abandoning the 64,000V cables) and the provision of additional substations with rotary convertors at Norwood, Streatham, Sutton and Purley – all to be supplied by further extensions to the 11,000V network. An amount of £18,000 was allocated for extensions to existing substations at Chapel Junction and Norwood. For the South London line, substations were to be provided at Brixton and South Bermondsey – but both would also reinforce the former SECR electrified routes to Holborn Viaduct and London Bridge. Existing substations originally provided for the LSWR and SECR routes electrification were available to be used and reinforced with extra plant, for example at Clapham Junction and Victoria substations.[68]

The first stage of the plan included installing the conductor rails on former LBSCR lines that had not previously been electrified such as the original main line from London Bridge to Purley which had been jointly owned with the SER and the two former SER branches from Purley to Caterham and Tattenham Corner. Stage One services commenced on Sunday, 25 March 1928 resulting in 8,530 less steam train miles and an additional 16,674 electric train miles being timetabled. The railway's traffic managers were told that 15 less steam locomotives were required 'saving' 52 enginemen but requiring an additional 36 motormen.[69] The new electric train services were to prove popular and convenient with the travelling public. The journey from Caterham to London Bridge had taken an hour and a half by bus and 46 minutes by the superseded steam trains; now, with the new electric service it took only 34.5 minutes.[70]

As part of the second stage of the programme, on Sunday, 17 June 1928 services with direct current trains commenced running on the original South London line between Victoria and London Bridge and on the Victoria to Streatham Hill route.[71] As part of the third stage the overhead wires were displaced to Sutton, but the third rail continued beyond to join the LSWR electrification at Effingham Junction. The third stage also included the conversion of the Balham to Crystal Palace line from alternating current to direct current. It was estimated that this third stage resulted in a decreased steam train mileage of 11,502 miles, a decrease in alternating current electric train mileage of 7,029 miles and an increase in direct current electric train mileage of 39,922 miles. This resulted in 18 less steam locomotives being used and 92 less 'steam' carriages. Seventy-two less enginemen were needed but 40 additional motormen were trained to operate the new timetable.[72]

The very last 'elevated electric' train left Victoria for Coulsdon North at just after midnight on the morning of Sunday, 22 September 1929. The trains were returned to the workshops at Peckham Rye to have all the electrical equipment removed. Most of the carriages began a new life being converted into direct current trains as per Alfred Raworth's April 1928 report. In total, 185 former alternating current vehicles were to be reused in direct current sets. Six of these converted carriages had three separate existences having also been originally conversions from 'steam' stock.[73] The 'milk vans', after the removal of their electrical equipment, had a second life as guard's vans on express goods trains.[74]

Much has been made by commentators of the frugality of the Southern Railway in 'recycling' the rolling stock, but what about the other plant to be disposed of? True, most of the plant scrapped had been paid for by the LBSCR and not the new Southern Railway, but the wastage included almost 50 route miles of the massive steel overhead gantries, 125 miles of copper conductor, 70 miles of single-phase 6,700V cables and 28 miles of the unique 64,000V cables from New Cross Gate to Gloucester Road. The electrical plant to be disposed of included the four 5,000kW transformers and 35 switching cabins containing the complex switchgear.[75] A few of the overhead gantries saw a useful afterlife supporting signals. The actual body of the carriages was not particularly valuable, and all the relatively new and expensive equipment inside – the transformers, control gear and motors – was scrapped.

Did this all make economic sense? In 1924 H.F. Trewman's contemporary textbook *Railway Electrification* examined the relative costs of alternating current traction and direct current traction for suburban trains.[76] He revealed that direct current energy cost 30 per cent more 'at the train' than a single-phase alternating current supply, but the alternating current trains were much heavier and more expensive to build. This is because each alternating current train had to carry on board its own 'substation' – that is some isolating switchgear and the step-down transformer. Trewman 'in his book' set out his full calculation process to determine the respective cost per train mile by considering the initial capital costs and the running costs. He also took due account of all the variables such as the route length, the intensity of the service to be provided and the losses in the electrical circuit. He also added into his formula a factor predicting the potential for increased traffic – the likely 'sparks effect'. Depending on the values applied to these variables, Trewman concludes that an intensively operated direct current service will be a few pence per train mile cheaper than a single-phase alternating current service, thereby giving credence to the Southern Railway claim that it could anticipate an annual operating cost saving of £59,580. It must be noted Trewman's 1924 calculations specifically applied to suburban lines; he provided completely different calculations for a main line electrification.

But did the Southern Railway's electrical engineers – Alfred Raworth and Herbert Jones – agree with Mr Trewman? Raworth, characteristically remaining loyal to his employer, was much later to declare, after Sir Herbert Walker's retirement, that he believed that the decision to standardise on the use of a third-rail direct current system was correct, as it had been firmly based on the established facts such as the known initial capital costs and the ongoing running costs. Furthermore, he claimed that his view was not only based on purely theoretical or economic grounds but was reinforced by the railway's practical everyday experience of running both systems.[77] To support this view, one former Southern Railway officer did not have any fond memories of operating the LBSCR system, claiming that the motor vans were

Having just left Victoria station, an N15 (King Arthur) class 4-6-0 with a Pullman train carrying travellers to France runs over newly laid conductor rails, while in the background the LBSCR overhead lines are still evident. (John Harvey Collection)

too heavy, the acceleration away from stations was poor and the overhead voltage fluctuated alarmingly.[78]

During the late 1920s it was to become obvious to the Ministry of Transport that little progress was being made in railway electrification throughout the nation, apart from the efforts of the errant Southern Railway, which was blatantly going against Ministry of Transport policy as expressed by Sir John Pringle in 1923. The Southern continued to incrementally extend its third rail system beyond the 20 miles from Charing Cross – the mileage from Waterloo to Guildford on the New Line was 31 – and furthermore was making no attempt whatsoever to standardise on 1,500V direct current electrification. Accordingly, the Ministry in 1927 deemed it necessary to form another committee to review the recommendations of the 1921 Electrification of Railways Advisory Committee.[79] This time the chairman was chief inspecting officer of railways, Sir John Pringle. The committee members included many experts from the days gone by – Sir John Aspinall, Sir Philip Dawson, Charles Merz, Sir Philip Nash and Sir John Snell. But there were some fresh faces at the table including Nigel Gresley from the LNER and Herbert Jones from the Southern Railway. Among those giving their evidence was Sir Herbert Walker.

The conclusion of the Pringle Committee was that for all future electrification direct current should be used either at the *higher* voltage of 1,500V or at the *lower* voltage of 750V. The use of 750V implies at least an appreciation that the Southern Railway was to obstinately continue to use the third rail. Later in 1932 however, the then Secretary of State for Transport, John Pybus MP, in pursuance of powers conferred upon him by the 1921 Railways Act, issued his Railway (Standardisation) Order which allowed the railway companies to electrify their lines using conductor rails charged at *up to* 750V direct current or overhead lines *up to* 1,500V direct current. This subtle change of emphasis was a significant concession to the Southern Railway as it allowed it to continue to use its 660V electrification.

Less contentious was the Pringle Committee's recommendation that both overhead and third-rail systems were permitted to use the earth connected running rails as a return path; but operators such as the London Underground would be permitted to continue to employ insulated return rails. The bulk of the final report concerned itself with the updating of standards in respect of conductor rails and overhead structures and regulations to ensure that any electric locomotives that might traverse the nation's railways were built to a standard 'loading gauge' profile. Many were to view the report as a crucial step towards a national electric railway network. A decade later, comment in the *Railway Gazette* lauded the Pringle Report as a step towards an integrated electrified railway across the nation.[80] Ironically, the editorial was written at the time of Herbert Jones' retirement and in the same issue the magazine complimented the Southern Railway's retiring chief electrical engineer for his part in the preparation of the Pringle Report. Ironic because, despite the document not being particularly contentious as it was an attempt at a difficult compromise to accommodate the Southern Railway – it was signed by all the committee members except for Herbert Jones. Perhaps the Ministry thought that by bringing the Southern Railway to the table the company could be 'brought on side'. But as in 1922 when Theodore Stevens had pressed the LSWR case with a minority report, Herbert Jones too was to do the same, appending a short and somewhat confusing statement thereby avoiding having to sign the main report and agree with any 1,500V proposal.

When the Pringle Report was published it was evident that the committee member representing the Southern Railway had dug in his heels and refused to line up with the aristocrats of the electrical engineering world. So, perversely, the only company electrifying its railway at the time was to walk away from a national committee set up to encourage the expansion of railway electrification.

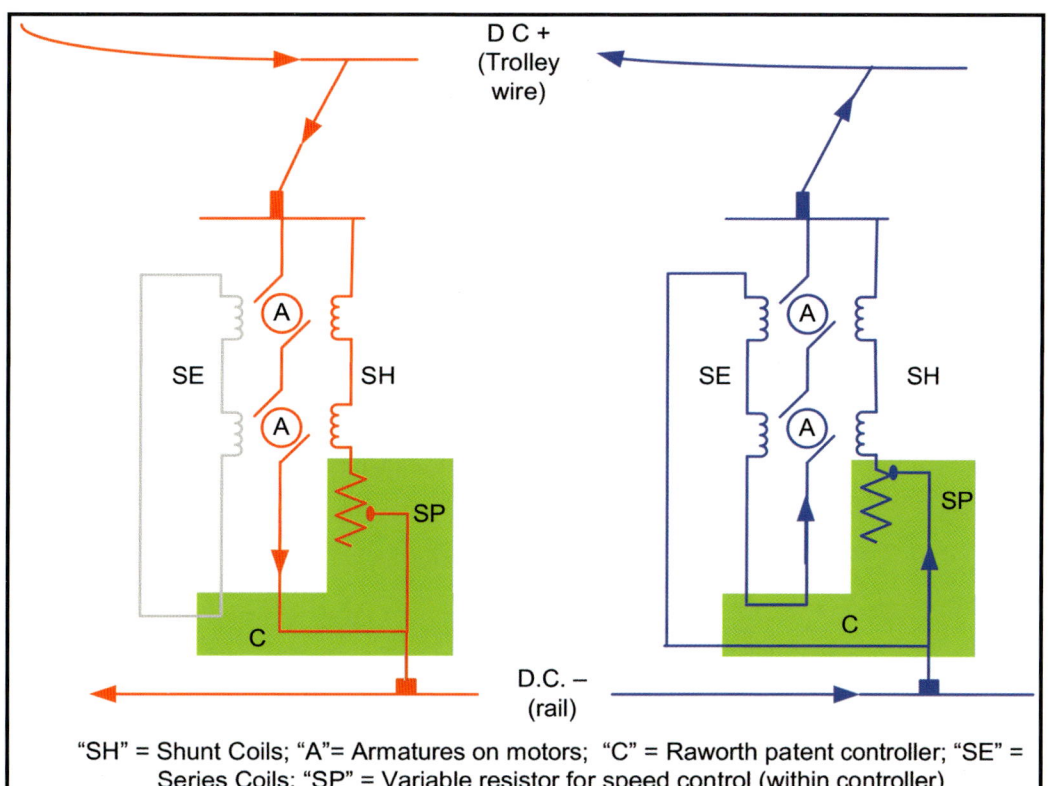

Simplified diagram showing details of John Smith Raworth's 'regenerative control'. Left - red lines- traction motors being used as motors, and right - blue lines - traction motors operating as dynamos (Author)

"SH" = Shunt Coils; "A"= Armatures on motors; "C" = Raworth patent controller; "SE" = Series Coils; "SP" = Variable resistor for speed control (within controller)

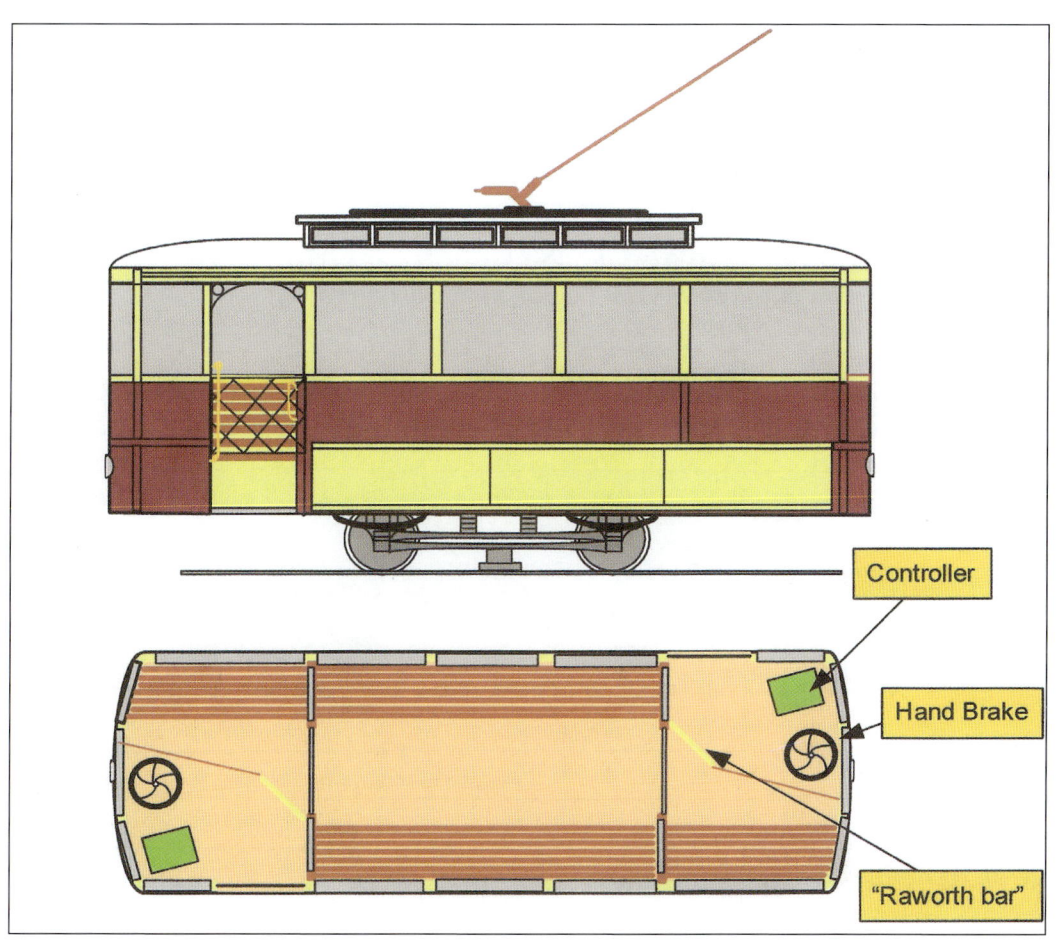

Typical Raworth demi-tram. (Author)

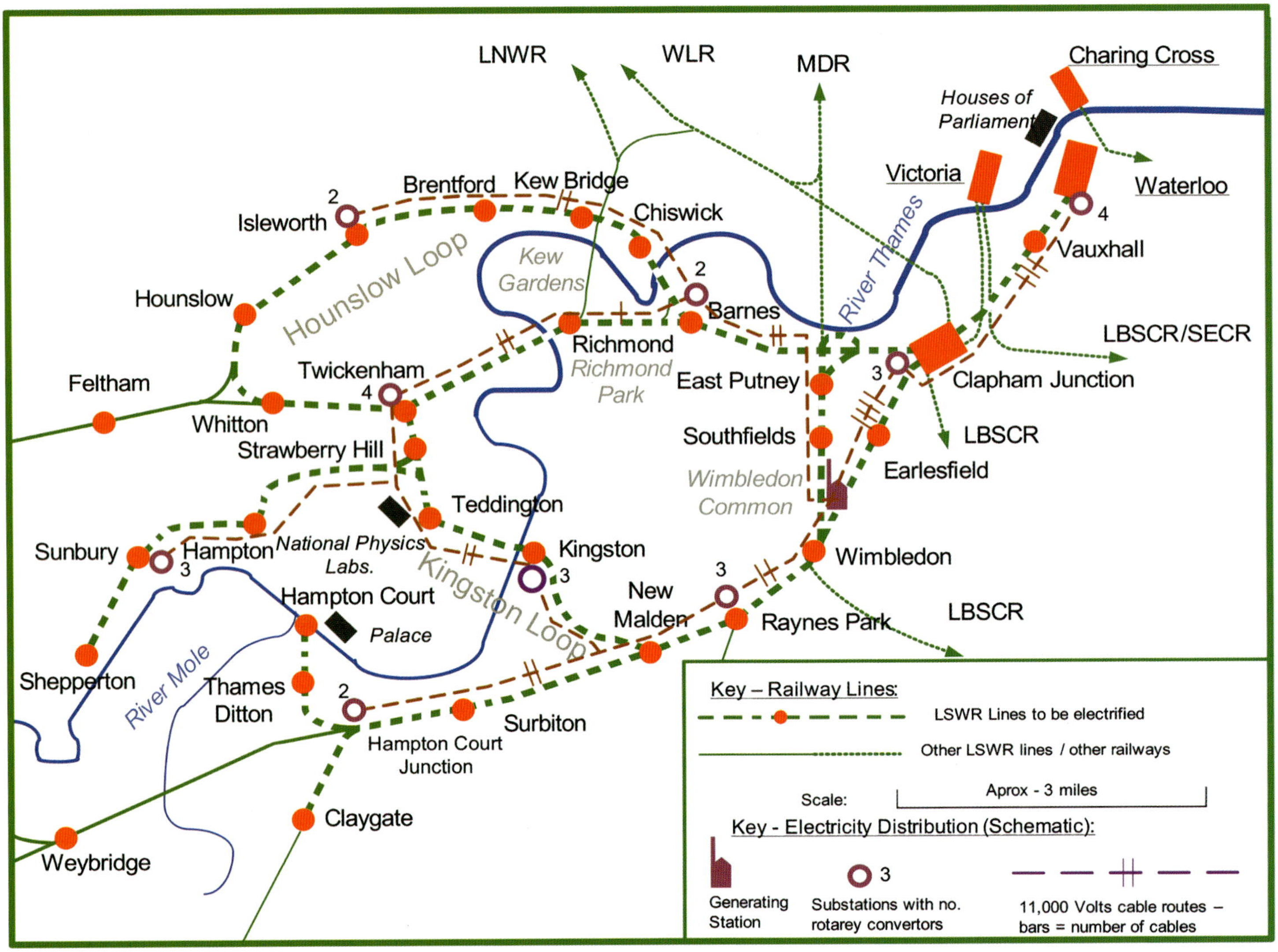

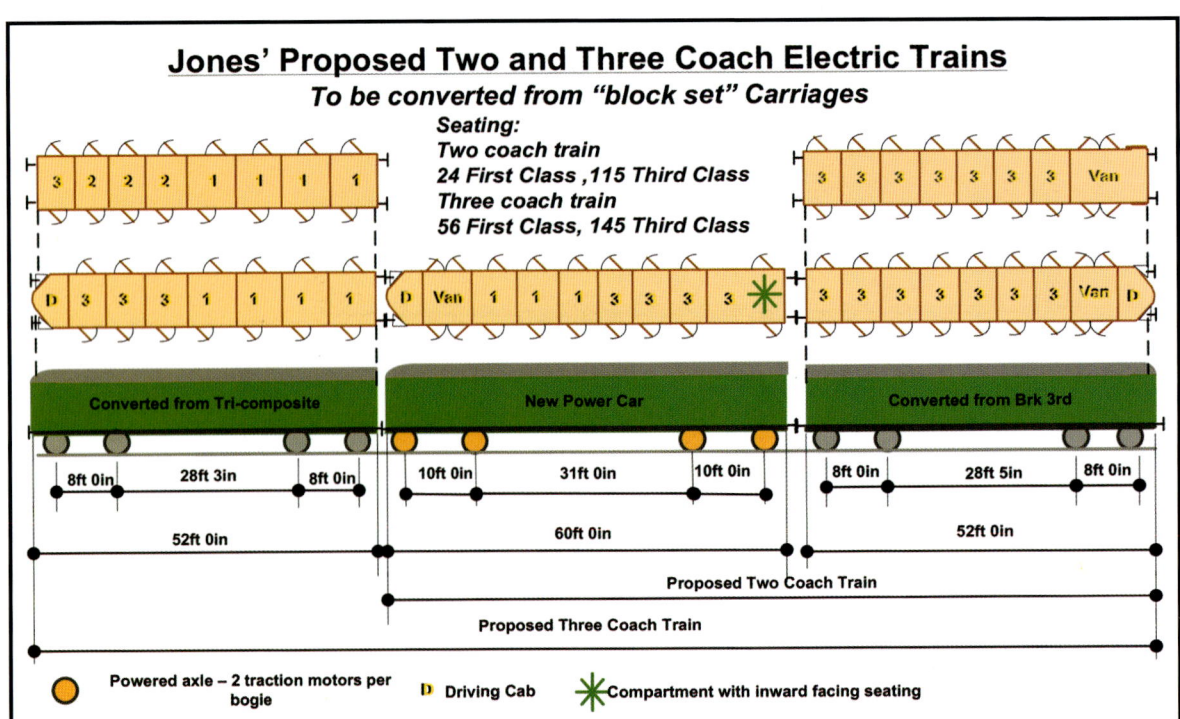

The LSWR electrification scheme. (Author)

Jones' proposal for LSWR electric trains. (Author)/Southern Notebook (Author/Southern Notebook (Southern Railways Group))

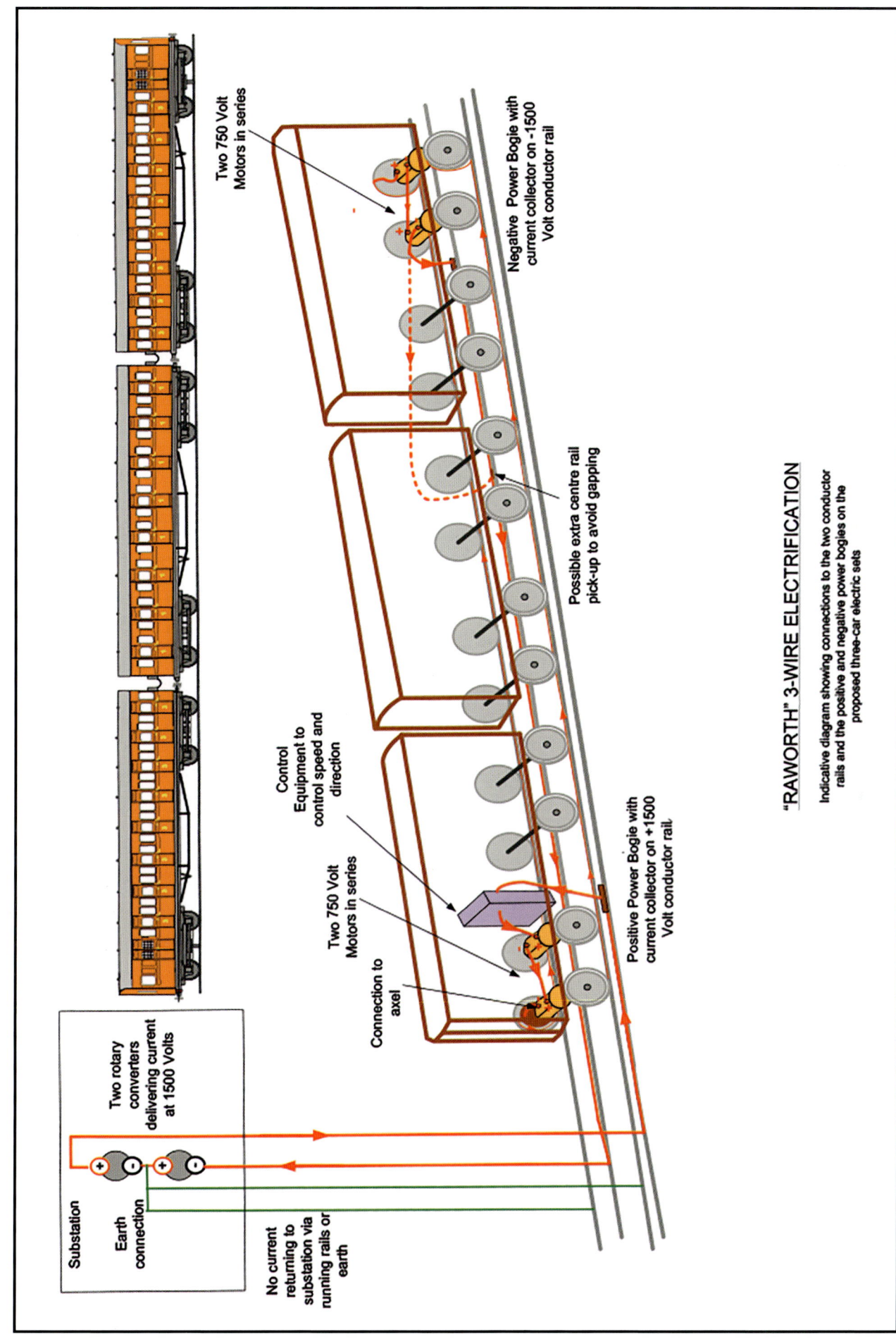

Alfred Raworth's 'three-wire' electrification proposal for the SECR. (Author)

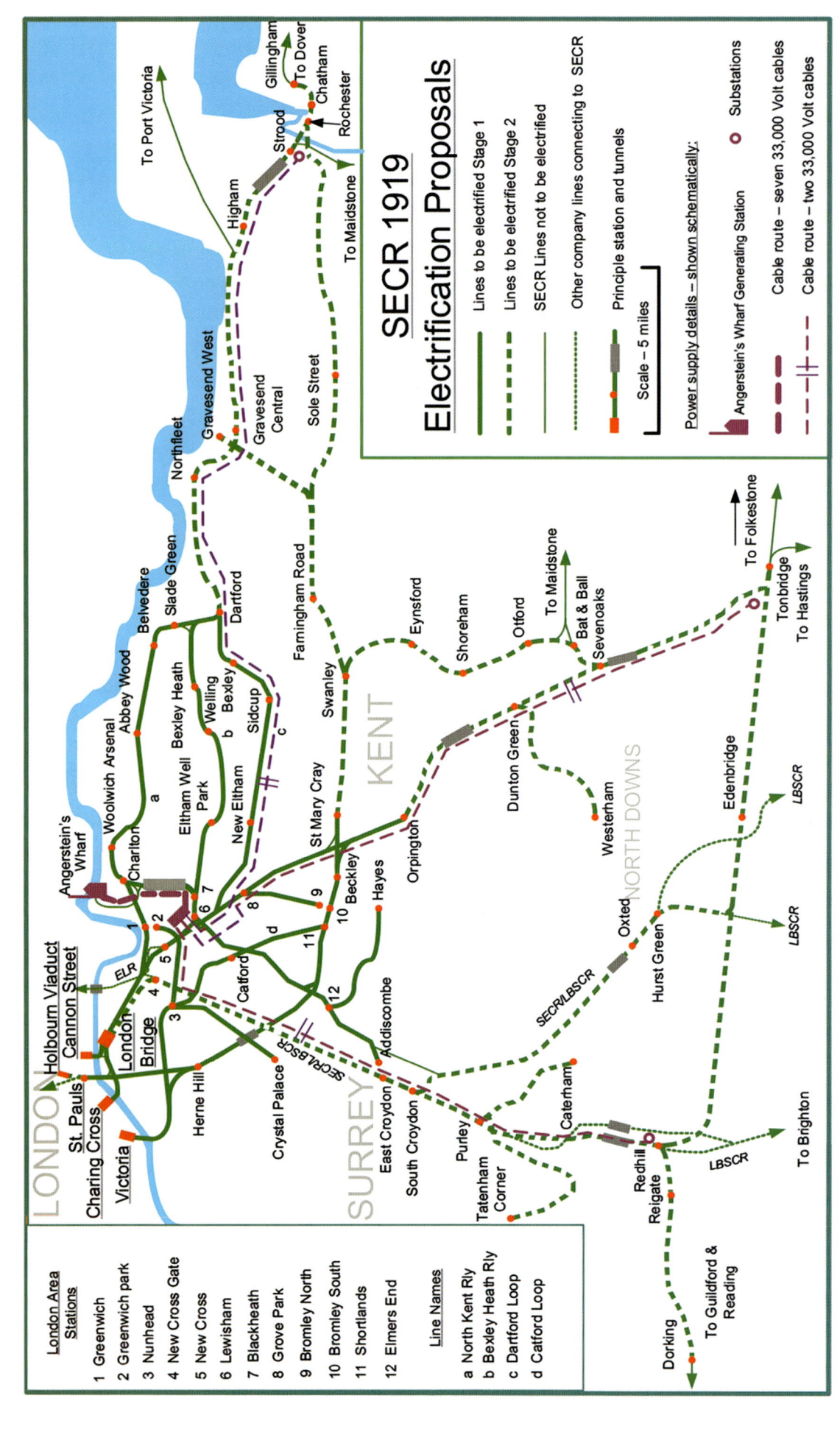

Alfred Raworth's proposals for SECR electrification. (Author)

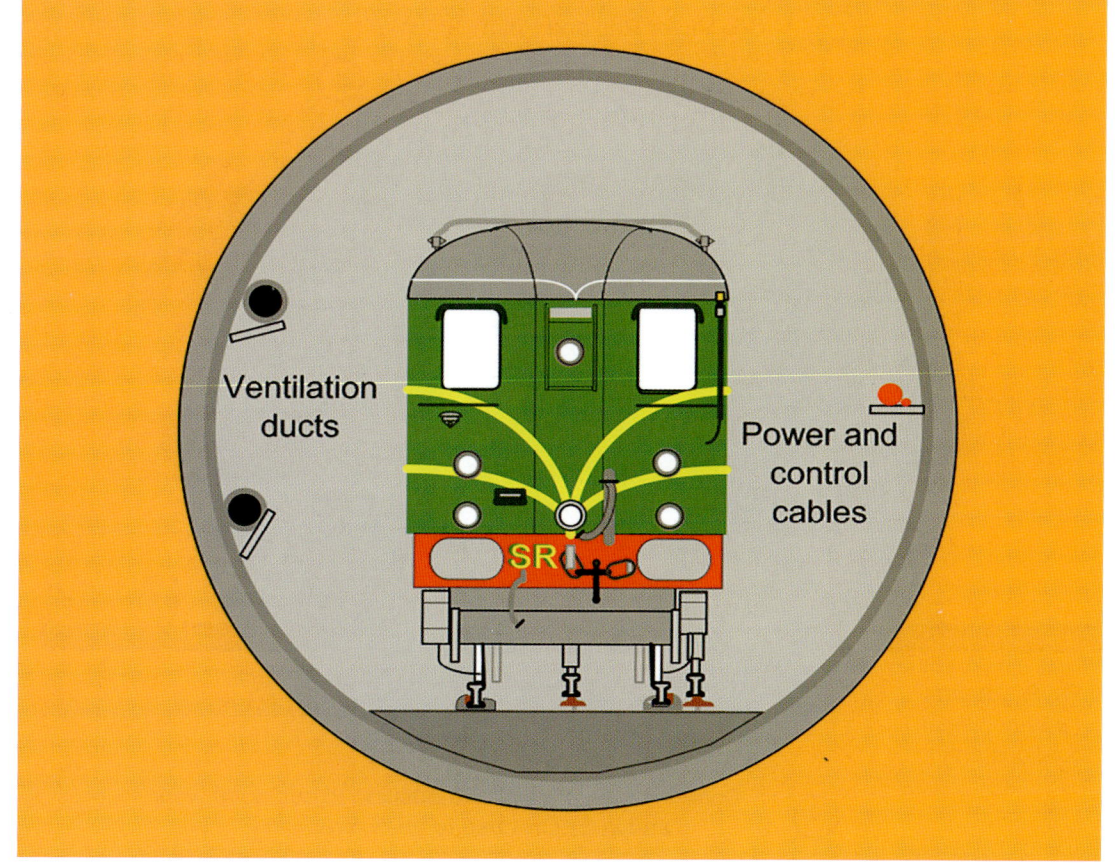

Comparison between the actual Channel Tunnel equipped with overhead wires and the reduced size bore that would have been required for the SECR scheme using conductor rails. The Raworth scheme would not however have been able to support the level of traffic on the actual Channel Tunnel as opened three-quarters of a century later on 6 May 1994. The Channel Tunnel mock-up (top), is at the National Railway Museum. (Author)

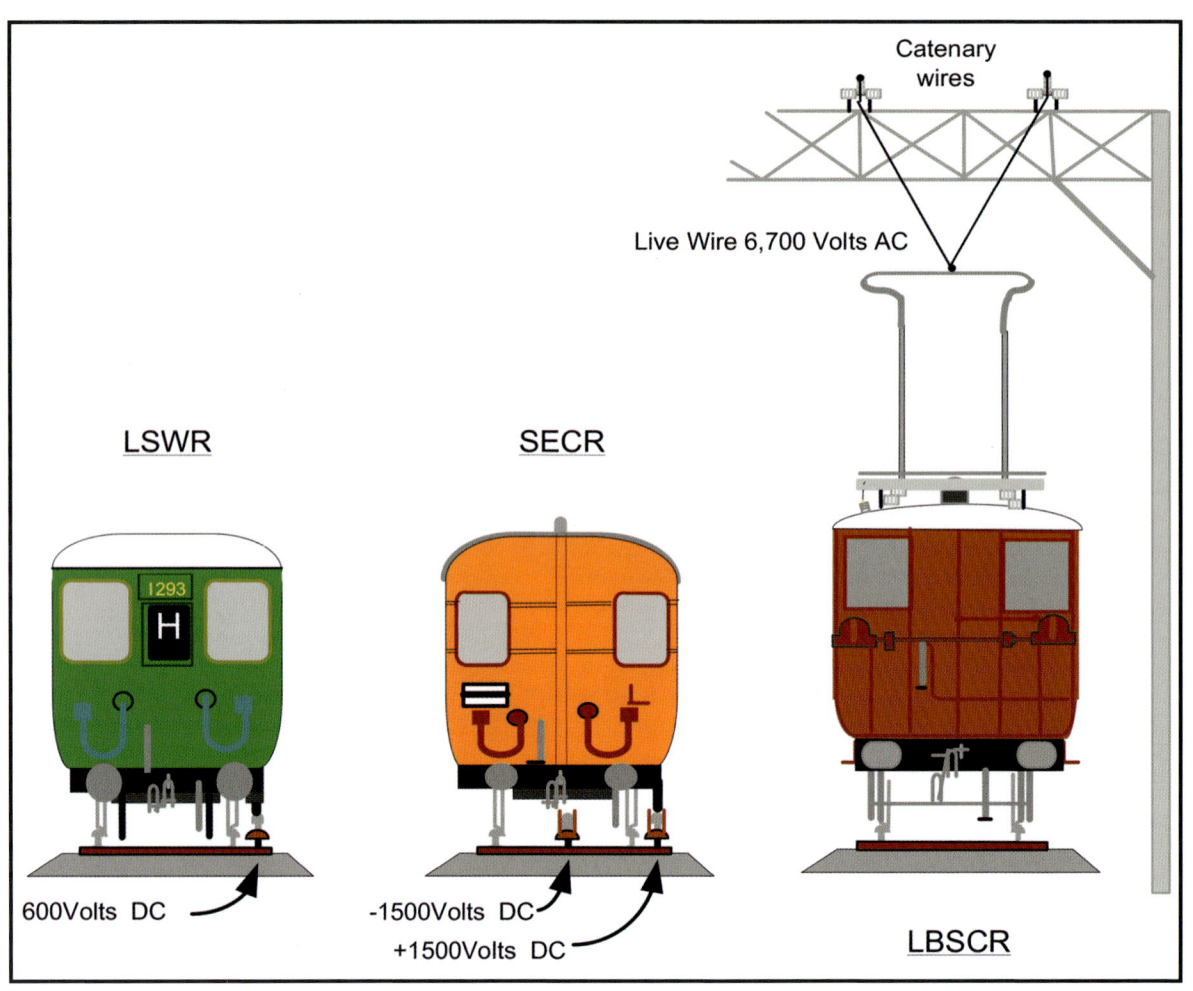

The three rival electrification schemes. (Author)

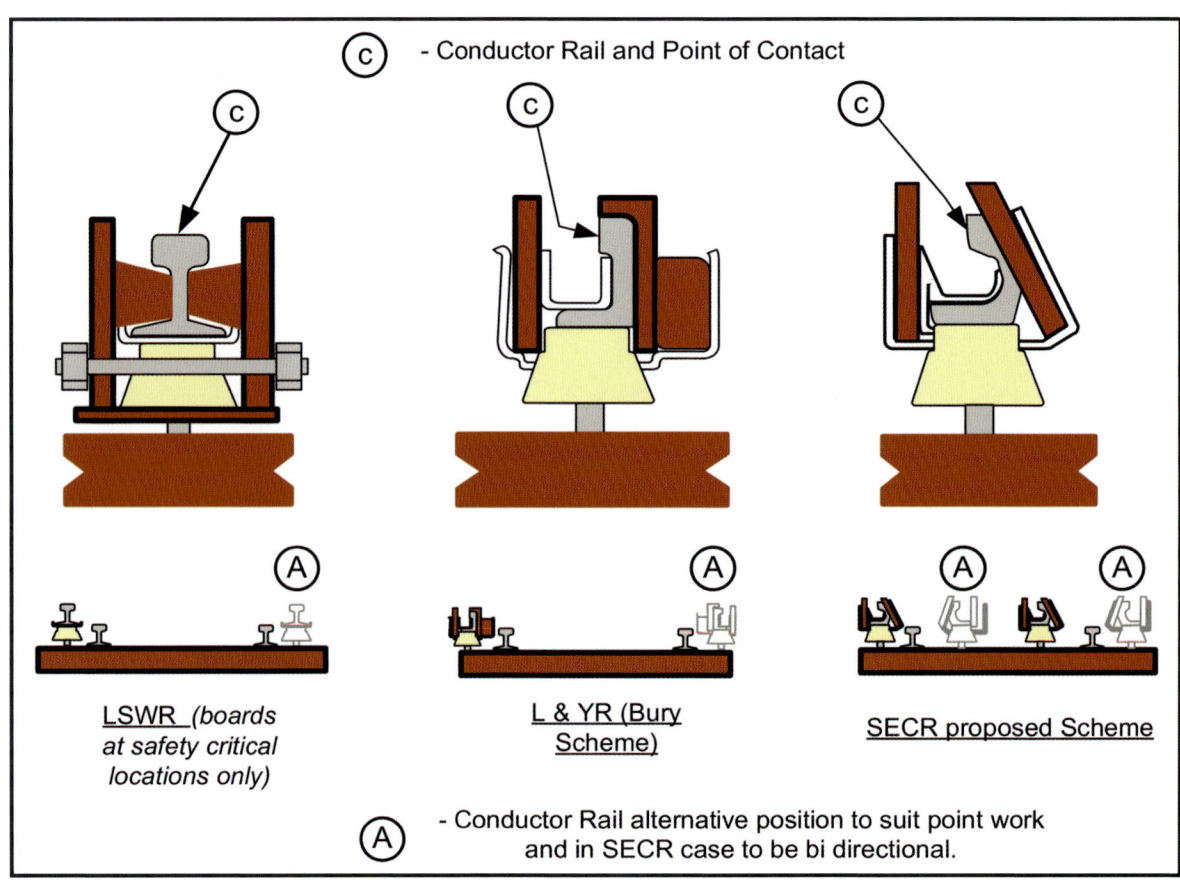

Comparison of the 'rail protectors'. (Author)

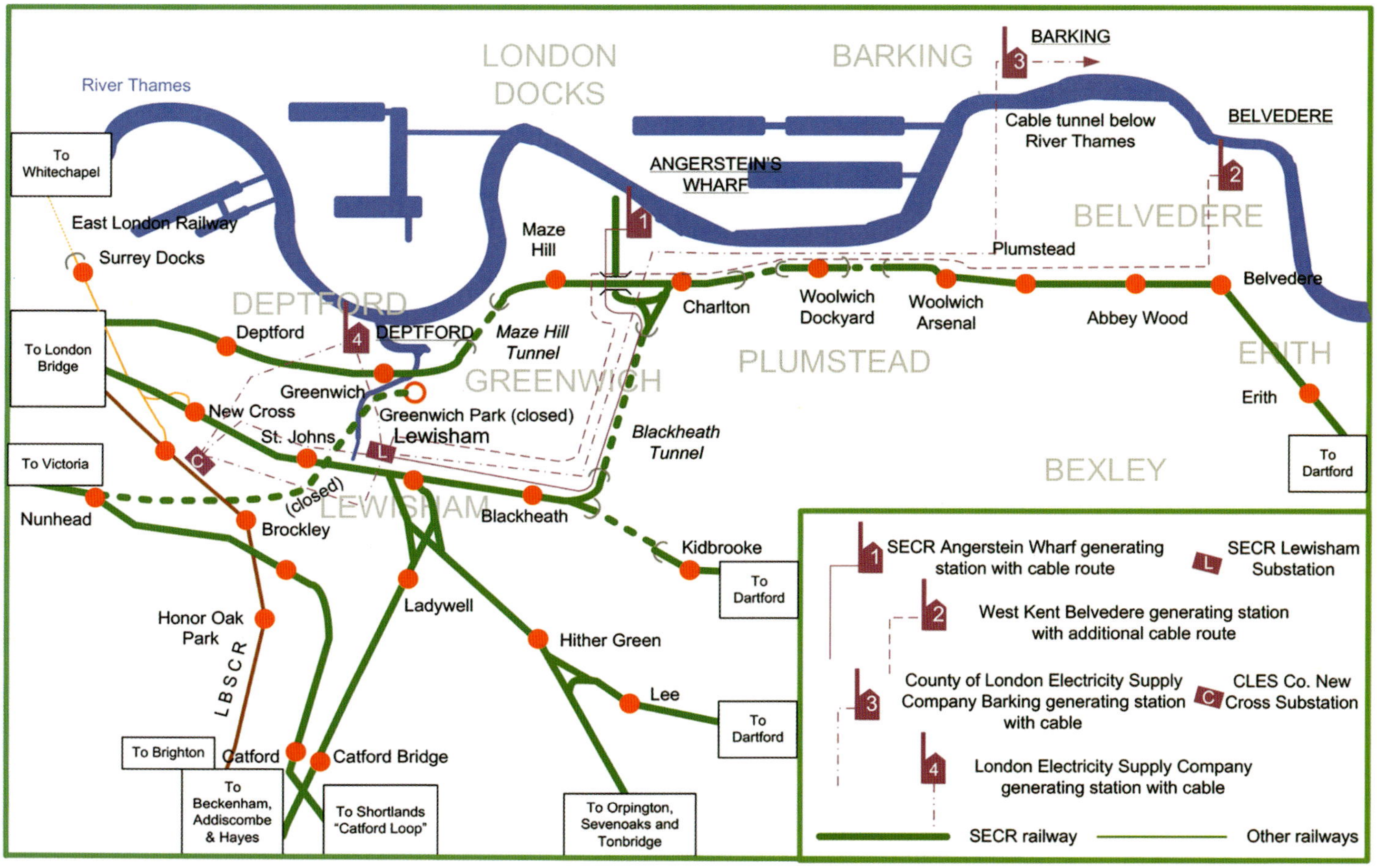

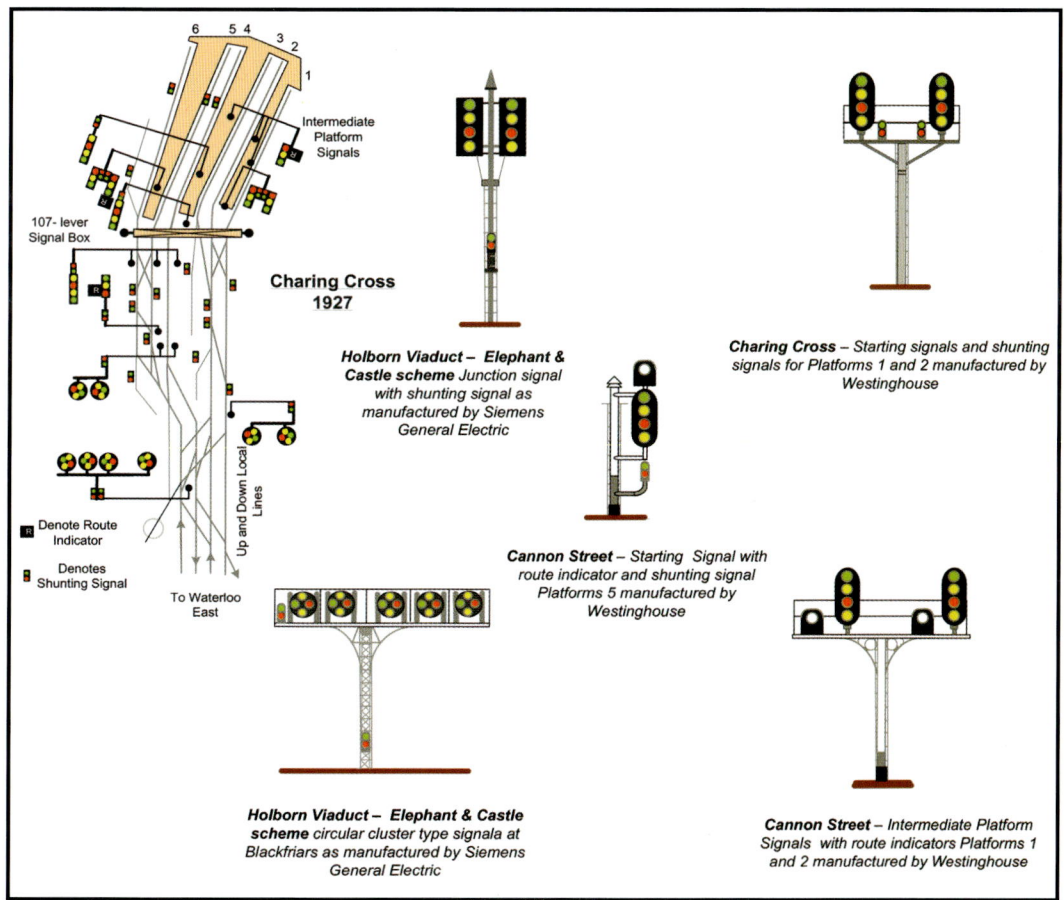

The various schemes to provide the SECR with electricity for its proposed electrification. (Author)

Examples of track layout and signals for Eastern Section colour light signalling scheme. (Author/Southern Notebook (Southern Railways Group))

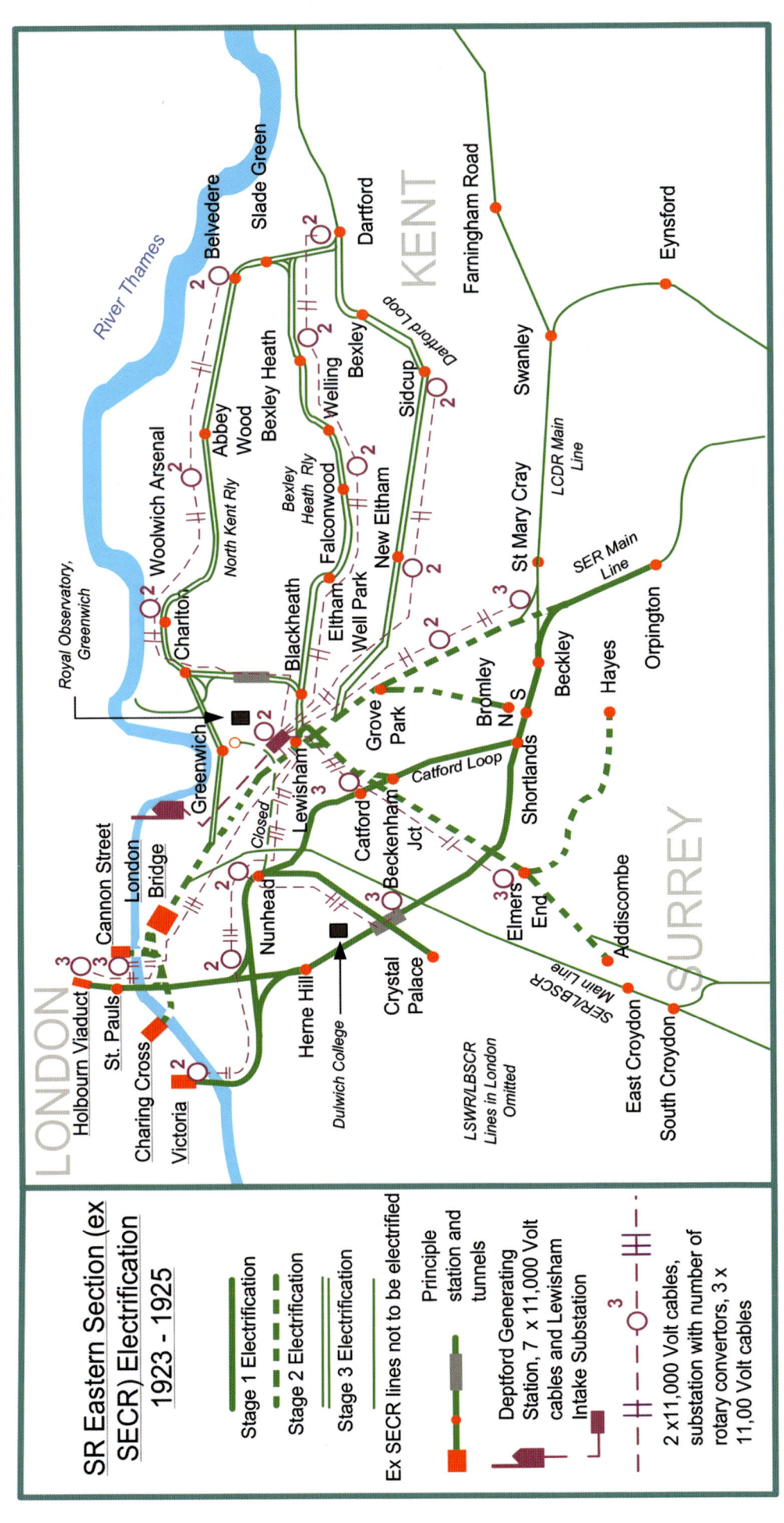

Southern Railway Eastern Section (ex-SECR) electrification. (Author)

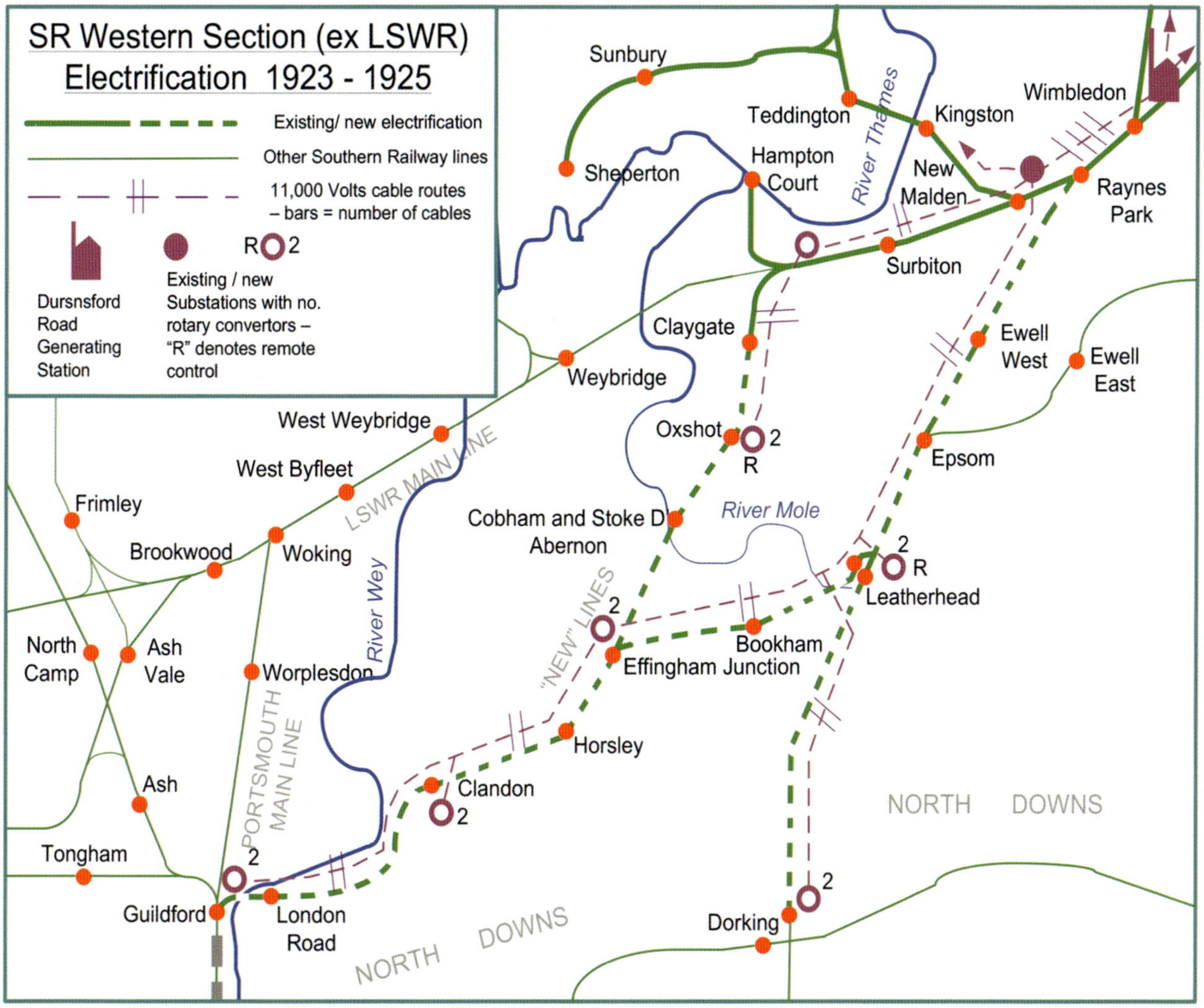

The second phase of the LSWR electrification scheme – the Southern Railway Western Section electrification. (Author)

A 1925 LSWR design (but built by the Southern Railway) motor brake third from set 1293 is here shown on display at the National Railway Museum at York on 29 October 2013. (Author)

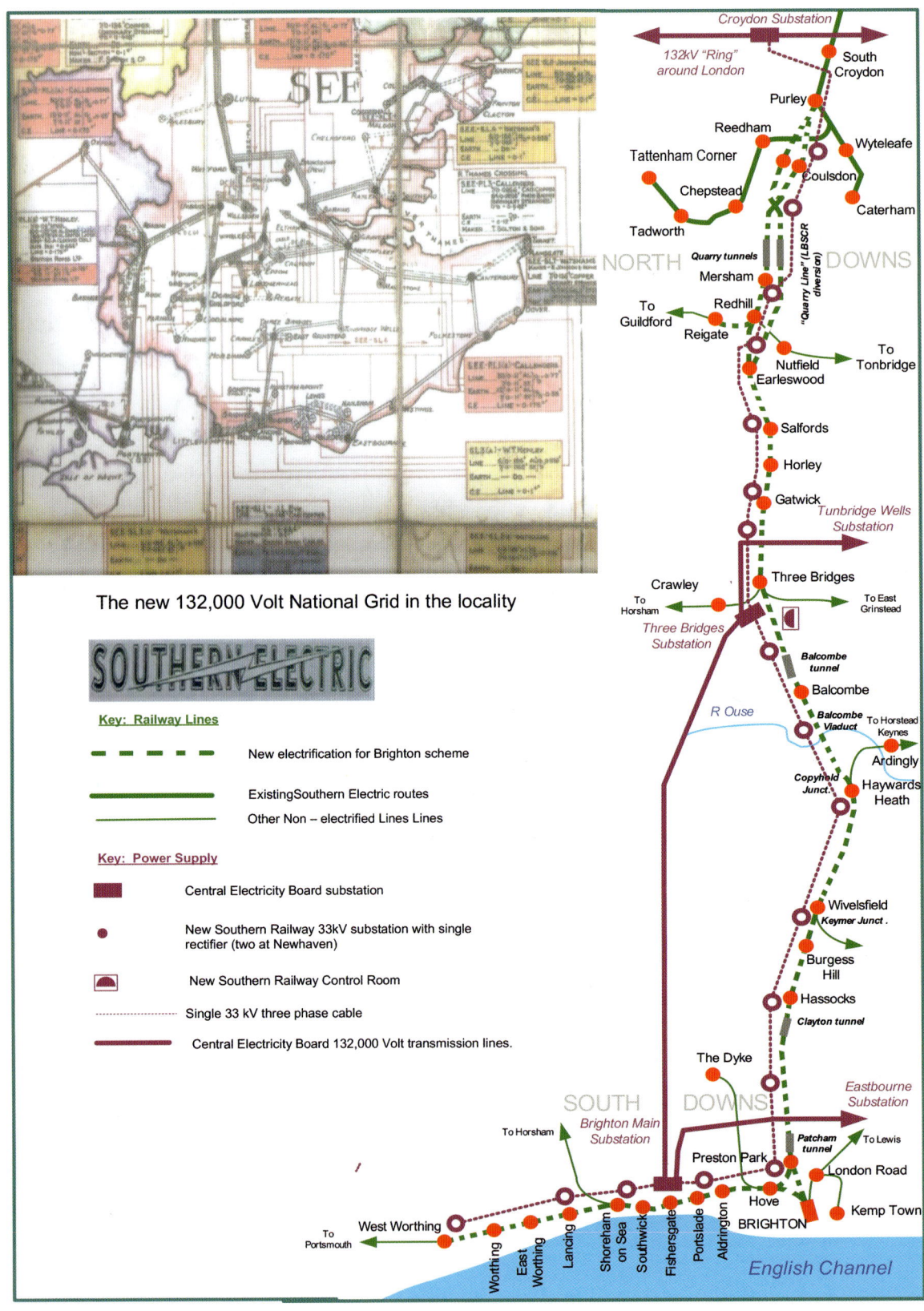

Map of the Brighton electrification scheme. (Author)

Inset: Central Electricity Board system diagram showing the extent of the original National Grid network in the Southern Electric area. (Martin James Collection)

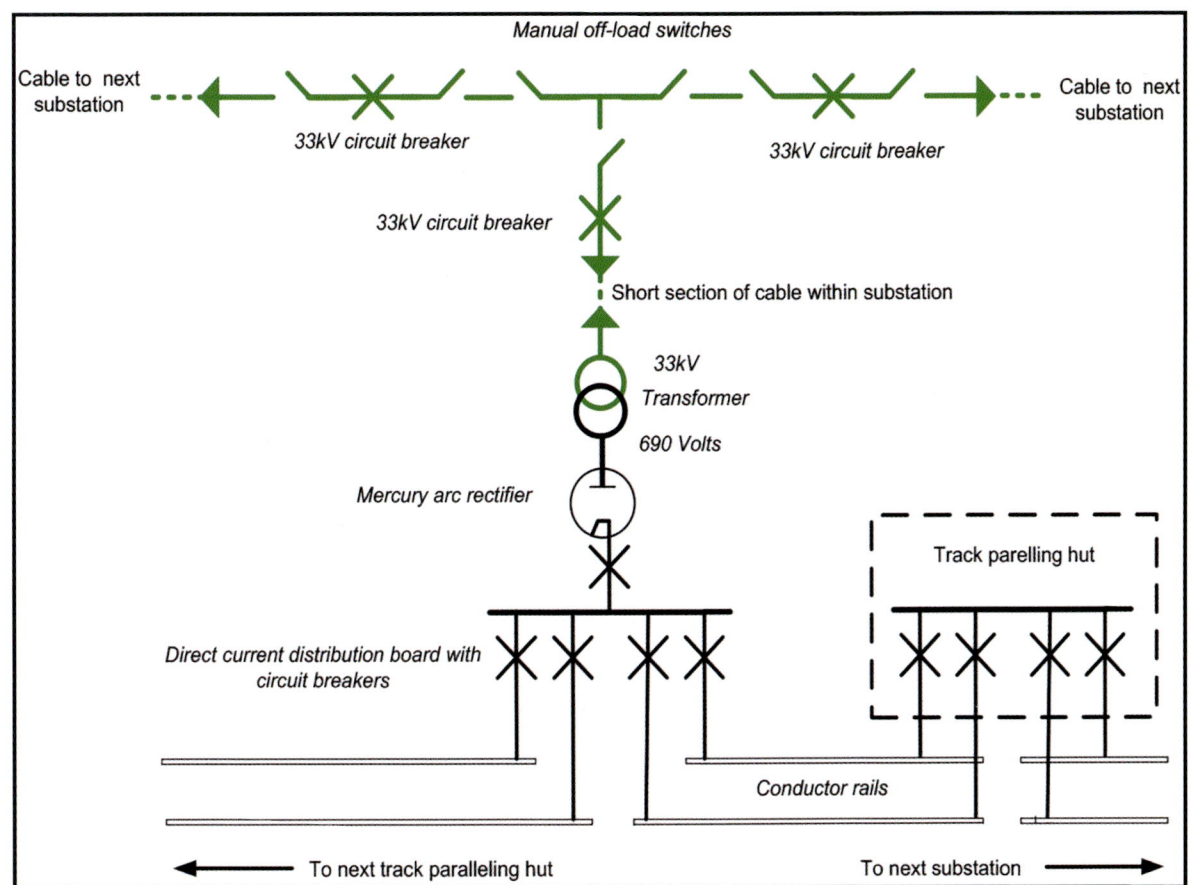

'Single line diagram' for typical Southern Electric substation. The three-phase equipment is represented by a single line; the direct current 'return' is not shown. (Author)

Above left: Siemens Desiro No. 450 006, at the rear of the 11.41 Waterloo to Reading train formed of two four-car sets, passes the 1939 built Wokingham substation on the 30 July 2017. Note the 'spare' concrete structure on the left intended to support a feeder cable for the electrification of the Reading to Guildford (ex-SECR) line. There have been plant renewals over the years, for example, the physically smaller transformer's low voltage connections necessitated that the original higher cable entry point into the building be blocked off. (Author)

Above right: A track paralleling hut in the twenty-first century. This one – part of the later Reading electrification scheme – is between Bagshot and Camberley substations. Over the years it has suffered from the work of graffiti artists who have trespassed onto the line. (Author)

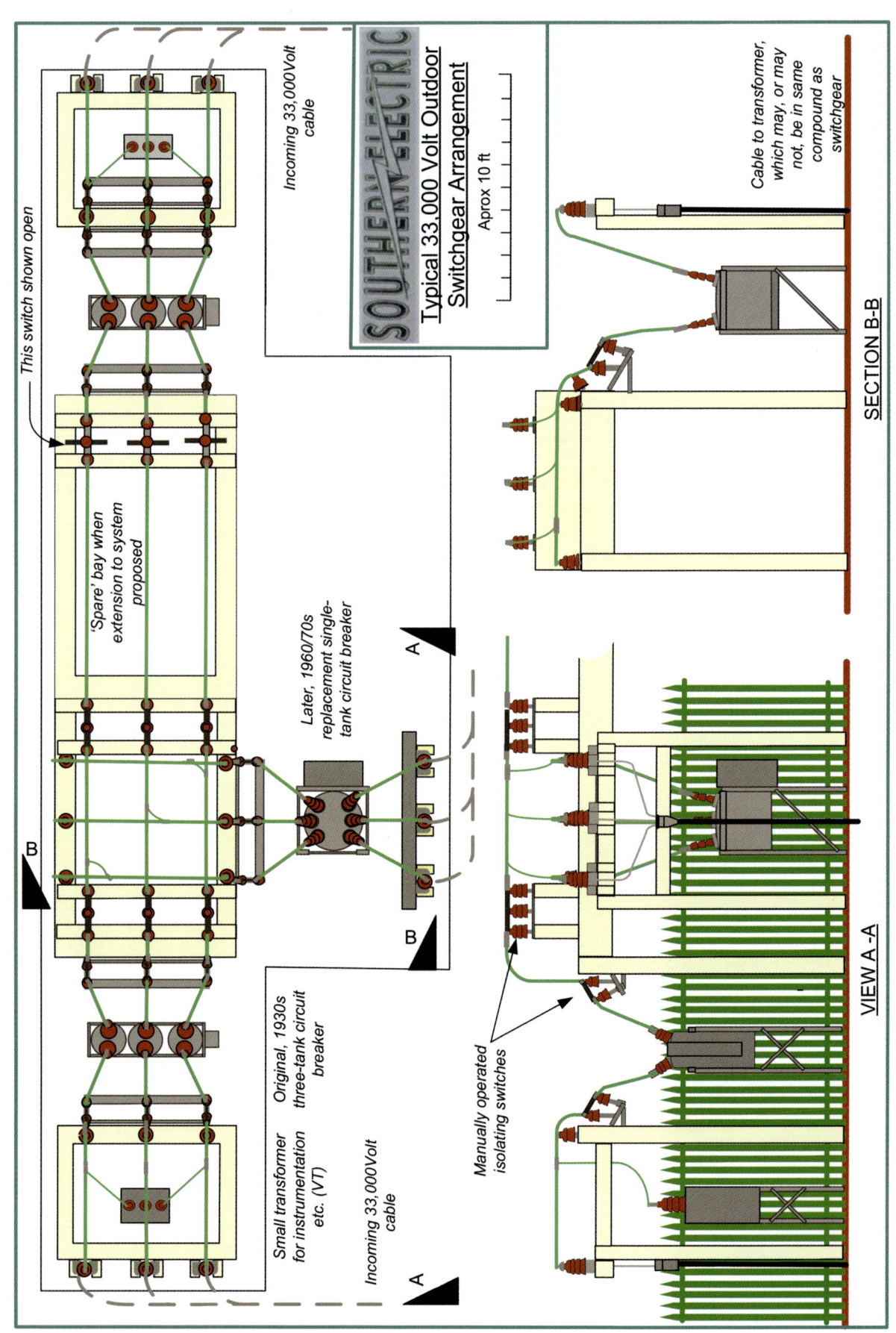

Typical 1930s Southern Electric substation – the outdoor switchgear. (Author)

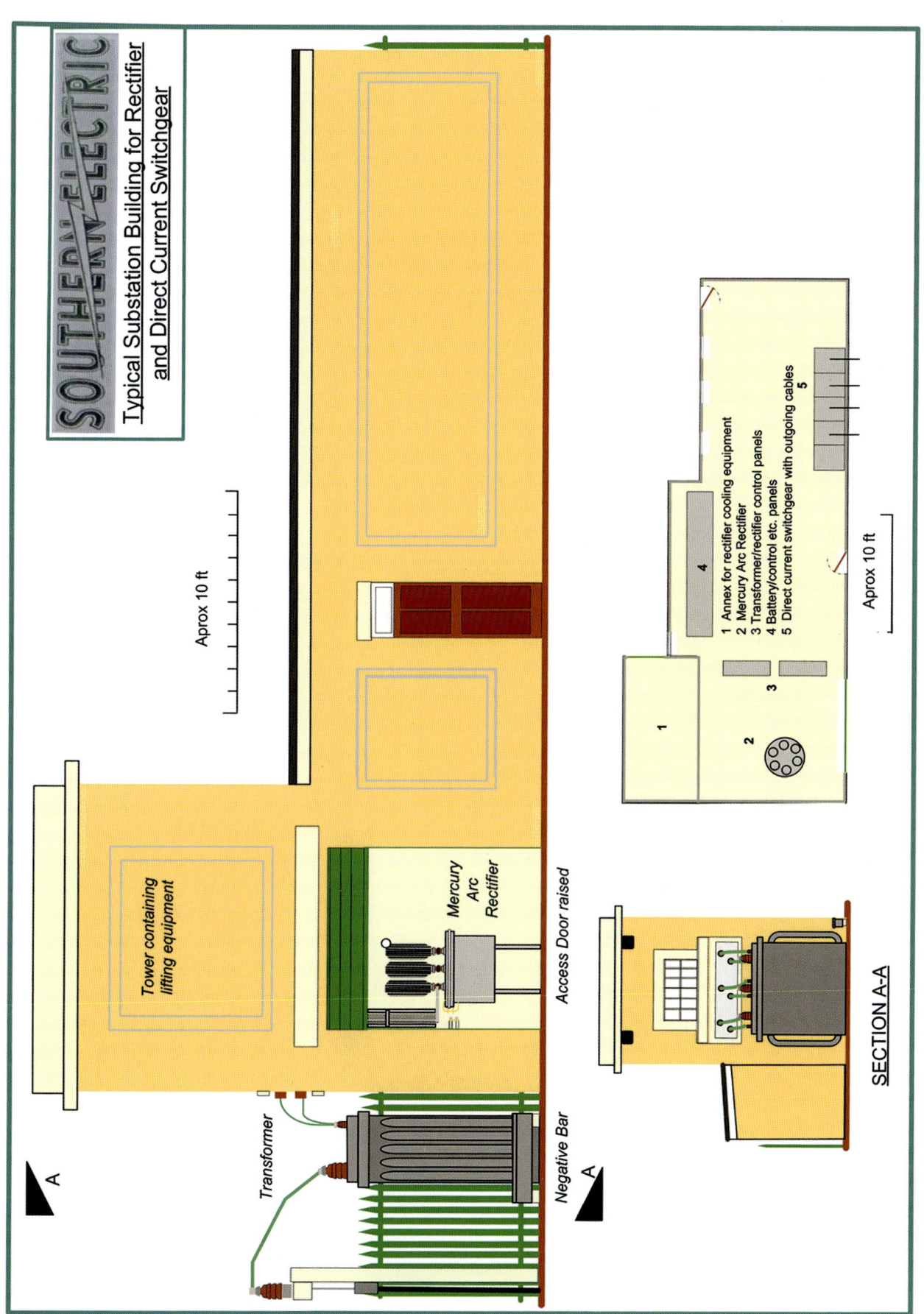

Typical 1930s Southern Electric substation – the substation building. (Author)

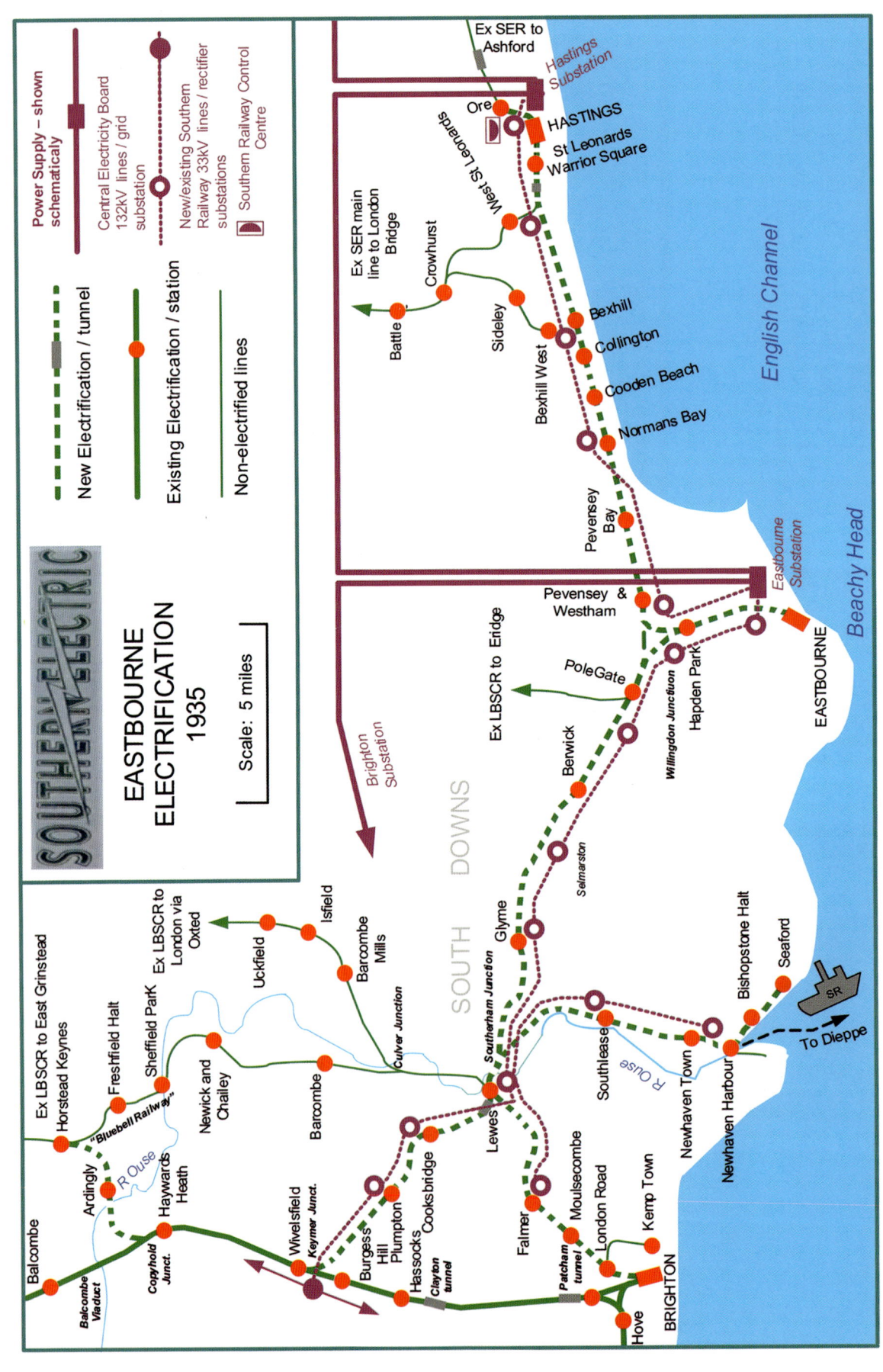

Map of the Eastbourne electrification scheme. (Author)

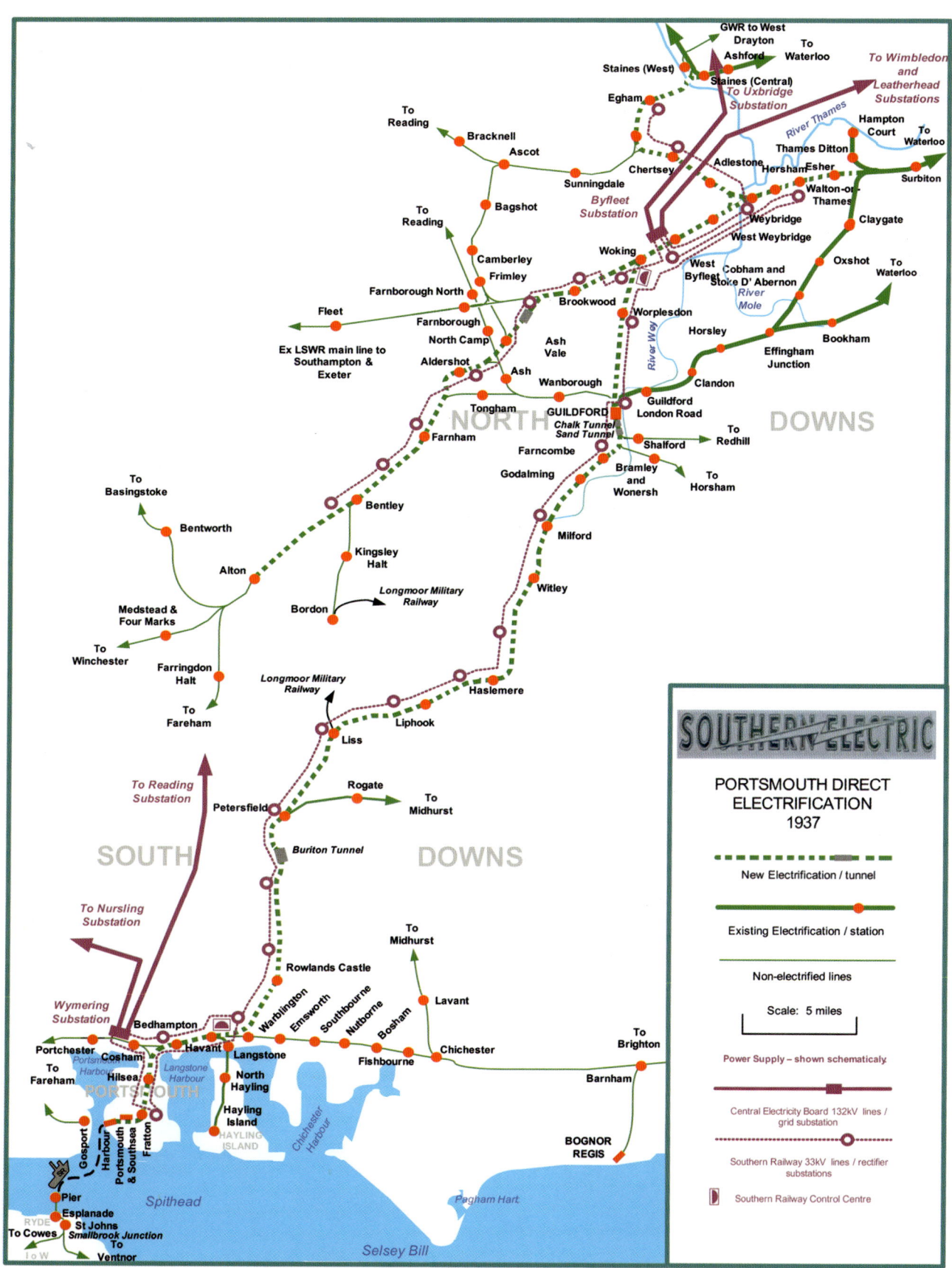

Map of the Portsmouth Direct electrification scheme. (Author)

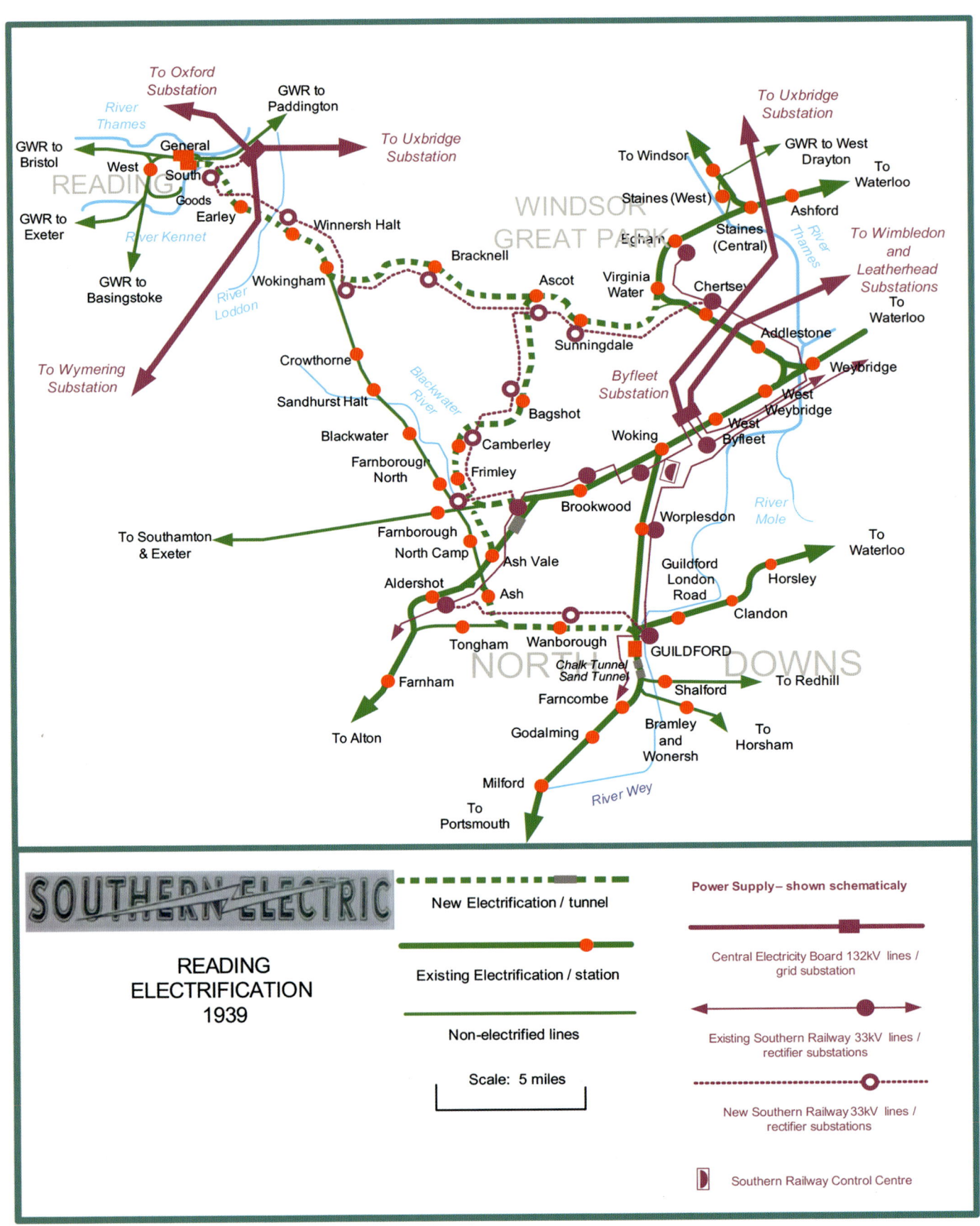

Map of the Reading electrification scheme. (Author)

PART 2
SOUTHERN ELECTRIC

CHAPTER 14

'THE WORLD'S GREATEST SUBURBAN ELECTRIFICATION'

In January 1925 when the Southern Railway was completing its three main electrification schemes and consequently causing much inconvenience to regular passengers, a young man named John Elliot was summoned to Waterloo station to be interviewed by Sir Herbert Walker.[1] During a convivial club lunch Sir Herbert had spoken with his friend Lord Ashfield – the chairman of both the United Electric Railways Company and the London General Omnibus Company – about his concern that, while the travelling public appreciated the new electric services, they were being less than tolerant of the necessary disruption to achieve this end. The discontent was being fuelled by unhelpful and inaccurate reporting in the London and provincial newspapers. Lord Ashfield advised that Sir Herbert should employ a trustworthy person to keep the passengers aware of what was being done to ensure that the press and public fully understood, and hopefully become sympathetic with, the railway's relatively short-term difficulties. The best person to do this was somebody who knew and understood the workings of all the great newspapers which then resided in Fleet Street, and Lord Ashfield knew the very man for the task: John Elliot.

John Elliot was certainly available, as at the time he was unemployed, after taking the blame for the *Evening News* being fined £10,000 for contempt of court. Elliot claimed that while he was standing in for the holidaying editor, it was the proprietor – Lord Beaverbrook – who ordered that the offending lurid murder story be splashed onto the front page. Elliot can be believed; he was the scapegoat and was given £400 as a parting settlement. Elliot's father was R.D. Blumenfeld, the editor of the *Daily Express*, another 'Beaverbrook' title. Given the anti-German sentiment of the time it was expedient not to be seen to be too German – for even the royal family had changed its name from Saxe-Coburg-Gotha to Windsor – so John Elliot Blumenfeld anglicised his Germanic name simply by dropping the Blumenfeld.

After the initial pleasantries, Sir Herbert bluntly said to Elliot: 'Lord Ashfield tells me that you can manage the press,' to which Elliot replied, 'Nobody can *manage* the press.' His advice to the Southern Railway general manager was that if he could be permitted to report directly to Sir Herbert, all the Fleet Street reporters would know that any information passed on by him was accurate and completely trustworthy. Elliot's suggestion was that he should set up an office at Waterloo to receive the reporters and editorial writers and there brief them as fully as possible. He knew from his own experience at the *Evening News* that it was difficult to get reliable and complete information from the staff at Waterloo, and hence 'journalists will do what journalists will do' – make things up themselves. This proposal met with Sir Herbert's approval. After being introduced to the chairman, the question of a job title for John Elliot was discussed. He was to be an assistant to Sir Herbert Walker – but an assistant what? Elliot had heard the term 'public relations' used when he had been in New York, so 'assistant for public relations' was what he duly became.[2]

Only six days after his appointment, Elliot was responsible for an advert appearing in the press with the heading 'The Truth About the Southern', extolling the valiant efforts of the Southern Railway's constituent companies during the Great War during which there had been a heavy toll on the system, and since the railways had not immediately been passed back into private control until 1921, there had been three years without any of the much needed capital investment required to rectify the effects of the war effort. After its formation, the advertisement proclaimed, the Southern had invested £10 million in electrification, new locomotives and main line carriages.[3] Furthermore it was made clear that the railway was vital to technological progress and wished to be an essential element in the post-war reconstruction. The advertisement became part of a series that reinforced this theme. This was a change in strategy – Elliot went on the offensive rather than simply reacting to misinformed criticism. The Elliot method was to ensure that the press was accurately briefed with reports that he had prepared with the assistance of the appropriate railway officer, and he was therefore able to fend off any adverse publicity. John Elliot would consult Alfred Raworth on matters appertaining to the electrification programme. Over the years, Raworth may have acquired the reputation of being 'difficult' with his fellow officers but was amenable to Elliot's non-technical efforts in the public relations sphere. Elliot accordingly found Raworth to be most cooperative and grew to like and admire him and later, after the pair had been colleagues for over two decades, he was to describe him as a 'friend to treasure'.[4] Elliot's efforts paid off as the Southern Railway began to receive a degree of admiration for its efforts and less criticism.[5]

After a year working as assistant for public relations, Sir Herbert agreed that Elliot's role should be widened to include advertising as well as publicity.[6] So, while in the electrical engineering sense the electrification projects justifiably belonged to Alfred Raworth, when it came to the image of the electric railway it is John Elliot who can claim some of the credit for something more abstract and ephemeral: the concept of a clean, efficient and modern mode of transport. He had been impressed by the United Underground Railway's 'bull's eye' logo that located the tube stations for passengers struggling along crowded pavements across London. What the Southern Railway needed to do to update its image was something just as smart and modern to advertise the new electric trains. Sir Herbert Walker feared that it might all become rather expensive, but nonetheless agreed to the production of striking new green enamel signs with white lettering to be erected on railway overbridges across London. The signs declared: 'Southern Electric – Frequent Trains to London Bridge' (or whichever terminus was appropriate to the location) – 'Cheap Fares Daily'. The words 'Southern Electric' were in bold capital letters standing out large above all the other lettering, below which was a large arrow directing any prospective traveller towards the station entrance.[7] Later versions of these signs included a dramatic 'lightning flash' intertwined with the 'Southern Electric' wording, creating an image of power and excitement.

John Elliot never claimed to have invented the term 'Southern Electric'. As part of his campaign, a press announcement in the form of an advertisement appeared in July 1926. This advertisement was possibly the earliest use of the description 'Southern Electric'. In that year, while Elliot was leading the railway's publicity endeavours the term became a 'brand' which seems to have simply emerged rather than being formally created. This simple, and somewhat obvious, description of what the company had to offer appears to have arisen from the zeitgeist rather than from any management decision. The logo and description 'Southern Electric' began to be used in Southern Railway advertising over the succeeding years, including many stylish art deco advertising posters. Despite the apparent lack of planning this new logo became familiar to the travelling public and for those who are fascinated by the railways of southern England the term 'Southern Electric' continues to resonate to this day. Even British Railways, Southern Region, inserted the wording 'Southern Electric' into its posters as late as the early 1960s. At no time was the wording 'registered trademark' discreetly placed alongside the logo.

Lord Ashfield's organisation provided another source of inspiration for the Southern Railway – the new modern art deco stations. Contemporary with the birth of the Southern Electric in 1926 the southward Northern line extension was completed. For this extension the architect Charles Holden had designed impressive modern station buildings which represented an early commercial flowering of modernism. Any description of modernism would have to include the phrase 'form follows function'. When the Southern Railway rebuilt many of its stations – often but not solely to accommodate the increase in traffic due to electrification – the chief architect J. Robb Scott, obviously inspired by Holden's work, was to be credited with stylish modern designs which fitted in perfectly with the Southern Electric image.[8] The term 'art deco' has been used to describe these new station buildings, but there is a slight paradox here, because this style included the application of decoration to the buildings using hard-edged, low relief, geometric designs, not quite the starkness of modernism. In short, the new Southern railway buildings were functional but embellished by a few bold geometric features.[9] What might previously have been the mellow honey colour of individual bricks in pre-grouping days was now clean white rendering or stonework with, at high-level in large, green block capital letters, the wording: 'Southern Railway'.

James Robb Scott was born in 1882 at a house in the Gorbals, Glasgow, and found employment with architectural firms in Edinburgh and London before he joined the LSWR where, as chief architectural assistant between 1909 and 1923, he worked on the rebuilding of Waterloo station. At the grouping in 1923 Scott became the new Southern Railway's chief architect. Detractors have questioned Scott's input into these designs, but as the chief architect he would have had a team of architects to do the designs and as the 'chief' he received all the credit. Scott's deputy Edwin Maxwell Fry was not an admirer of his superior and later claimed that it was he, not Scott, who had weeded out all the department's 'dead wood' and recruited the many able young architects who were to dramatically update many of the Southern Railway's stations.[10]

On completion of the legacy schemes the Southern Electric network continued to spread out across the suburban area. As per the reports from Jones and Raworth, third-rail electrification was installed on the whole of the LBSCR's suburban network. This was followed by further incremental advances such as from Dartford to Gravesend Central in the former SECR territory and, in July 1929, the railway announced that the LSWR electrification was to be extended from Whitton (where the Hounslow loop returns to the Windsor line) on through Staines to Windsor and Eton Riverside.

To plan and supervise the incremental extensions, Sir Herbert instigated a steering committee chaired by his traffic superintendent Edwin Cox, which included senior officers from every department.[11] Following a recommendation to electrify a line, every detail of the scheme was examined by this committee,[12] but it was Alfred Raworth who prepared a summary report for Sir Herbert, for presentation to the full Southern Railway board of directors. His report had to gather together many aspects of the railway's business – traffic, rolling stock, civil engineering, as well as the electrical power engineering, and it is significant that Sir Herbert entrusted this task to Raworth. With an enthusiasm reminiscent of his father's attempts to electrify tramways at a minimal cost, Alfred had now found favour with his general manager for planning and executing electrification schemes with minimal expenditure. In many ways, the professional relationship that Raworth had with Sir Herbert must have been similar to that with his late father at Raworth's Traction Patents. Alfred Raworth was entrusted with the electrical engineering and final assimilation of the report material, while the older man was the figurehead who took the ultimate responsibility. This use of trusted individuals epitomised Sir Herbert's style of management, for he had a talent for identifying those in which he could put his complete trust and allowed them to get on with the work without hindrance. This does not mean that Sir Herbert took no interest in what Raworth was doing on his behalf – he most certainly did, but he did not interfere – which is quite different. Given the trust that was placed in Raworth and the fact that he was asked to make persistent incursions into non-electrical engineering aspects of the railway's business, it would have been surprising if his position relative to the general manager was not envied by some, particularly as to most of the other railway officers he remained an 'outsider'. Furthermore, given that he came across, initially at least, as being stand-offish, it is hardly surprising that many departmental leaders found Raworth 'difficult'.

An early example of one of the relatively small expansions of the Southern Electric network was the extension from Dartford to Gravesend Central. Raworth's report stated that the route length to be electrified was to be almost seven miles, and the existing steam service resulted in 156,000 train miles per annum, which was to increase to 247,700 train miles per annum with a service comprising of three electric trains per hour in each direction throughout the day. To work the service, eight-coach trains would be used in the rush hour period (the Southern Railway at the time continued to use the clumsy arrangement of two standard three-car electric sets with two unpowered trailers sandwiched in between), and during the 'slack hours' only a single three-car set would be used. A new Southern Railway rectifier substation at Northfleet was required and this would be equipped with two rotary convertor sets. For every new extension, it was always recorded in the minutes that the new works would be built to the same electrical standard as the rest of the network, a detail that the concerned directors needed to know to allay any fears that the engineers, and in particular one suspects, Alfred Raworth, who had a well-known alternative scheme in mind – were not heading back to what they would have considered to be the uncharted and financially dangerous waters that constituted a dreaded 'non-standard' proposal. While new rolling stock was provided for each extension, it was not of course intended that these trains would forever be used on the new line, but most electric sets did spend most of their working lives on the services for which they had originally been built.

Alfred Raworth's estimate for the Dartford to Gravesend Central scheme was that the electrical equipment would cost £97,821, new rolling stock £18,750, station alterations £35,000 and £8,060 was allocated for 'other works' resulting in a grand total of £159,631. Sir Herbert Walker would then inform his directors that he proposed to allocate about two thirds of this expenditure to capital and the remainder to the renewal budgets or a suspense account. He would then present the overall financial appraisal. In the case of the Gravesend Central extension the total annual running costs (including the 4 per cent interest charge on the capital element) were expected to be £24,818 which was £2,238 more than the then present cost of operating the steam service. But,

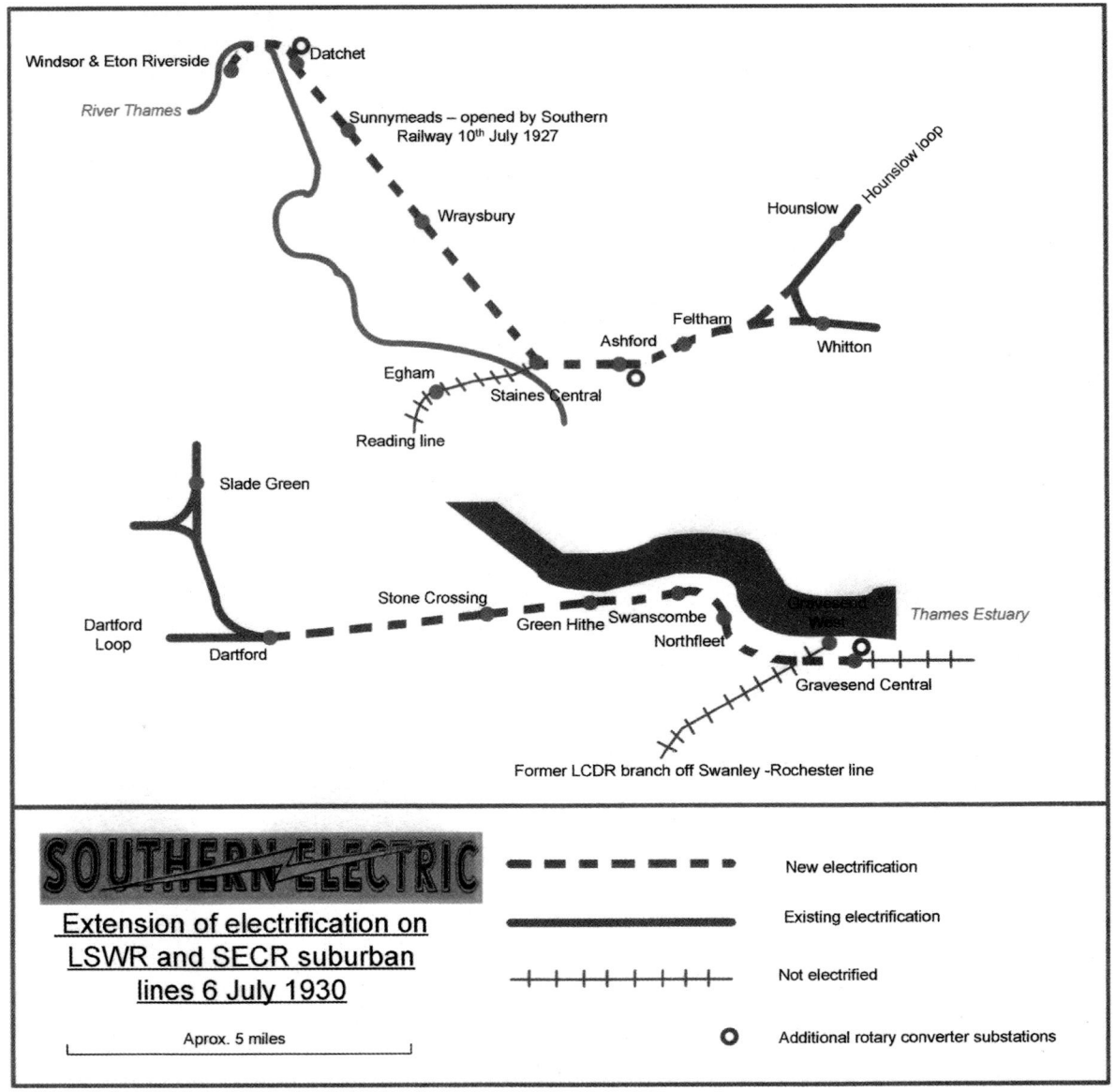

Most of the lines electrified 1925–1930 were on the former LBSCR lines, but the electrification was also extended to Windsor and Gravesend. (Author)

of course, there was the expectation that the 'sparks effect' would dramatically increase receipts and the 'deficit' would be transformed into a healthy surplus.[13]

The optimism for a significant 'sparks effect' was well founded, for traffic statistics for the first six months of 1929 compared to the previous six months showed a continuing increase in passenger journeys. On lines electrified during 1928 and 1929 there had been a total of 10 million journeys – an increase of 37 per cent. For lines electrified during 1925 and 1926 there had been a 3.5 per cent increase to 51 million journeys, and even for lines electrified back in 1924 there had been a 2 per cent increase over the same period, to 35 million journeys – an increase of 2 per cent. Overall the increase had been 8 per cent.[14]

So successful were the new Southern Electric services that complaints of overcrowding were being received. In 1928, three trains an hour during 'business hours' on some services into Charing Cross were proving to be inadequate and had to be increased.[15] Further complaints of overcrowding were received in 1930 resulting in an additional 326 electric train miles a week being added to the timetable. Extra morning trains were run to Cannon Street from Barnehurst, to London Bridge from Sutton and to Holborn Viaduct from Wimbledon. Extra evening trains were run from Cannon Street to Dartford via Greenwich and from London Bridge to Wallington.[16]

Early in 1927 the Southern Railway once again invited tenders from General Electric, English Electric and British Thomson-Huston for the supply of rotary converters for the expansion of the network. All three companies were British, even if two of them were effectively subsidiaries of American corporations. Two of the manufacturers joined forces, offering to produce only 14 of the 28 machines that the Southern Railway required, and the third company declined to provide a price for the work, offering no explanation for this refusal. After some negotiation, the '14 only' offer was revised to the full number required but the order for the last nine, the manufacturers insisted, must not be issued until November 1927.[17] Alfred Raworth was convinced that the British suppliers were unashamedly operating a 'ring' by fixing prices and deciding among themselves which of them was to have the business. Furthermore, he believed, they did not all have the competence to design the plant or sufficiently stringent manufacturing standards. Raworth put these points to Sir Herbert advising him that with tens of thousands of pounds worth of contracts in the offing, this situation could not be tolerated. Raworth emphasised that he had a duty to provide new plant that would give reliable performance under traffic conditions and that he was not going to delegate this responsibility. Sir Herbert respected Raworth's opinion and presented it to the directors who agreed to support this standpoint. From then on, Raworth, in almost every instance, was granted complete freedom to choose the plant he required from whoever he thought could manufacture it to his exacting standards and charge an acceptable price.[18]

The manufacturers' views on their design and build competence was not recorded but as far as any suggestion of a 'ring' was concerned they seemed not only to concur with this view but to praise the system they used to control their prices and sales. Given the lack of orders for heavy electrical plant the manufacturers were closing ranks in a combined effort to keep their industry intact. A director of the British Electrical and Allied Manufacturers Association was unrepentant in a press statement published in the *Railway Gazette*: '… such a system is practically an economic necessity, having regard to present-day conditions. There is no secret to it – we have always been quite open in our dealings …'

The Southern Railway lost patience with the British suppliers. Upon Sir Herbert and Alfred Raworth's advice, the chairman, Brigadier-General Everard Baring, called together several of his directors and subsequently gave the verbal instruction that a formal order be placed with the Swedish manufacturer Allmanna Svenska Elektriska Aktiebolaget (ASEA) for all 28 rotary converters[19] at a contract price of £120,000. The railway was to publicly argue that it had made a considerable saving by buying from ASEA and reminded the British suppliers that the Southern had a duty to its shareholders and to its 70,000 staff. The British manufacturers were not impressed and complained that continental manufacturers worked together internationally making efforts to secure British business. They also claimed to be busy – despite the depressed economic climate – and condemned the Southern for not approaching other British manufacturers such as Metropolitan Vickers, Bruce Peebles and Mather & Platt.[20] Walker and Raworth knew that the response from the other companies would have been much the same as had been the case with the original three firms and Raworth's informal approaches would have determined their ability to produce the rotary converters on time and to the railway's stringent specification.

For Raworth, the ASEA order necessitated him locating his passport and making the trip to Vasteras, a city 62 miles west of Stockholm in Sweden, to witness the testing of the rotary converters and from there he reported back to Sir Herbert that he was satisfied with his dealings with the Southern Railway's new supplier.[21] Far from this being a one-off order it was the start of a new and fruitful relationship between the Southern and ASEA. A month later a further 20 rotary converters were ordered for delivery 18 months hence. Another 30 were ordered for delivery by the end of 1929 at the same price, but ASEA was obliged to pay any import duty charged by the British government and the deal was subject to the 'political and economic' circumstances not altering 'appreciably'.[22]

This contract to buy Swedish equipment did not go unnoticed outside the railway industry and there were questions in the House of Commons. The parliamentary secretary to the Board of Trade, Sir Burton Chadwick, refused MP's request for an inquiry, stating that all parties had made full and frank statements, so nothing could be gained from further investigation by the Board of Trade.[23] The anti-competition stance of the manufacturers and the apparent tacit support of the government seems strange to us today, but the contemporary thinking was that in the poor economic climate – and drawing on the spirit of cooperation encouraged by Lloyd George that had been evident during the Great War – manufacturers should work together to minimise waste, and the grouping of the railways into four large companies had been carried out in this same cooperative spirit. The Southern Railway did not seem to have suffered from any damaging bad publicity as all the blame seems to have been directed towards the manufacturers.

The Swedish rotary converters were duly delivered, but the British manufacturers unashamedly continued to 'price fix'. In January 1928, the Southern Railway tendered for six more rotary converters together with the associated power transformers. ASEA tendered £1,590 for each transformer and £2,940 for each rotary converter. Blatantly, and almost mischievously, British Thomson-Houston, English Electric and General Electric all tendered the same prices: £1,782 for each transformer and £3,312 for each rotary converter. This amounted to an

At the time Alfred Raworth first met with ASEA its logo incorporated a swastika. Political sensibilities soon removed this controversial symbol due to the rise of the Nazi Party in Germany. This nameplate is not off any railway owned equipment but a County of London Electricity Supply Company's transformer. (Alan Booth)

176 • ALFRED RAWORTH'S ELECTRIC SOUTHERN RAILWAY

additional £564 for each transformer-converter combination compared with ASEA's offer. Despite Alfred Raworth's concerns, continuing to buy abroad was not politically acceptable in the then current economic climate and the railway did need to maintain cordial relationships with the British manufacturers. Accordingly Sir Herbert Walker discussed this issue with his chairman, Everard Baring, and a deal was subsequently struck to pacify the British manufacturers and public opinion. ASEA was to supply the transformers only, at a cost of £9,400 – as tendered, but with a slight reduction as two of the transformers were to be of reduced power rating – and British Thomson-Houston would provide the rotary converters plus the direct current switchgear and connections to the track at a cost of £13,063 for each site.[24] Later, in part to further overcome the 'political' difficulties, ASEA was to form a British subsidiary company, ASEA Ltd, with a registered office at 5 Chancery Lane, London, and opened workshop premises (mainly used to assemble rather than fully manufacture equipment) in Walthamstow, North London.[25]

Alfred Raworth was now a powerful and influential man in the Southern Railway hierarchy. His enhanced status and increased salary had in 1929 enabled the Raworth family – Alfred, Ruby and the still unmarried Marjorie – to move to a new house, leaving their relatively small 'Meadow Cottage' in Stoke D'Abernon and taking up residence in a much larger and grander house, 'Ladywood', situated in the more upmarket locality of Silvermere to the west of Cobham. Throughout the 1930s the Raworths usually employed up to three live-in servants to attend to their domestic needs.[26] The house with its spacious and secluded grounds was set well back off the main road. From the rear garden of Ladywood, Raworth could view the lush

Purley substation under construction. Note the 11,000V above-ground cables attached to posts, and the low voltage direct current cables that will be connected to the conductor rails. (Bluebell Railway Museum)

Inside Purley substation, three of the British Thompson Houston rotary converters that were the subject of the 'price fixing' allegations. (Bluebell Railway Museum)

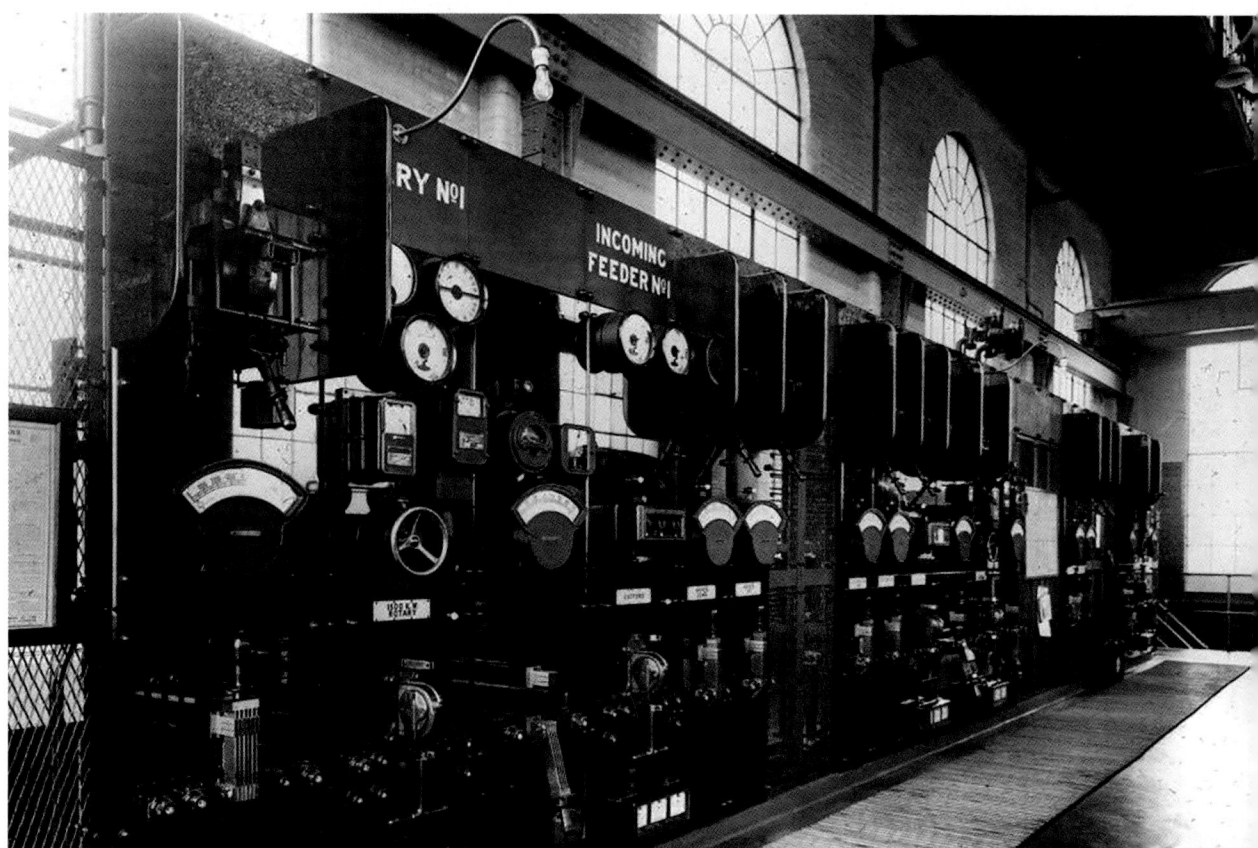

The switchgear control panel in Purley substation. (Bluebell Railway Museum)

golf links and the prestigious St George's Hill residential area.

Following the demise of Raworth's Traction Patents, Alfred's brother, Basil, had left England to work for the Hong Kong branch of the General Electric Company, where he rose to the rank of branch manager. As was the custom with families working in the British colonies, his two sons were sent to England for their education. Basil's elder son – christened John Smith Raworth – studied at the prestigious Repton public school in Derbyshire before going to Trinity Hall, Cambridge, while his younger brother was later sent to the equally famous Marlborough College.[27] During the school holidays the two boys would stay with their aunt and uncle at Ladywood and since their mother and father were on the other side of the world, they were effectively brought up by Alfred and Ruby who were substitutes for their parents.

At the secluded and tranquil Ladywood, away from the stresses and strains of the Southern Railway, a more amiable Alfred Raworth enjoyed relaxing with his family and friends, but many Southern Railway staff continued to find him hard-headed and difficult to work with.[28] John Elliot agreed that Raworth's 'acid' approach did not endear him to his fellow officers, but those who had close dealings with him respected his 'outstanding technical efficiency' and his personal loyalty to Sir Herbert. The final costs of the electrification schemes that he engineered were regularly well below budget. For example, the cumulative savings from the short extension to West Croydon, the Whitton to Windsor via Staines scheme and the extension from Dartford to Gravesend amounted to nearly £170,000.[29] The cynic might suspect that since Raworth produced the estimates in the first instance, he would always be able to present himself in a good light at the end of any project, if he had deliberately overestimated the original project cost. But what any estimator must do is to provide the customer – in this case the Southern Railway board of directors – with a realistic price that will not be exceeded. If these electrification projects had cost more than had been allowed for in the budget, the additional costs would have destroyed Walker's business plan by eroding into, eliminating, or worst of all exceeding the value of the additional traffic receipts. And naturally, the directors, who were always reticent to spend any money at all and desired that every project be carried out at minimum cost, would be happy to receive the bonus of a small underspend.

The success of Southern Electric contributed to the financial case for new railway construction in the form of the Wimbledon and Sutton Railway. This railway had been authorised by Parliament as long ago as 1910. The Southern Railway directors agreed in October 1927 to spend £270,000 to purchase land and a further £822,000 to be allocated to electrification.[30] The line was opened in stages between July 1929 and January 1930. The six intermediate stations were of the 'single platform' type, with a central 'island' platform accessed by subway or pedestrian bridge. Naturally, the Southern Electric modernist concrete style of architecture was used for the new stations. This was very much a passenger line, only St Helier station situated in the middle of the line had a goods yard, which closed in 1963, but there was also the Express Dairies milk depot in use from 1954 until 1978 which did provide regular steam or diesel-electric locomotive hauled trains of milk tanks, to relieve the monotony of the steady stream of electric trains.[31]

A look at the Southern Electric route map of 1931 demonstrates the immense progress since 1923. Apart from the original LSWR electrification which had been carried out around the time of the outbreak of the First World War, the electrification programme had been completed in seven years, a remarkable achievement. The Southern Electric was now being described as the 'world's greatest suburban electrification'. The use of the word *greatest* is interesting for it implies an element of subjectivity and can mean different things

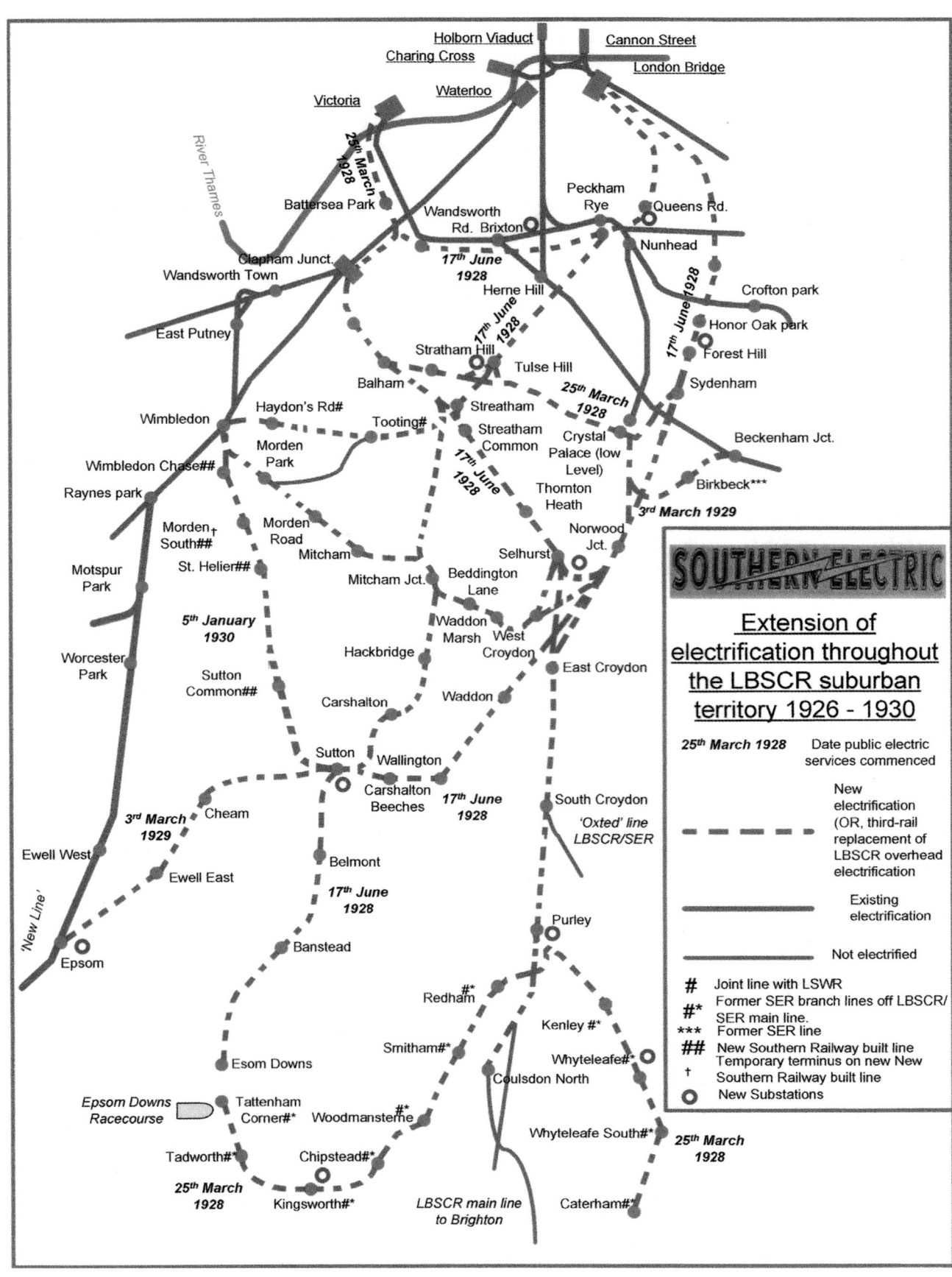

The expansion of the third rail into LBSCR territory. (Author)

A pair of 2-NOL sets 'Nos. 1310 and 1311' approaching Mitcham with a train from Wimbledon on 6 June 1949. Note the single electrified track and the 132,000V Grid line running between Beddington and Wimbledon. The Wimbledon to Mitcham line was electrified before the construction of the National Grid which was later to play a vital part in the extension of the Southern Electric to the south coast. (Meredith/Transport Treasury)

to different people, but in June 1926 the *Railway Gazette* had concurred with this claim,[32] agreeing that it was possible to cite electrified railway mileages longer than the Southern Electric network, but it was the traffic density on the lines operating in a confined, but economically very important, geographic area that sufficiently singled out the Southern Railway's system for the accolade 'greatest'. By 1931 the Southern Electric suburban lines served six major London termini and radiated out across all South London and the outer suburbs, often offering alternate routes between stations and serving all the important towns within this expanding suburban area.

The character of the lines may have changed, but the Southern Electric had been laid over rather than completely superseding the traditional railway network. The Southern Railway continued to use steam locomotives for the freight traffic for domestic coal, general merchandise and mail, including parcels, and at the main termini the long-distance express and stopping services continued to be hauled by steam locomotives. This was all a long way from Raworth's vision of an 'electric railway', when steam with all its smoke, dust, and necessary equipment such as turntables, water columns and coaling stages had been eliminated. But

Raworth, and probably Sir Herbert, saw the realisation of this futuristic ambition as an incremental process, as bit by bit the full electric railway could be achieved.

The 1929 general election produced a Labour government led by Ramsey Macdonald. He did not have a parliamentary majority, so was unable to carry out the Party's plan to nationalise the railways. This may have been good news for the Southern Railway directors, but what was even better news for them was that just before the election the Conservative government had abolished the passenger duty charged on railway ticket sales. It was the then Chancellor of the Exchequer, Winston Churchill, who had instigated the removal of this tax, on condition that the railway companies capitalise the amount paid in tax in 1925 and spend 90 per cent of this money on schemes to 'modernise'. This gave the Southern Railway a pot of money containing over £2 million to start to spend from 1929 onwards.

Some money was to be spent on 200 new 'steam' carriages for main line use, but most of this new money was to be spent to further extend the Southern Electric. So, with invigorated finance following this windfall from the government, Sir Herbert Walker confidently stood up at a chief officer's meeting and declared: 'Gentlemen, I have decided to electrify to Brighton.'[33]

CHAPTER 15

A NEW DESIGN FOR AN ELECTRIC RAILWAY

Before making such a bold pronouncement Sir Herbert needed confidence that such a scheme was not only financially viable but also technically possible. The existing Southern Electric infrastructure was not suitable for a longer distance application such as the 'main line' extension to Brighton. The 660V direct current electrification required many manned substations and the distance from London to the south coast prohibited the simple extension of the 11,000V cables. If a similar electricity distribution arrangement were to be used, a new power station on the south coast would be required. Fortunately, the history of Southern Electric was now to run in parallel with the history of the new National Grid, and Sir Herbert's electrical engineer (new works) had some innovative ideas that would reduce the amount of electrical plant required and the necessity to have the substations manned at all times.

Winding back to 1919 again, Sir Eric Geddes' Electricity (Supply) Act had created the Electricity Commissioners whose major task was, after negotiations with existing statutory electricity suppliers, to establish regional joint electricity authorities that would acquire generating stations and any interconnecting transmission lines and operate them as a unified whole.[1] This proved to be a difficult task and in 1925 a new committee, given the title 'Committee Appointed to Review the National Problem of the Supply of Electrical Energy', was established with the Rt Honourable Lord Weir of Eastwood as chairman.[2] The 'problem' indicated in the title was that relative to other countries, Great Britain had low electricity consumption, and this was perceived to be due to an inadequate public electricity supply. This first Weir Report showed Great Britain twelfth in a list of countries rated by 'consumption of units of electricity per head of population', with only 110 units per head, while the table leader, California, scored 1,200 units. Other nations – Tasmania, Norway, Sweden and the USA as a whole used around 500 units per head.[3] Part of this apparent discrepancy was due to the availability of hydroelectric power in these other countries, but the main reason was probably that it was the abundant British coal that supplied most of the nation's energy. Coal was used to make coke and 'town gas' to provide both heat and light and was also the fuel for industrial stationary internal combustion engines. Lord Weir argued that Great Britain had for several decades lagged behind other countries – notably the USA and Germany – in the sphere of electrical engineering, and furthermore it was only by using more electricity that it would be possible to make economic and social progress in the still new twentieth century.

Lord Weir believed that if the railways made more use of electricity to power their trains this would be a catalyst to initiate an expansion of the electricity supply network. He proposed what he termed an electricity 'gridiron' to facilitate further railway electrification. His report was reviewed by the Cabinet in May 1926[4] and he argued that the proposed 'gridiron' must be constructed as a physical entity and that the government must pass the necessary legislation to set up a central electricity authority specifically to build and operate the proposed nation-wide asset. This was accepted by the government, resulting in the Electricity (Supply) Act of 1926, which created the Central Electricity Board (CEB). The CEB was not strictly a nationalised organisation but a 'public corporation' like the British Broadcasting Corporation (BBC), and the required capital was raised through the City of London without government guarantee.

The National Grid took only five years to build. The CEB contractors commenced construction of the first transmission tower (or 'pylon' as the public will always call them) in Central Scotland in 1928. The final tower was built on the outskirts of the New Forest in September 1933 when there were 4,000 miles of transmission lines, of which 2,894 miles were the 'primary lines' operating at 132,000V. From the Grid substations there radiated over 1,100 miles of 'secondary lines', usually at 33,000V (but some operated at 66,000V) and these lines were in appearance miniature 132,000V lines. Usually, each tower carried two three-phase lines or circuits, resulting in six wires being provided with an extra earth wire strung from tower peak to tower peak. In the early days, some routes had only one three-phase circuit attached, creating a lopsided appearance until the second set of three wires was attached later. The total number of towers built was 26,265 – working out at about seven towers per mile. Each tower weighed over three tons and was on average 75 feet (23 metres) high.[5] The balance of the route mileage consisted of underground cables. Most of the 132,000V lines remain in service today but, following completion of the 400,000V 'Super Grid' in the early 1970s, these original transmission lines now perform a secondary 'distribution' function like the original 33,000V lines.

The new National Grid established substations along the railway route to Brighton. The first, at Croydon, was built at Beddington Lane – about two and a half miles away from the railway. A Southern Railway-owned 33,000V cable was laid by the CEB at the railway's expense, but the total cost was shared, because a 132,000V cable had to be laid along the same route.[6] Three Bridges Grid substation had a common boundary with the railway (as did many of the original Grid substations). Also, the Grid substation at Fishergate was built alongside the Brighton to Portsmouth line just to the west of Brighton. (See plate 10)

This enabled Raworth to design an electrical distribution system that was different to that in the London area. In place of duplicate high voltage cables radiating from a single supply point – Durnsford Road or the LESC's Deptford generating stations – a single high voltage cable route was to be installed running between the Grid intakes – the availability of *two* supply points avoided the need for duplicate cabling. These cables would operate at 33,000V – the same as the Grid's secondary voltage. Furthermore, rather than duplicating rectifiers at the substations, Raworth proposed only one rectifier at each substation, security of supply being afforded by interconnecting the substations using the conductor rails. With this arrangement three single-rectifier substations would feed the same length of railway that would previously have been supplied by two double-rectifier substations. Therefore more, but smaller, substations would be required.

In the past the substations would have had to be manned, but new innovative technologies would remove the need to employ substation attendants. First, the labour intensive rotary converters were to be replaced by mercury arc rectifiers which required only minimal maintenance and, as a bonus, could safely use the new national frequency of 50 cycles per second. These were huge thermionic 'valves' – often massive versions of the glass valves which were once universally used in domestic radio and television sets before the transistor or silicon chip eras. The Southern Railway however used steel tank rectifiers as there were some early difficulties in producing such large glass containers. The usual six transformer connections were connected to the rectifier's anodes. A pool of mercury was the device's cathode.

The device could only conduct electricity in one direction, therefore for one half of a cycle an arc would flash from each anode in turn down into the mercury, while during the other half cycle it would not. By this means a smooth direct current supply could be produced by combining the six 'half waves' taken from the cathode. The intense heat generated by the repeated arcs evaporated the mercury to form a gas which condensed on the inside of the casing to drip back into the pool. By this means the cathode – which would quickly burn away if it had been made of a solid metal – was regenerated.

The other reason why the substations need not be manned was because Alfred Raworth had worked closely with ASEA to develop a pioneering control system which today would be termed a 'supervisory control and data acquisition' (SCADA) installation. Today, with digital technology, these are computer-based systems using video screens, but Raworth had to develop a mechanical system to remotely operate the switchgear and monitor the loads and voltages.

The LBSCR main line had been electrified as far as Purley and this would continue to be suppled from the LESC at Deptford using the railway standard frequency of 25 cycles per second. The difference in frequencies was not an issue as there was no intention to interconnect the LESC's alternating current supply with that from the CEB. Due to the anticipated increase in traffic expected to follow the Brighton electrification scheme, an additional feed at 11,000V, 25 cycles per second was to be taken from the former LSWR generating station at Durnsford Road to reinforce the line from Victoria through Clapham Junction towards Balham. Some substations on the original electrical network required additional rotary converters.

At each of the Grid intake points the CEB was to own the circuit breakers controlling the 33,000V connection to the Southern Railway substations. These circuit breakers required auxiliary equipment (current transformers and voltage transformers) to measure the load current and the supply voltage from which the electricity meters calculated the units consumed. The 33,000V cable route was trackside and connected a string of Southern Railway substations. At one of the substations along the route a circuit breaker would be open creating two 'radial' feeds from each direction. This 'split' or 'open' point could be 'moved' to another substation by a sequence of switching, involving closing the open circuit breaker before the opening of another at the new open point. It was not possible to leave the two Grid supply points in 'parallel' for too long, as the railway's 33,000V system might transfer a massive amount of power from one end of the country to the other if the Grid load took a 'short cut' courtesy of the Southern Electric cables rather than travel around the Grid as intended.

So, confident that his extension of the third rail to the south coast was practical, Sir Herbert presented his proposals to the directors on 21 January 1930. He tabled a report written by Alfred Raworth which was presented in the usual format as another 'incremental' extension to the Southern Electric network. A 36-mile extension of the conductor rail south from Purley to Brighton was proposed. Additionally, from Preston Park – just outside Brighton station – the LBSCR line along the south coast towards Portsmouth was to be electrified as far as West Worthing, as were the first two miles of the Redhill to Guildford line as far as Reigate. Despite speculation that this 'main line' scheme was to make some use of electric locomotives, Raworth confirmed that only multiple unit electric stock was to be used. Raworth's report stated that the 1930 tally of 1,971,983 train miles per year using steam-hauled trains would increase to 4,921,200 electric train miles per year. The project would cost £2,700,600. Within this total spend was £260,000 for high voltage cables, £450,000 for substation equipment, £260,000 for conductor rails, £430,000 for the electric trains, £130,000 for a repair workshop and £162,000 to re-signal the main line from Coulsdon to Brighton with colour light signals.[7] This was duly approved by the Southern Railway directors and so commenced the electrification to Brighton which was a significant step change in the history of the Southern Electric.

The whole project was to take only three years from board approval to completion. Alfred Raworth was to be responsible for the design, layout and installation of all the electrical equipment,[8] and it was he, not Herbert Jones, who was to be given the credit for the design of the electrical equipment aboard the new rolling stock.

Meanwhile, Lord Weir produced his second report specifically investigating to what extent the new National Grid could be used by the railways. He was less enthusiastic about the Brighton electrification than might have been expected. He saw Sir Herbert's plan as merely an extension of the suburban network, driven by a 'commuter' traffic business model – not a 'main line' model.[9] Not surprisingly, Merz and McLellan were commissioned to provide advice on main line electrification and recommended two possible main line schemes as exemplars, one for part of the LNER, the other for part of the LMSR. The schemes would comply with the Ministry of Transport's policy by using 1,500V direct current overhead electrification and it was proposed that electric locomotives would be used exclusively within the electrified areas. The LNER scheme was to electrify the former Great Northern main line from King's Cross in London to Leeds. This would cost £5,646,323 and reduce operating costs by £624,630 a year.[10] Similarly, for the LMS it was proposed to electrify the West Coast Main Line from Crewe to Carlisle, with an electrified 'branch' from Weaver Junction to Liverpool Lime Street station. This would have cost, according to Merz and McLellan, £5,123,370 and would have accrued an annual saving of £127,766.[11]

The other railways were not impressed. Despite agreeing that the magnitude of any main line electrification project should not be a deterrent to the potential benefits, they showed little conviction that the level of electrification proposed was viable. They exhibited a degree of defeatism, opining that the two schemes for part electrification would not reap sufficient benefit and that full electrification would eventually cost the railways nearly £341 million of which £261 million would have to be raised for capital expenditure. They should not have taken too much notice of the final figure but followed Sir Herbert Walker's example and electrified using an incremental, one small step at a time approach. But it would be difficult to raise the capital as the Stock Exchange did not like uncertainty especially when raising money for long-term projects. Also, the railways were concerned that Weir's report was based on out of date traffic figures which, due to the current economic depression, were falling. What if more trade was lost to road haulers? There were also those who retained faith in steam traction, claiming that there was scope to improve the efficiency of steam locomotives, a not unreasonable stance, for over the next 20 years improvements were to be made. These included the universal application of superheating – the process of putting more energy into the 'saturated' steam by passing it through super-heater flues – and many detailed improvements such as providing improved bearings that did not 'run hot'. Other forthcoming design changes were intended to facilitate the easier maintenance of locomotives.

This lack of enthusiasm for electrification was to continue while the Southern Electric forged ahead. Sir Herbert's advice should have been followed by, for example, the LMSR which was led by Sir Josiah Stamp whose railway was experiencing a decline in profitability.[12] In 1933 he publicly stated that he believed that the suburban problem would not necessarily be solved by electrification as the important advances in steam locomotive design 'could not be brushed aside' and the LMS had spent £7 million on electrification in North London and £3.25 million on electrification around Liverpool – but these were their legacy schemes, begun before the grouping by the LNWR and LYR. He believed that any increase in revenue due to electrification could not be guaranteed.[13] This complacency was too much for the long-time advocate of railway electrification, Sir Philip Dawson, who urged Sir Josiah to follow the lead of the Southern Railway and not to procrastinate ' … until traffic drops still further'.[14] Throughout the 1920s and 1930s Alfred Raworth and Sir Herbert Walker had to combat the general prejudice that electrification was too expensive, but they showed that with good planning, good engineering and a rigorous control of material and labour costs, electrification projects could be economically delivered.[15]

CHAPTER 16
THE BRIGHTON ELECTRIFICATION

Brighton was one of the most important Southern Railway destinations outside London. The town has an ancient history dating back to the Domesday Book when it was known as Brightelmstone. After becoming a health spa in the eighteenth century, there followed a massive growth of the town due to the popularity of sea bathing. From 1780 onwards, rows of Georgian terraces were built, completely changing the character of Brighton from a fishing village into a fashionable resort frequented by the gentry. Brighton's reputation was further enhanced by the regular and prolonged visits from the future King George IV while he was Prince Regent, during his father King George III's long period of mental instability from 1811 onwards. In the late eighteenth century, the Prince Regent had a new palace – the Royal Pavilion – built in the town and designed by the architect John Nash.

Before the coming of the railways, the wealthy, when seeking the pleasures of Brighton, would have travelled by road from London in horse-drawn coaches. The road to Brighton also played host to early steam-powered coaches. One of these, *Infant*, designed by Walter Hancock, made the run from London to Brighton on 1 November 1832. With the arrival of the railway in 1841, the town became an ever more popular destination for day trippers from London. The population of Brighton rose from 7,000 in 1801 to 120,000 in 1901. The influx of so many people devalued the town and it consequently came to be regarded by many as a 'Marine Metropolis'.[1]

The LBSCR could boast in 1888 that its express trains were the heaviest in the world with 360 tons behind the locomotive. Relations were strained between the SER and the LBSCR due to the two companies having to share the line between London Bridge and Redhill. This bottleneck impeded the Brighton express trains and the subsequent delays brought much adverse press criticism directed towards the LBSCR. To alleviate this situation the LBSCR was forced to build what amounted to a new main line parallel to the existing route to avoid Redhill. With parliamentary approval granted in 1896, the 'Quarry Line' was opened in 1900 and ran from Coulsdon North to Earlswood.[2]

The LBSCR was to pioneer the use of Pullman coaches – or 'cars' as we must call them with respect to their American origins. The conservative English travellers did not take to the Pullmans at first, being suspicious of the open saloon interior, preferring the seclusion of the traditional small compartments.[3] In 1905 the Pullman Car Company was taken over by Lord Dalziel, who based his new business at Preston Park works – just outside Brighton. Some cars were built at Preston Park, but new stock was mainly built by other rolling stock contractors, with only maintenance being carried out at the company's own works.[4] The LBSCR's development of the Pullman service culminated in the introduction in 1908 of a new seven car 'train deluxe', 'the most luxurious train in the world'. This new service was soon to be officially named the *Southern Belle*.[5]

The use of Pullmans on the Brighton line was to be continued by the Southern Railway. While most of the railway's operations were 'in house' the Southern did contract the Pullman Car Company to provide most of the on-train catering either in Pullman vehicles or railway-owned restaurant cars. In 1931 there were 20 trains daily from either Victoria or London Bridge to Brighton that had at least one Pullman car attached, plus another 15 to other destinations, such as Portsmouth, Newhaven, Eastbourne and West Worthing. Most of these trains had a single first and third class car, but the 11.05am and the 3.05pm departures from Victoria were the all-Pullman *Southern Belle*.[6] Most of the Pullman cars in use in 1930 were now rather old – and some were very ancient indeed, being American imports complete with clerestory roofs.

The Southern Railway was not going to use the euphemism 'modernisation' as a pretext to reduce the quality of comfort and amenity on its proposed new electric services, therefore the new electric multiple units would have to include some Pullman cars and stock for the all-Pullman *Southern Belle*. But the Pullman cars were not the property of the Southern Railway, and the Southern began negotiations with the Pullman Car Company expecting them to provide 38 new vehicles. The Pullman Car Company was taken aback, saying that it could not raise the required £205,000 to purchase the new cars. It was suggested that the Southern Railway build the new Pullmans at its own works and provide them free of charge; alternatively, the Southern should provide the money for the Pullman Car Company to purchase the vehicles from another contractor. Neither was acceptable to Sir Herbert Walker and after much arm-twisting the Pullman Car Company agreed to raise the capital themselves, but the railway would take over the debt and effectively purchase the vehicles if the company was unable to meet its repayments.[7]

The Southern also had to build more of its own new rolling stock for the proposed express and fast trains on the electrified Brighton line. So far, it had only built suburban electric rolling stock and it had often been expedient to obtain most of these by placing redundant wooden coach bodies onto new steel underframes. The Southern however did not have any suitable 'old' main line carriages – indeed it had an ongoing programme to build new main line carriages for steam services. The LBSCR, unlike the LSWR and the SECR, had not been building new corridor coaches in recent years. As late as 1922 the LBSCR continued to build non-corridor carriages for the Brighton services.

The passenger carriages built by the Southern for its electric services provided a standard of comfort and amenity that was to be equalled in later years but never to be exceeded. There was a caveat: the high un-sprung weight of the motor bogies did gain the motor coaches a reputation for giving the crew and passengers a rather rough ride.[8] All the coaches in these new electric trains were built to the same length as the new stock built for the SECR line electrification. There were two types of six-car train built specifically for the Brighton electrification. Each had gangwayed 'corridor' connections between the vehicles but no connection was provided between the sets. Therefore, the usual 12-coach trains were strictly two separate trains coupled together. Between the motor coaches in each set were four unpowered trailers, one of which was a Pullman car providing the catering services – a splash of umber and cream amidst the line of Southern olive-green carriages.

The new motor coaches continued the curved front-end styling that had first been used on the new suburban trains introduced for the SECR suburban electrification. The driver's cab was the full width of the coach and had two forward-facing windows either side of the route number display board. Behind the cab was a 10-foot (3 metre) long luggage and guard's compartment. The passenger accommodation in the motor coaches was in two open saloons comprising 28 seats for smokers and 24 seats for non-smokers.[9] The earlier suburban trains had their electrical control gear mounted above floor level, directly behind the driver, but the Brighton line express stock had a new design of electro-pneumatic control equipment specified by Alfred Raworth, which was tucked tidily away below floor level within the underframe of the motor coach. Another new 'Raworth' feature was that low voltage power for carriage lighting and control systems was obtained using a 5kW motor-generator set, and this too was mounted below floor level. The electrical equipment for the motor coaches was supplied by two British manufacturers, both then part of the AEI group: British Thomson-Houston provided the traction motors, while the rest of the electrical equipment was made by Metropolitan Vickers.[10] Both bogies on the motor coaches were powered. The electric motors were mounted within the bogies and drove the axles through gearing. The total power of all the motors on a six-car set was 1,800 horsepower (1,325kW). The design top speed for the motor coaches was 75mph.

The Southern Railway trailers were all side-corridor vehicles and were of three types: full third class, full first class and a first and third class composite coach. Richard Maunsell was the chief mechanical engineer, but his assistant, Lionel Lynes, took charge of the rolling stock design and development.[11] These carriages followed Southern Railway practice, contrasting with the other main line companies in that an access door to every compartment was provided on the 'compartment' side of the carriage and four or five doors on the 'corridor' side. This aided entry and egress for passengers on busy routes, which enabled station stops to be timetabled to take as little as 20 seconds.[12] Unlike the motor coaches and the Pullmans, which were described as 'all steel', the trailer bodies were of timber-framed construction making use of oak, teak, mahogany and pitch pine. Sheet steel attached to the framing was used to clad these coaches. The interiors of the first class compartments were of American walnut and the third class compartments were of African mahogany and birch plywood. The hammock seats were comfortable and there were rugs on the floors of the first class compartments, but only plain linoleum for the third class passengers. Armrests were provided for the third class passengers, but these had to be folded up if the full seating capacity of the compartment was required.[13] The Pullman car in each set – which were all given ladies names such as 'Anne' or 'Ruth'– had a small kitchen and seats for eight diners in the third class saloon and a further eight in the much more spacious first class saloon. There was also a more private first class compartment in which four could dine in comfort.

Twenty of the six-car sets – which with the imminent introduction of the 'telegraphic' designations would become known as the 6-PUL sets – were built for the Brighton electrification. To specifically accommodate the high number of first class season-ticket holders who journeyed to their City offices via London Bridge, the Southern Railway built a further three six-car sets – originally known as the 'City Limited' stock but soon to be classified 6-CIT. These sets had three of the first class only trailers enabling the city gents with their first class season tickets to sit behind their copies of the *Financial Times* three abreast, six to a compartment. To provide them with a full English breakfast should they so desire, a Pullman Car built to the same specification as those in the 6-PUL sets was included.[14]

Even these 6-CIT units were not to be the best trains on the Brighton line, for three five-car all Pullman sets were built for the new all-electric *Southern Belle* service. All 15 cars were constructed by the Metropolitan Carriage and Wagon Finance Company. The Southern Railway provided all the electrical equipment and naturally the train crew, as the driver and guard had to be railway staff, but the Pullman Car Company provided the chefs and attendants. The daily maintenance was the responsibility of the railway, but programmed maintenance, repairs and modifications were carried out by the Pullman Car Company at its base at Preston Park, Brighton.[15] These five-car

trains were comprised of a third class motor coach at either end of the set (these also contained the guard's compartment as per the other main line sets) and the three intermediate cars were a single third class car and two first class cars which also contained the all-important all-electric kitchens. The vehicles were built to the traditional Pullman standard, with open style interiors complete with sumptuous upholstery, fine curtains, hardwood panelling with elaborate marquetry and many ornate features and fixtures. Externally they presented a smooth and uncluttered appearance with large windows to each seating bay. As was the Pullman Car Company's usual practice the third class cars were numbered: the motor coaches were 'Car No. 88' to 'Car No. 93'; the parlour thirds were 'Car No. 85' to 'Car No. 87'; and the first class cars were all given ladies' names: 'Doris', 'Hazel', 'Audrey', 'Vera', 'Mona' and 'Gwen'.[16] These female names may seem a bit staid today, but probably many Pullman Car Company managers' wives, daughters and mistresses were duly honoured by the naming. The electrical specification was as for the 6-PUL and 6-CIT sets.

For the intermediate stopping services to Brighton, the Southern Railway built 33 four-coach sets. These were soon to be designated 4-LAV. The 'LAV' in the description denotes that there were lavatories on board – but only in one of the carriages – the other three vehicles had full width compartments without side corridors giving a 'suburban' look to the train, but internally these new coaches were built to the same high standard as all the coaches then being built by the Southern Railway. So that the 4-LAVs could run as multiple units with the older suburban sets, they were equipped with the 'old' electrical control equipment. Alfred Raworth had designed a system for running 'old' and 'new' control systems together, but the complete lack of any

Southern Electric 6-PUL No. 3018 photographed on the Ouse Valley viaduct. The date is unknown, but the BR logo is the later post-1957 version. The headcode '4' denotes the non-stop Victoria to Brighton service. (Mike Morant Collection)

THE BRIGHTON ELECTRIFICATION • 191

The *Brighton Belle* in Southern Railway days. Trains could be composed of two 5-BEL units or, as here, a single five-car unit. (John Scott-Morgan Collection)

photographic evidence of trains made up of both types of stock suggests that this arrangement was either never used or was unsuccessful.[17]

Passengers in the splendid new rolling stock were completely unaware of the electricity supply infrastructure which powered their trains. (See plate 11) This included the 33,000V, three-phase cables supplying the railway-owned substations. The cables were installed in wooden troughing supported above the ground by ferro-concrete posts every 3 feet.[18] From the CEB's Fishergate substation, one cable route was installed westward through substations at Shoreham and Lancing to West Worthing – where for the time being there was to be no alternative Grid supply from the Portsmouth direction. A second cable route from Fishergate ran at first eastward and then northwards alongside the main line, connecting substations at Portslade, Brighton, Preston Park, Pangdean, Hassocks, Keymer Junction, Haywards Heath, Ouse Valley and Balcombe Tunnel to reach the CEB's Three Bridges Grid substation. From Three Bridges Grid another route headed north linking substations at Gatwick, Salfords, Redhill, Merston and Star Lane and hence to the third Grid substation at Croydon[19] – where the CEB gave the Southern Railway circuit breaker the plant number 'SR 1'. All the original 18 Southern Railway substation sites are still in use today, but with all their original equipment renewed.

The 33,000V cables were supplied by the Southern Railway's neighbour at Eastleigh – Pirelli General. What is surprising is that Raworth, who was always an innovator, did not use Pirelli's latest technology for his early projects, namely oil-filled cables. Within these cables a thin mineral oil flows through small, perforated ducts between the phase conductors. The oil is an insulator forced under pressure into the cable's paper insulation. By this means a better insulation is achieved with less paper insulation and the cable can thus be smaller overall for its designed load and hence easier to install in confined locations. The oil is kept at a pre-defined pressure and as it expands and contracts with heat during the load cycle a constant pressure must be maintained. This is achieved by cylindrical header tanks which have an internal bellows-like mechanism which keeps the pressure constant. Later cable installations on the Southern Electric network were to use oil-filled cables and the lineside oil tanks that can be spotted from trains today give witness to this. Today oil-filled cables are obsolete – modern polymeric insulation is deemed superior and the oil-filled cables are now considered environmentally unfriendly as they are liable to leak into the ground. Alfred Raworth was to remain loyal to conventional cable which was manufactured with more oil impregnated paper. Possibly for such a high-profile project, with so many new and novel features, taking the risk of using the newly developed oil-filled cable technology was just one innovation too far. Pirelli General was contracted to supply 50 miles of its three-phase cable at a price of £112,000.[20] For later electrification schemes three-phase cables were not used (apart from the short sections leading into the substations), instead within the lineside wooden troughs the cables were of the single-core type – that is individual single-phase lead-covered cables.

At each substation along the cable routes, the alternating current high voltage switchgear was in the outdoor part of the installation enclosed by an 'unclimbable' fence to keep intruders away from the high voltage conductors which were bare copper bars supported on porcelain insulators and fixed well clear of the ground by reinforced concrete frames and posts. The 33,000V switchgear was arranged to suit the 'ring' concept. Three circuit breakers per substation were usually used, each connected on one side to a common copper busbar. Two of the circuit breakers were connected to the incoming feeder cables and the third was connected to the transformer. Additional 'off-load' knife switches operated on site manually were provided to enable plant to be safely isolated for maintenance or repair. Alfred Raworth, following receipt of all tenders, recommended to the directors that ASEA's tender price of £62,560 be accepted for the supply of the 33kV switchgear.[21] The single transformer was also in the outdoor compound but placed directly against the substation building with the six low voltage conductors passing through the wall to reach the mercury arc rectifier inside. (See plates 12 and 13)

The flat-roofed substation buildings housing the direct current equipment had (or rather have – for many remain in situ) a pronounced tower at the end next to the high voltage compound and at the base of this tower, safely enclosed within an expanded metal screen, was the mercury arc rectifier – the additional height of the building was necessary to house the hoist to lift all or part of the rectifier for maintenance or replacement. The substation buildings were originally of plain brick construction but later they often had a white rendering applied in line with the 1930s modernist style.

Once again it had been necessary for Raworth to look beyond the British electrical plant manufacturers to procure the mercury arc rectifiers. The rectifiers were purchased from Brown Boveri, a Swiss manufacturer which had pioneered the use of these rectifiers a decade earlier, installing three units each rated at 800kW with an output voltage of 580V for the Berne tramway system in Switzerland.[22] Raworth reported to the directors that prolonged negotiations had been necessary to agree terms with Brown Boveri for the complete rectifier units (the mercury arc rectifiers with the transformers), which resulted in the Southern Railway paying £100,654 to the Swiss company.[23] The substation transformers reduced the 33,000V to 690V alternating current to feed into the rectifiers. The rectifiers were rated at 2,500kW but were so designed that, if necessary, they could operate at 4,000kW for five minutes.[24] At the output voltage of 660V direct current 2,500kW equates to a massive 3,790 amps. Cooling of the units was essential, and this was achieved using a closed water system situated in an annex behind the main building.[25]

A typical Southern Electric substation – precise location not known but probably part of 'Brighton' scheme. (Author's Collection)

Next to the mercury arc rectifier cell was the direct current switchboard which would have typically five high speed air-break circuit breakers at double track locations or nine circuit breakers at quadruple track locations. The contract to supply this direct current switchgear was awarded to a British company – British Thomson-Houston, at a tender price of £24,956.[26] One circuit breaker (rated at 4,000 amps) would be the incoming feed from the rectifier and there was a circuit breaker (rated at 2,500 amps) controlling each outgoing feed.[27] There were twice the number of outgoing feeder cables as there were railway lines as each conductor rail was interrupted outside the substation and a direct current feeder cable was connected to either side of the break to supply current in both directions.

ASEA was once again successful, when after prolonged negotiations it won the bidding to provide the direct current connections to the track, on condition that it carry out all the fabrication work at its Walthamstow premises – the railway thereby ensuring the use of some local British labour.[28] The conductor rails did not continue to the next substation, for at a halfway point they finished outside what was termed a 'track paralleling hut'. Inside these 'huts' (actually a small but substantial building) there was another switchboard equipped with high-speed air-break circuit breakers (rated at 2,000 amps) – one for each conductor rail in each direction and all connected through a common busbar – effectively paralleling all the conductor rails.[29] These switches were normally closed but did provide the means of isolating individual sections of rail should this be necessary.

The first of the Southern Electric control rooms using the ASEA control system was built at Three Bridges to control the whole of the new electrification south of Purley. This was a modernist single-storey building with art deco overtones. Rising above the main building was a 'D' shaped superstructure which was the main control room containing the panels with the electrical system's mimic diagrams.[30] These mimic diagrams, positioned on a line of panels which swept around the curve of the 'D', were schematic representations of the electrical system, with appropriately positioned lamps to indicate the status of operable plant (a green lamp for open, a

A partially dismantled direct current circuit breaker. These were mounted on wheels and were rolled into the switchgear cubicles at the substations and track paralleling huts. (Author)

red lamp for closed) and switches, again located appropriately on the diagram, used to operate the remote equipment. Instrumentation too was indicated on the mimic panel – voltmeters and ammeters positioned against the circuits being measured.

The Three Bridges control room was designed to exude the quiet dignity that any good control room should have. The wide-open space was kept clear of any furniture save a rather scruffy and slightly out of place second-hand desk in which the two men on duty would sit. Lighting was by elegant art-deco uplighter lamps. The steel mimic panels were painted a glossy black, with all the diagram symbols in a light cream colour. Along the top of the panels were painted the names of the substations and track paralleling huts. The 33,000V cables were represented by a single cream line running the length of the diagram with the switchgear, transformer and rectifier symbols – with the operating switches and indicator lamps for each – located directly below the substation name, and for each of these there was placed a voltmeter and an ammeter. Below the rectifier was a representation of the direct current switchboard – again with indicator lamps – leading onto the conductor rails. Along the lower part of the mimic diagram the parallel conductor rails were shown and between the substations the paralleling hut switchboards with indicator lamps and control switches were displayed.

It would of course have been impractical to individually 'hard-wire' every control

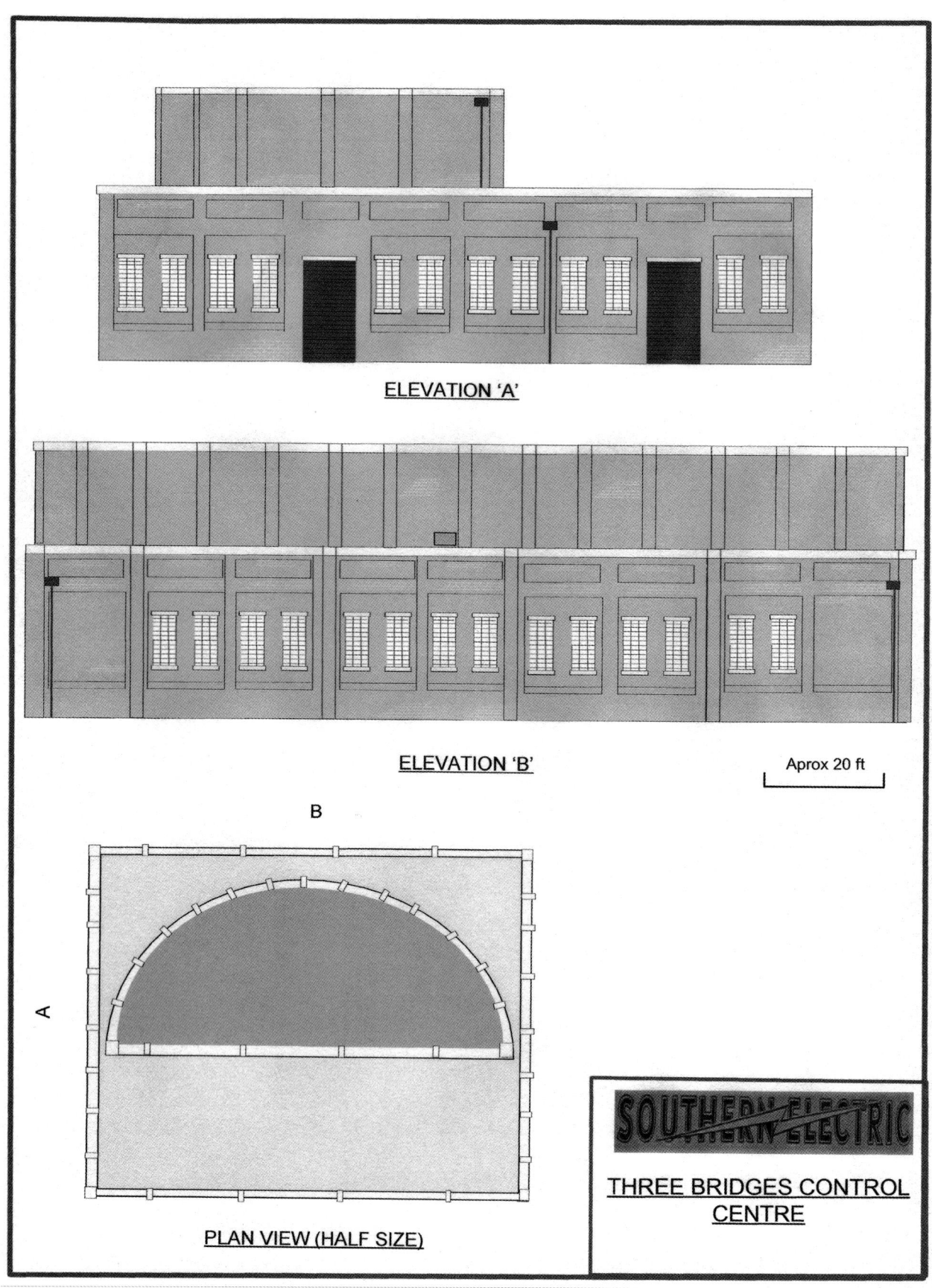

The Southern Railway's first control centre at Three Bridges. (Author)

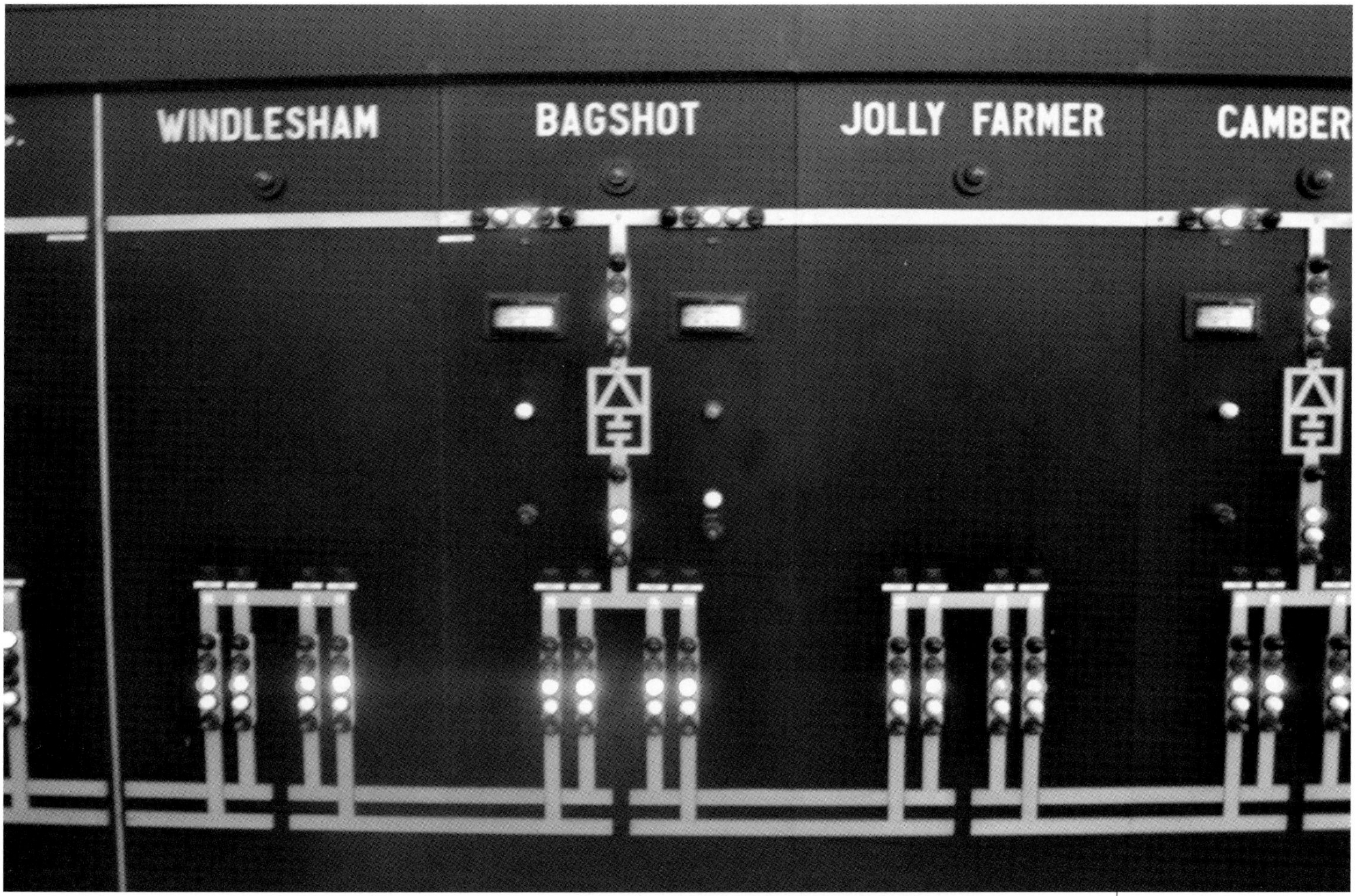

Typical control panels. These are part of the later installation at Woking. (Author)

switch to every substation item of plant, so an electro-mechanical 'polling' system had been devised by ASEA and Raworth. Mounted out of sight behind the mimic diagram panels in the control room were the transmitter cabinets, each of which controlled two substations. At the substations there was a partner panel. Between the control room and the substations, a 'pilot cable' was installed with the 33,000V cables. The Eastleigh-based Pirelli General Company supplied these small cables.[31] Each control room selector panel needed four cores (wires) within the cable in addition to the return wire that was 'common' for all the other wires. By a system of electrical pulses, the selector disc in the transmitter cabinets would rotate until it was aligned to the switch contacts on the control board and through the control cable pulses would simultaneously drive the corresponding

ASEA transmitter unit positioned behind the control panels at the Woking control room. Each of these cabinets operated two substations. (Author)

THE BRIGHTON ELECTRIFICATION • 197

disc at the substation end so as to become aligned with the appropriate plant control contacts. Each control room selector had 50 switch positions. If it was required to, say, open a circuit breaker then the method of operation was for the operator to depress a 'calling key' by the circuit breaker symbol on the mimic diagram to activate the selector disc. To confirm the completion of the synchronous rotation to align the sending and receiving discs, a white lamp was to illuminate to show that there was direct electrical contact between the switch on the panel and the operating relays for the equipment at the substation. The operator would then be able to carry out the switching operation. Similarly, it was possible to 'call up' voltage or current readings for the substation selected.[32]

In the event of a circuit breaker tripping (automatically opening) due to a fault on the system, the control room staff were to be alerted by the ringing of a bell and the dropping of a mechanical shutter to display a 'fault' sign board. To determine at which substation the fault operation had happened, the appropriate 'fault' lamp positioned at the top of each substation panel would light up. To find out which circuit breaker had operated, the system had to be interrogated by depressing the 'check key' on the mimic panel and the control discs would rotate until a position was found at which the 'change of state' had occurred and this would be indicated by the red lamp (the circuit breaker 'closed' indication) going out and the green circuit breaker indication lamp illuminating (circuit breaker 'open' indication).[33]

The Southern Railway also needed to be able to operate the circuit breaker associated with the Southern supply at the Grid substations. To do this it entered into an agreement with the CEB to have these circuit breakers integrated into the Southern Railway ASEA control system,[34] thus setting the precedent – still applied today – that the supply authority and the railway have joint control responsibility of these infeed breakers.

The Southern Railway and ASEA were justifiably proud of the control room with its pioneering supervisory control equipment, which was the first of its kind to be built anywhere in the world.[35] The control room was almost complete on 5 May 1932 when a party of directors visited and were duly impressed by the technology and the architecture.[36] Alfred Raworth also hosted a press visit to the new control room and the nearby substation at Salfords in July 1932 and following the visit the impressed *Railway Gazette* journalists were taken out to lunch by John Elliot.

There was another significant visitor to the new control room. Given ASEA's involvement in the project, a visit by Crown Prince Gustaf Adolph of Sweden during his state visit to Britain was inevitable. The prince, later to be King Gustaf IV and fresh from his honeymoon after marrying his second cousin Princess Sibylla,[37] was accompanied by the Prince of Wales, later Edward VIII. The royal party arrived from Victoria in one of the new six-car electric trains to be greeted by Sir Herbert Walker, Alfred Raworth and Edwin Cox. Alfred Raworth conducted the tour of the control room and Prince Gustaf demonstrated his skills as a control engineer by opening and closing a circuit breaker at far away Lancing substation. Following the tour, the party returned to the new train which was berthed in the nearby north sidings. Luncheon was taken in the new Pullman car 'Ruth'[38] – which identifies their train as the 6-PUL unit number 2017.

The London to Brighton main line beyond the suburban area was only double track – one line in each direction. The electrification of the LBSCR's Quarry Line diversion, as well as the original line through Redhill, effectively made a 16-mile four-track section and to further increase capacity additional tracks were installed for about a mile between Copyhold Junction and Haywards Heath where the station was provided with extra platforms. To accommodate the proposed

12-coach trains the new standard length for platforms was 800 feet (246 metres). This involved lengthening platforms at Brighton Central and Hove stations. Some track and signalling modifications were made at Hove and Redhill to facilitate the splitting and joining of separate train portions. The track layout at Brighton was rationalised allowing the removal of 17 crossovers and a double junction. To service the new trains, a three-road carriage shed was installed at West Worthing. At Brighton, the former paint shop at Brighton works was converted into a 12-road carriage maintenance shed.

The most significant Brighton line improvement was the installation of full colour light signalling from Coulsdon North to Brighton. This 36-mile installation was at the time the largest in the country and was supervised by the assistant engineer ('signals and telegraphs'), Lieutenant Colonel G.L. Hall, who was formerly in the Royal Engineers and had been an inspecting officer for the Ministry of Transport. Much of the colour light signalling was automatic and was introduced in stages allowing the closure of signal boxes at Cane Hill, Quarry, Worsted Green, Earlswood station, Tinsley Green and both the North and South boxes at Three Bridges.

The removal of the semaphore signals and commissioning of the colour light signals was completed in stages overnight. The long section between Coulsdon North and Balcombe tunnel was completed in six hours between 2am and 8am on a wet and miserable Sunday morning. Commissioning tests began as early as 3am and as dawn broke the semaphore posts were on the ground and the new colour light signals standing in their place. The seven manual signal boxes were abolished and the new box at Three Bridges was in service, replacing three old boxes. Track circuiting was an integral part of the new signalling – particularly for the automatic signals – and this too had to be commissioned within the very tight time slot.[39] To successfully carry out this massive operation considerable detailed planning had been necessary, together with much preparation work, such as the installation and testing of new signalling circuits before the changeover.

The final installation of the new four aspect signals between Preston Park and

Standing in front of the new control panels at Three Bridges, HRH The Prince of Wales is flanked by Crown Prince Gustaf Adolph of Sweden on his left, and by Sir Herbert Walker to his right. Alfred Raworth, on the left of this group, maintains a respectful distance, smoking a cigarette. (ABB)

Brighton also took place in a few hours on the night of Saturday, 15 October 1932. This included the commissioning of the new, all electric, signal box at Brighton, which had 225 'levers' replacing a total of 582 mechanical levers at six of the local signal boxes at Lover's Walk, Montpelier Junction, Brighton South, Brighton West, New England and Holland Junction.[40] Brighton signal box was not one of the elegant 'glasshouse' designs, but a rather brutal modernist brick building perched inelegantly above the tracks on reinforced concrete posts, set back off the main line, straddling sidings associated with the former LBSCR's locomotive workshops.[41] The box commanded a clear view of both the 10 platforms at Brighton terminus and the locomotive sheds and carriage sidings. Within the box, the signalmen worked from three large diagrams set above the block instruments using the 225 stainless steel levers, each only 2 inches long (about 50 millimetres), to operate the complex circuits requiring 13,346 electrical contacts within the box. There were 115 miles of signalling wire and 10 miles of telephone wires inside the building.[42]

The electrification of the Brighton line proceeded smoothly, enabling the first electric train to appear at Redhill station on 21 March 1932. On Sunday, 17 July 1932 at the start of the summer timetable, a temporary service of electric trains to Three Bridges commenced, and on the same day an electric service was inaugurated to Reigate.[43] Electric trains ran from Victoria and London Bridge to Three Bridges – one train an hour in each direction. This, combined with a revision of services on the Reading to Tonbridge route, reaped the Southern Railway early benefits from the Brighton electrification by reducing the number of steam locomotives required by six (three on Sundays), thus requiring 15 fewer pairs of footplate men (six on Sundays). Twenty-six additional motormen were added to the staff of the electrical department.[44]

A test train was run to Brighton on 1 November 1932. This was an ad hoc set which included a power car from each manufacturer (the Birmingham Railway Carriage and Wagon Company and the Metropolitan Carriage Wagon and Finance Company) marshalled at either end. This was a test run in preparation for the full timetable which was due to commence on 1 January 1933. This first electric train to enter Brighton station made the journey from London in just under an hour, exactly 100 years after the first steam coach – *Infant* – had arrived in Brighton, having made the gruelling seven-hour

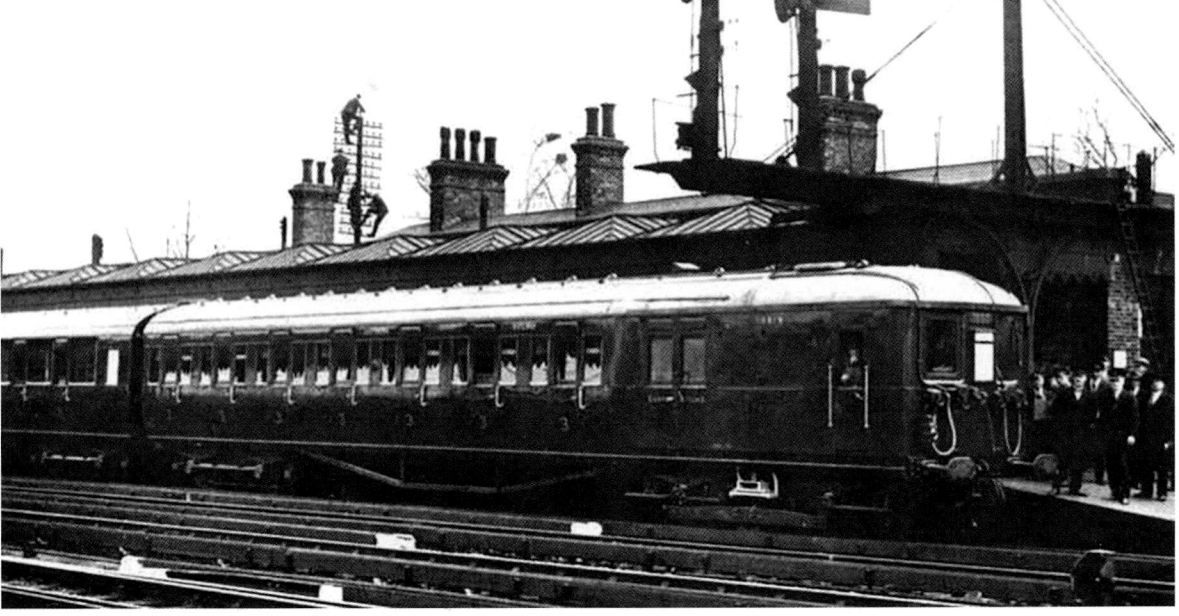

The first electric train arrives at Redhill station on the 21 March 1932. This was a 4-LAV unit. Note linesmen working on the telegraph lines. (Author's Collection)

road journey from the capital.⁴⁵ As was to be expected there was a great deal of press interest in the Brighton electrification. The *Railway Gazette* congratulated the Southern on its courage in embarking on its first main line electrification scheme – particularly courageous in its use of multiple unit trains.⁴⁶ Knowing now how successful the Southern Railway's main line electrification was to be, it is difficult to appreciate just how bold a step the company had made in 1930.

The Southern Railway's Brighton electrification was a rare piece of good news for a nation gripped by the worldwide Great Depression. The railway's achievement contrasted with the news from Glasgow where work on the new ocean liner (to be named *Queen Mary*) had been suspended and hundreds of men laid off because Cunard did not have the funds to complete the project. This chronic economic downturn had its roots in the American Wall Street stock market crash of 1929, exacerbating the already sorry state of the British economy. Britain's trade with the rest of the world fell by a third. As the first test runs of the new electric trains were made late in 1932, there were 3.5 million Britons claiming unemployment benefit and many more forced into part-time employment.⁴⁷ Without any heavy industries, such as steel production and widespread coal mining, the south of England suffered less, but the Southern Railway was unable to fully escape the consequences of the reduced economic activity. It is in this context that, Winston Churchill's 'give-away' budget encouraging investment by abolishing passenger duty earlier in 1929 – before the October Wall Street crash – must be viewed as being particularly prescient.

On 30 December 1932, the Brighton electrification was officially opened by the Lord Mayor of London, Alderman Sir Percy Greenaway. The ceremonial party – which mysteriously did not include Sir Herbert among the Southern officials – took the special electric train from Victoria station to West Worthing. The journey from Victoria to Worthing was scheduled to take 74 minutes but was completed in 72 – a leisurely pace. The maximum speed was 74mph through Haywards Heath and speed was reduced to only 56mph through Merstham Tunnel – a location where the steam-hauled trains rarely exceeded 45mph. At Shoreham, a town between Brighton and Worthing and a beneficiary of the new electric services, each dignitary in the official party received the gift of a commemorative medal from Shoreham Urban District Council. These medals depicted on one side an electric train, and on the reverse St Mary's church, Shoreham.

The party received a civic reception at Worthing led by the mayor, Councillor T.E. Hawkins, and the town clerk, who greeted the train amidst a forest of flowers, foliage and bunting. There was a reception in the station dining room where 'Worthing Sunshine' cocktails were offered. After the festivities at Worthing the party re-boarded the train to make the short journey to a flag-bedecked Brighton station, where all the steam locomotives in the vicinity paid their due respect to the upstart electric train by sounding off their whistles with shrill and sustained blasts. This time it was the mayor of Brighton and other local dignitaries greeting the train. Perhaps to make the point that this first main line electrification represented a sea change in railway development, the speeches were made in front of an exhibit representing a similar leap in railway technology from just over a century before – a full-size replica of Stephenson's *Rocket* which the Southern Railway had borrowed from the LMSR. Following the speeches, the party adjourned to the Royal Pavilion for lunch.

This was not the end of the celebrations, for that evening the Worthing Corporation hosted a dinner at the Burlington Hotel in Worthing. The new Southern Railway chairman, Gerald Loder, who succeeded the late Everard Baring, was present. Appropriately, Loder had been the Conservative MP for Brighton between 1889 and 1905 and had been a director of the LBSCR, succeeding Charles Macrae as chairman during the dying days of the LBSCR's existence in December 1922.[48] Arguably Loder had been elected chairman purely because it was now the former LBSCR's director's turn to take the top job, for he was not highly regarded by his fellow board members.[49] In his speech he seemed to see the occasion as the celebration of the completion of the LBSCR's electrification proposal. He even gave reference to the SECR's desire to electrify to Redhill. He obviously lacked any understanding of the engineering issues – this scheme was unlike anything advocated by either the LBSCR or the SECR. But Loder did praise Sir Herbert and many other officers including Alfred Raworth who had made their contributions to the endeavour. One of the honoured guests was Sir William Forbes who had been ousted from the Southern management team in 1923 by Sir Herbert Walker. Perhaps this was why Sir Herbert absented himself from what was to become an 'LBSCR celebration'. His prepared speech was read by his assistant general manager, the dapper Gilbert Szlumper. The speech at first emphasised the Southern Electric business case by explaining that the electrification enabled the railway to vastly increase train services – and hence its revenue – without a proportionate increase in operating costs.

Sir Herbert Walker now saw Alfred Raworth as an essential participant in the Southern Electric expansion due to his skills in planning and managing the schemes and for his tough dealings with suppliers during contract negotiations. Later in the speech Szlumper read out it became clear that Raworth's engineering expertise had gained the general manager's wholesome praise as more details were given of what the chairman had just lamely described as the 'technical and more difficult' aspects of the electrification. Sir Herbert described the Three Bridges control system as

> a masterpiece of ingenuity and the last word in electrical practice. The idea and all the enormous amount of detail in connection with it was worked out by Mr Raworth in conjunction with the contractors who gave every assistance during his endeavours. The system of remote control of substations had taken them a long way forward in the electrification of railways …[50]

The next day, 31 December, the last steam-hauled *Southern Belle* departed Victoria station hauled by the LBSCR war memorial engine, the Baltic tank *Remembrance*. The last steam through train to Worthing was hauled by one of the King Arthur class locomotives *Sir Durnore*, departing from Victoria at 9.05pm, just as the empty electric *Southern Belle* arrived at the station to be berthed there overnight. The following day, Sunday, 1 January 1933, the full electric timetable began. To celebrate the new service and to prove how smooth the running of the new electric trains was, the Southern Railway arranged for the London Madrigal Society to sing aboard the *Southern Belle*. The society members performed from Victoria to Brighton, to the apparent delight of the passengers.[51]

The new electric service on the Brighton line was six trains an hour with some additional trains at peak periods. The six were: a non-stop from Victoria to Brighton scheduled to take one hour exactly; a Victoria to West Worthing semi-fast; a Victoria to Brighton semi-fast; a London Bridge to Brighton semi-fast; and two all-stations trains, one from Victoria and one from London Bridge. All of these trains ran to the strict 'Walker' principle of consistent minutes past each hour throughout the day so that the regular passengers need never consult a timetable.

The new electric trains caused much excitement with the local residents and proved to be popular with the travelling public. Ticket sales at Brighton rose considerably during the first week and the weekend takings at one smaller station on the coast rose dramatically when ticket sales jumped from 1,800 to 2,500.

At Aldrington Halt, ticket sales doubled in the first week of electric operation. At West Worthing, third class return tickets to Worthing were sold out as many children were taken on the round trip to experience the new electric trains. On the first Saturday, 7 January, 7,500 passengers arrived from London to watch the Brighton versus Chelsea football match.[52] Throughout January 1933, the trains on the services to Brighton and Worthing carried an additional 73,000 passengers increasing receipts by 5 per cent – a result which pleased the Southern Railway, as the success of this 'off-season' trade augured well for the summer traffic to these seaside destinations. The Southern also hoped that more Londoners would be tempted to reside on the coast and take advantage of the improved services.[53] The new timetable required the electric trains to run an additional 83,800 miles a week, with a reduction in the steam train mileage of 30,160. The use of the electric trains resulted in 120 passenger coaches (including 19 Pullmans) being withdrawn or moved to other services. Twenty less steam engines (12 on Sundays) and 44 less pairs of footplate men (24 on Sundays) were required.[54]

Soon after the start of the full electric services on the Brighton line, further electric trains were built for the stopping services. If the 4-LAVs were a step back in evolutionary terms compared to the new main line stock, then the two-car 2-NOLs were a further step back. In the newly introduced 'telegraphic' nomenclature, the 'NOL' appellation depressingly stands for 'no lavatories'. The men from Maunsell's mechanical engineering department scoured the former LSWR network looking for old wooden coaches that were surplus to requirements and suitable for cobbling together to form bodies to be placed on the now standard 62-foot underframes. Obviously more than one old body was needed for each new coach body and at times when the length did not come to the full 62 feet an empty dummy section was used to fill the gap. The 2-NOLs, like the 4-LAVs, had the 'old' control equipment and comprised of a motor brake third sitting 73 passengers in seven normal-sized compartments plus a 'coupe' compartment having seats on one side only; a driving trailer 'composite' having seats for 24 first class ticket-holders in three compartments; and 60 third class passengers in six third class compartments.[55] The last of the 2-NOLs were to be retired in 1959 when it became apparent that these venerable old wooden bodies just could not take any more use. Despite their Edwardian appearance, the 2-NOLs had a lot of character and were very much part of the Southern Electric scene, and their passing was much mourned by many enthusiasts.

Regretted by many, in 1934 the *Southern Belle* was controversially renamed the *Brighton Belle*. The renaming was timed to coincide with the auspicious occasion when Brighton Corporation opened its new covered seawater swimming pool.[56] Nonetheless this iconic train was much loved by its regular passengers and continued to run until 1972.

Sir Herbert Walker had been eventually authorised to spend up to £2,854,082 on the London to Brighton electrification scheme. The final cost was £2,290,160 putting it well within budget. Alfred Raworth was not responsible for all the expenditure of course, but his input was by far the riskiest, requiring the implementation of many novel features, often after tough negotiations with suppliers.

CHAPTER 17
MAINTAINING MOMENTUM

Sir Herbert Walker has justifiably been given the credit for the electrification programme and he was viewed by his contemporaries as being the railway manager who was second to none.[1] He was the driving force behind the whole project, persuading the directors to authorise expenditure and then ensuring that the completion of each incremental extension was on time and within budget. He could not have done this unaided; he needed talented assistants and he later confided to John Elliot that 'without Cox and Raworth it could not have been done'.[2] By the time of the historic completion of the Brighton electrification, Walker, Cox and Raworth had become a formidable team. While they had earned respect and some admiration, they were not necessarily liked by the Southern Railway staff. Walker may have had his 'common touch' when meeting lowly staff on his travels around the railway, but the savings from electrification meant some job losses, particularly at the steam motive power depots – but nonetheless priority had been given from the earliest LSWR days to the redeployment of displaced engine drivers as motormen. Walker would also have upset the average railwayman when he described the statutory reduction of the working day to only eight hours as 'iniquitous'. He based this assertion on the premise that a signalman at a remote country station with an 'easy' eight-hour working day had the same conditions of service and pay as a signalman working flat out all day at a main junction in London.

The 1930s was a period when, despite the increase in passenger revenue due to the success of the Southern Electric, the railway's overall receipts were falling due to the Great Depression and road competition. Sir Herbert had to instigate an overall pay reduction of 1.25 per cent across all grades of employee (including himself) and this followed a 2.5 per cent reduction back in 1928.[3] The serious loss of goods traffic was exacerbated by the restrictions on rail freight charges imposed by the Rates Tribunal, resulting in the railways being unable to compete with the unrestricted charges made by the road hauliers. To rub salt into their wounds, the railways were one of the largest local 'rates' ('council tax') payers in the country and were thus contributing towards the upkeep of the nation's roads – the public assets used by their main competitor.[4]

John Elliot's considered opinion of Sir Herbert was that despite being from the older Edwardian generation of railway managers, he had dragged the Southern Railway into the twentieth century by creating a modern transport system. Elliot observed that Walker ensured that the railway's activities were simply organised without over-elaboration and that the railway was run safely, punctually and – not forgetting the shareholders – profitably. Walker expected every one of his chief officers to ensure that all their staff, either at headquarters or down the line, performed their duty to meet these three requirements.[5]

Following a rather over-ambitious programme of excursion packages which had not shown a profit, John Elliot had an uncomfortable interview with Sir Herbert in which he was given a profound insight into his general manager's commercial focus. He was told that: 'You must learn what we are here for. It is not just to please the public but to enable us to pay ordinary dividends each year and so be able to raise capital for further electrification.'[6]

Sir Herbert's commercial expertise was beyond question, for he understood what the travelling public required and ensured that they were provided with regular and frequent train services. There does however seem to have been a blind spot in his thinking. This was a failure to fully appreciate the potential threat from the private motor car. Walker had been born into an age without any motorised road transport and would hardly ever resort to the use of the Southern Railway chaufeur-driven Daimler, even to the extent of using the rival trams to visit an out of the way location on one of his frequent inspection visits. He could see no reason why anybody should ever feel the need to travel by motor car.

Sir Herbert and his traffic manager, Edwin Cox, seem to have been 'joined at the hip', both being physically tall and fiercely loyal to each other. Cox was a man who had by his 'force of character, drive and competence'[7] risen from being an office boy to chief operating officer. Walker certainly appreciated Cox's organisational skills and his unbounded creative energy and when Cox wanted to retire in 1936, he tried to persuade him to stay on so that they could both retire at the same time. Not all the staff however were so impressed with Cox – many viewed him as a good talker with a keen operating sense but on occasions, he was apt to make too hasty a decision. He was ambitious and confident in his own abilities but somewhat opinionated. One of his closest colleagues declared that 'Cox sits in his room surrounded by mirrors in which he sees only himself.'[8]

John Elliot described Alfred Raworth as 'a cool and dedicated technician' in whom Walker had complete confidence.[9] Herbert Jones, as the chief electrical engineer, should have taken the full credit for the Southern Electric electrical engineering and not Alfred Raworth, as he was merely the 'electrical engineer new works'. In 1936 Raworth's salary was £3,000 a year, only £200 less than was paid to Jones,[10] and given Raworth's privileged position working so close to Sir Herbert Walker, the hierarchical structure had become blurred. Jones and Raworth kept separate offices at Waterloo and London Bridge respectively and each had their own clerical and technical staff. It has been stated that there was animosity between the two men, and it has been inferred that Sir Herbert deliberately kept them apart in their remote offices to avoid conflict. A kinder interpretation might be that this allowed each to concentrate on his separate bailiwick, for Jones had a large and dispersed department to run, and Raworth had a busy design office with tight project management commitments.

There is another explanation, in that Walker operated a 'divide and rule' policy, creating a situation of rivalry as each department strove to be 'the best'. With his limited technical knowledge he had to have trust in each of his senior officers and allow them a free rein to run their departments with minimum interference from himself – the only criteria being that his set objectives were met. Each 'chief' had total control and Walker, needing their loyalty, would provide them with unquestioning support. Arguably, if you were one of Walker's 'first lieutenants', you were above any criticism and in complete control of your domain. For example, the chief engineer, George Ellson (who had patented the 'rail protector' with the SECR), may have been a highly respected bridge engineer[11] and was later to provide 'inspiring leadership' during the forthcoming difficult war years,[12] but he exhibited bouts of uncontrollable temper and his bullying was so viscous that it was claimed that his staff could not carry out their work to the best of their abilities because they were constantly avoiding any contact with him. This was reported to the assistant general manager, Gilbert Szlumper, who argued that nothing could be done and that the staff should accept the situation.[13] The only explanation for such negligent inaction would be that Walker would hear no criticism of his chief officer. Gilbert Szlumper also had his issues with Sir Herbert. It will be noted that he was assistant, not deputy, general manager. Szlumper's role was to be no more than the general manager's 'bag-carrier', to be available to do his bidding – such as reading out his speech at Worthing when Sir Herbert deigned not to attend. When Sir Herbert took a long trip to South Africa, his assistant only learnt about it from picking up the office gossip and had to directly ask if what he had heard was true.[14]

Raworth's small and rather basic office at Station Approach must have seemed very appropriate to visitors given his reputation for austerity. E.P. Leigh-Bennet, when writing in the Southern Railway's magazine for its passengers, described Raworth's office as ' …a drab little building in the purlieu of London Bridge station which nobody wants to enter except railway contractors [where] all the multitudinous details of this vast electrical undertaking are put down with deadly accuracy on paper …'.[15]

Within this unpromising workplace there worked a small and intimate team. At least two of Raworth's staff, his chief clerk, W.G. Young, and his principal assistant engineer, W.C. Moore, were long serving and remained with Raworth throughout the hectic expansion years after their appointment by Percy Tempest in 1922.[16] The team suffered sorrow in September 1932 when Henry Young, the 16-year-old son of W.G. Young, was killed by a motorcycle which collided with him as he walked with his mother through their home village of Shipbourne near Tonbridge.[17] The boy's death was even more tragic as Mr and Mrs Young had now suffered the loss of all five of their children. To support the grieving parents at such a difficult time, staff from the Southern Railway 'new works' department, with Alfred and Ruby Raworth and W.C. Moore and his wife, attended the funeral at Shipbourne parish church.[18]

Raworth was able to keep the cost of his electrification programme low by good engineering and – due to the efforts of Sir Herbert – by maintaining a rolling programme of work to keep his small team of technical staff continuously engaged. A guarantee of continuing orders enabled plant and cable costs to be kept low following the negotiation of long-term contracts. For example, in 1936 Raworth negotiated a 10-year agreement with Pirelli General for the supply of 33,000V cables at a price that was over 20 per cent cheaper than the first order for the Brighton scheme in 1930.[19] Wherever possible the work was carried out in-house by Southern Railway operatives, avoiding the use of contract labour. This continuity of work meant that once the Brighton scheme had been completed it was possible to seamlessly move on to the Eastbourne electrification.

The Brighton electrification may have realised a large part of the old LBSCR's ambition, but that company had also intended to include Eastbourne and Newhaven in its electrification programme. Following the success of the Brighton electrification, Sir Herbert Walker now proposed further electrification not only to Eastbourne but also further east along the south coast to Hastings and Ore. The seaside resort of Eastbourne had earned a reputation for being more 'refined' than the over-popular Brighton. Newhaven, once a minor fishing port at the mouth of the River Ouse, had become the embarkation point for cross-Channel ferries to Dieppe, but the Southern Railway initially intended to continue operating the boat trains with steam locomotives, not electric multiple units. From Hastings the conductor rails would extend into former SECR territory, Ore being the first station after Hastings on the line to Ashford.

Despite the success of the Brighton electrification, Sir Herbert had to use all his skills of persuasion to convince the directors to sanction the scheme. On two occasions the unconvinced board deferred a decision until the next meeting.[20] Raworth and Sir Herbert had to revisit the cost estimates before the directors sanctioned the scheme.

This extension to the Southern Electric was from Keymer Junction on the former

LBSCR Brighton main line, to Lewes where the branch to Newhaven and Seaford was also to be electrified as was the direct line from Lewes to Brighton. The conductor rails continued down the main line to Willington Junction near Polegate where the route divided at a triangular junction. One arm of the triangle continued to Eastbourne, the other along the coast to Hastings. All electric trains ran to Eastbourne and then reversed to reach Hastings and Ore. The conductor rails were installed on the northern arm of the triangular junction, but today the track has been lifted and non-railway redevelopment has taken place. (See plate 14)

The budget finally agreed with the board was £1,750,000. The electricity distribution for the scheme was like that for the Brighton electrification with two more Grid connections made – at Eastbourne and at Hastings. The 16 new substations on the 33,000V network followed the same general design as those built for the Brighton electrification. An additional feeder cable with its associated circuit breaker was installed at the Keymer Junction substation connecting the new scheme to the existing network. At Lewes substation two extra circuit breakers were provided for the 33,000V cables to Falmer substation (on the Lewes to Brighton line) and to Southlease and Newhaven substations on the Seaford branch. At Newhaven substation – because it was, and always would be, at the end of the electricity distribution network – two transformer rectifier units were installed.[21] The only substation on the Lewes to Brighton line at Falmer was a 33,000V dead-end, but the conductor rails provided interconnection to the Brighton substation via a track paralleling hut at Roedale. Most of the AESA supervisory equipment for the Eastbourne extension had been installed at the Three Bridges control centre in advance of this scheme, but five substations at the Hastings end of the route were to be controlled from the Southern Railway's second new control centre at Ore. Space might have been found at Three Bridges for five more panels, but Walker, Raworth and Cox already had their minds set on another big electrification scheme: London to Hastings using the old SER route through Sevenoaks and Tonbridge when it was intended that more panels would be installed at the newer control room.

The electrification was, as usual, to be the catalyst for improvements along the route. At Lewes, the curving tunnel was realigned to accommodate the longer carriages of the new electric trains and this was achieved by widening the bore by 3 feet (0.9 metres) for 230 feet (71 metres). The curves through the station were also eased and the platforms were extended by adding 200 feet (62 metres) to their length. South of Lewes the railway crosses the navigable River Ouse on a bridge which had a central lifting section to allow the passage of tall vessels using the river. The conductor rail could be interrupted for such a short distance, but obviously this was not possible for the cables. It was decided to install the power cable, pilot cable and the pair of direct current 'jumper' cables in ducts laid directly onto the riverbed. To achieve this, it was necessary for the river to be Damned in sections and the water pumped out to lay the pipes, which were set in concrete, ready to install the cables.[22] At Eastbourne, the four platforms were lengthened to accommodate the longer trains. Signalling and track alterations were carried out at several locations and electric lighting was installed at 10 stations. At Ore a new four-road carriage shed was built to accommodate the electric trains.[23]

For the Eastbourne electrification, 16 more six-car sets were built to the same high standard of interior decoration and passenger comfort as before, but unlike the original 'Brighton' sets the Pullman car was substituted for a buffet car with a small pantry. While the Southern Railway referred to these carriages as 'buffets' the later, and better, definition as 'pantry' cars was more accurate as there were no sales counters, just two serving hatches. The rest of the carriage length was made up of five first class compartments. A space was provided to store portable tables to

Schools class 4-4-0 No. 30909 *St Paul's* at St Leonards West in the 1950s. In the background is the St Leonards West substation; 30909 is on the former LBSCR line and is probably not 'wrong line' working as it appears, but at the rear of an empty train from St Leonards depot to Hastings, from where it will haul the train to London Bridge. The former SER line to London is behind the substation which displays evidence of a spare bay to accommodate the proposed Hastings electrification on the alternative route to the capital. (R Caroll Collection)

enable the Pullman Car Company catering staff to provide refreshments in these compartments. The other trailer vehicles were two corridor thirds and a corridor first. The motor coaches at either end were open saloons as per the Brighton units, but the two power bogies had modified springing to reduce the rough riding of the earlier units. These trains were designed to run at a maximum speed of 90mph, but speed restrictions on the routes would curtail them to 75mph. Again, each six-car train had inter-carriage 'corridor' connections within the set but no access between the six-car sets. The corridor connections were of a new and improved type, having a larger cross section area providing easier through access for the portlier passengers.[24]

In 1935 more two-car sets were built which, unlike the 2-NOLs, were a genuine new build for the stopping and semi-fast services on the extended main line routes. These were designated 2-BIL and were formed of a driving motor third class coach with a guard's compartment and a first and third class trailer with a driving compartment. The 'BIL' referred to 'bi-lavatory' as there was a lavatory provided at one end of each carriage to which access was possible via the side corridor – but there was no corridor connection between the two vehicles in the set. The interiors were to the same standard as the new express main line stock. The motor coach had seven third class compartments seating a total of 56 passengers and the trailer had four third-class compartments for 32 passengers with 24 first class passengers in the four first class compartments. These new two-car trains were made up into two-, four-, six- or eight-car formations as required. One hundred and forty 2-BILs were to be built between 1935 and 1938 for use across the Southern Electric network.

The full electric service to Eastbourne and Ore began on 7 July 1935 following opening celebrations on 4 July with dignitaries from the City of London and from the principal new destinations of Eastbourne, Bexhill and Hastings in attendance.[25] Eastbourne and Ore business travellers were not to be deprived of their full Pullman comforts as the rosters for the 6-PUL sets were rearranged so that they

2-BIL unit No.2045 at Lewes takes the line to London in 1961. Behind the train Lewes substation can just be seen. This was a larger than usual substation having extra bays to connect the dead-ended 33,000V cables to Falmer and Newhaven. (Rail online)

were to be used on these new services as well as the original Brighton and Worthing routes. The new timetable provided an hourly fast service from Victoria with additional trains from London Bridge at peak times. The operational flexibility of the electric trains was demonstrated by the ease with which they could be divided to serve different destinations, such as the splitting of trains at Haywards Heath creating separate portions running to West Worthing and to Ore via Eastbourne. Every half hour, stopping trains ran from Brighton to Seaford and from Brighton to Eastbourne and Ore.[26] The trains from London to Ore all ran via Eastbourne – demonstrating the ease of train reversal, this only requiring the driver to walk the length of the train, step into what had been the rear cab and continue the journey.

The dividing and combining of the new multiple unit electric trains was generally less hazardous than the marshalling of the unpowered two-car trailer units – but not without incidents. For example earlier, on 8 November 1934 at Haywards Heath, one of the six-car units travelling from West Worthing collided with a stationary six-car unit which had arrived from Brighton. The driver of the West Worthing train had correctly stopped at the 'home' signal protecting the Brighton train before obeying the subsidiary 'calling-on' signal which permitted moving forward subject to extreme caution. Unfortunately, the motorman misjudged his speed and failed to stop the train when attempting to brake on a very greasy rail. Some passengers, mainly those who were standing and waiting to alight at Haywards Heath, received minor cuts and bruises. There was some damage to the ex-Brighton portion requiring the transfer of all passengers to another train.[27]

There was a curious late addition to the scheme, not mentioned in the board minutes at the time of the approval or in any publicity material used to announce the scheme. This was the electrification of the branch line between Haywards Heath and Horsted Keynes. Horsted Keynes station was a junction on the former LBSCR East Grinstead to Lewes line. The electrification was an inexpensive addition to the main project, as it was possible to electrify the 4-mile-long railway by simply laying out the conductor rail and bonding the running rails at the rail joints – no extra 33,000V cabling or rectifier substations were required. Rather unconvincingly perhaps, the official explanation for the electrification given later was that the service from Seaford was taken on to Horsted Keynes to avoid congestion at Haywards Heath.[28]

The only intermediate station on this line was at Ardingly, and three miles to the north of this tiny village is Wakehurst Place, which was at the time the home of the Southern Railway chairman, Gerald Loder. He was a keen gardener, a one-time president of both the Royal Arboriculture Society and the Royal Horticultural Society. Wakehurst Place planted out by Loder is now a National Trust property which is leased to the Royal Botanical Gardens based at Kew.[29] It is no surprise therefore that there was some critical comment. Journalists, behaving as predicted by John Elliot, were making up a story because they had no reliable information. They intimated that this minor line was electrified purely for the chairman's benefit. However, if Loder or his guests were to travel to Wakehurst Place, they would most likely, as did Southern Railway director Leo Amery in May 1932,[30] travel to Haywards Heath by train and then be chauffeur-driven to Wakehurst Place – a drive of just a little longer duration than that from Ardingly station.

On Wednesday, 19 June 1935 the conductor rails from Copyhold Junction on the main line to Horsted Keynes were energised. The live conductor rail was also extended for a further 100 feet (30.5 metres) towards East Grinstead to allow for shunting movements.[31] An hourly train from Horsted Keynes would connect at Haywards Heath with a Victoria to Brighton fast train allowing passengers on the branch train onto a fast service to Brighton; then the train would wait in the station for the Victoria to Ore train – which would follow to Lewes before branching off to Seaford, thereby providing passengers on the Victoria to Ore train with a service to Seaford and all the intermediate stations.[32] By then however, Gerald Loder, ennobled as Lord Wakehurst, had resigned as chairman in December 1934 due to ill health. Any benefit he might have had from the improved service to Ardingly was short lived as he was to die in April 1936. Traffic on the branch was never high, indeed between 27 September 1936 and 3 January 1937, the branch Sunday service was discontinued – a most un-Southern Electric occurrence.[33]

Did Gerald Loder press for the electrification of the line to boost his standing with the local community? As Southern Railway chairman he may have had some limited discretion to make day-to-day decisions. If he had a personal wish for the extra electrification, he needed a good reason to sanction the extra work. Edwin Cox would have argued that the service would allow better utilisation of the new electric multiple units by integrating the branch services with the main line services. But Walker, Raworth and Cox had a hidden agenda not to be shared with their sceptical directors. A distinct pattern was emerging: this trio, in a similar fashion to a skilled professional snooker player when building up to a high-scoring break, were not just concentrating on the present scheme, but on what was to come next, and what was to follow that. Events were to show that the Horsted Keynes branch electrification was part of a much longer-term development strategy.

Further evidence of the railway's officers always looking beyond any current project came in December 1933 when the Southern Railway announced that it would immediately commence work on an additional 23 miles of electrification to Sevenoaks.[34] This would extend the range of the suburban Southern Electric services and rather than extend the 11,000V supply – which should have been possible – two new Grid supplies were obtained from locations some distance away from the routes to be

Between August 1960 and October 1963 it was possible to see Southern Region electric trains and Bluebell Railway steam trains together at Horsted Keynes. At its opening, the Bluebell Railway did not use the Southern Region station, but soon afterwards it was possible to witness a scene such as this. On the left a Southern Electric train – pair of 2-BILs – and on the right a Bluebell Railway train with one of its diminutive P class 0-6-0s. (Southern Railways Group/John Sutton Collection)

electrified. Only seven rectifier substations were built for this scheme,[35] and – as a further indication of future expansion – these substations were controlled from a third new control room built at Swanley. The northern Grid intake was at Northfleet, and in the south at Tunbridge Wells. To reach Northfleet Grid the 33,000V cable was laid along the then non-electrified former LCDR main line and Gravesend West branch. A switching station was built at Swanley comprising of three 33,000V circuit breakers in the usual outdoor compound but without a transformer rectifier – this being substituted for the cable from Northfleet. To reach Tunbridge Wells Grid from the last substation at Sevenoaks, a 33,000V cable was laid for six miles alongside the non-electrified former SER main line to Tonbridge. At Tonbridge, the cable route crossed what had been the original SER main line to Folkestone and then continued down the SER Hastings main line for a further three miles to the Grid substation which conveniently has a common boundary with the railway a quarter of a mile south of High Brooms station. The fact that so many of the early Grid substations were adjacent to railway lines cannot be coincidental giving Lord Weir's desire for the railways to be a major user of his 'gridiron' concept.

Two routes to the Kentish towns received the conductor rails: an extension on the old SER main line from Orpington and an extension on the original LCDR main line from Bickley to Swanley and then down the branch to reach Sevenoaks via Otford. Also included for electrification was the spur installed by the SECR to

MAINTAINING MOMENTUM • 211

2-NOL No. 1843 awaits passengers in the bucolic calm of Horsted Keynes station. Note that notwithstanding the initial SR instructions to staff, by the time this photograph had been taken the line to Haywards Heath had been partially singled, and only one line in the station had the conductor rail. (Meredith/Transport Treasury)

interconnect the old SECR and LCDR systems between Chislehurst and St Mary Cray. The primary intention may have been suburban electrification, but this new scheme resulted in electrification creeping a little further down the former SER and LCDR main lines. It might have been expected that for a modest expenditure on conductor rail and rail bonding, the branch line to Westerham could have been served by electric trains as was Horsted Keynes. But the Westerham branch was a dead-end, without any potential to be used to further extend the Southern Electric network.

The Southern Railway management was very keen for the electric services to run to St Mary Cray in advance of the completion of the Sevenoaks scheme, to take advantage of the rapid expansion of housing in the locality.[36] Also, with the onset of summer, additional passengers were anticipated, visiting a new lagoon swimming pool nearby.[37] Services duly commenced on 1 May 1934. There were three trains an hour in each direction on weekdays and two every hour in each direction on Sundays.[38] Following the full electrification an improved timetable was introduced – for example, on weekdays the number of trains to St Mary Cray increased from 28 every day to 54 and trains to Otford increased from 20 to 54 a day. The service to Sevenoaks via Swanley was an extension of the existing service from Holborn Viaduct to Shortlands via Catford and the Charing Cross to Orpington service was extended to Sevenoaks. These trains stopped at every station along their route. To operate these extended services, 11 more three-car electric sets were provided. These 3-SUBs were not new, but, for the last time, there was a further rebuilding of old steam carriages onto new underframes. To supplement the 3-SUBs, four more two-car unpowered trailer sets were constructed. The trains normally consisted of a single three-car set, but at rush hour periods they were made up to eight coaches by using two three-car sets with one of the unpowered trailer sets in the centre of the formation.[39] These trains would be divided at Swanley or Orpington with only a single three-car set continuing to Sevenoaks.[40]

The 33,000V switching station at Swanley was a temporary arrangement as it was due to be relocated as part of the imminent rebuilding of Swanley station. Disaster struck when the switchgear was damaged due to a train crash. This incident followed close on the heels of another fatal crash on 2 April 1937 when a Southern Electric suburban train, after crossing the bridge over the LSWR main line on its way to Victoria, collided with the rear of another electric train which was stationary just outside Battersea Park station. The accident was duly investigated by the chief inspector of railways, Lieutenant Colonel Mount, who concluded that the accident was due to errors by a relief signalman. The violent collision caused the steel underframe of one of the wrecked carriages to strike the live conductor rail, resulting in the tripping of the direct current circuit breakers thereby safely disconnecting the electricity. There were 10 fatalities in the crushed wooden carriage bodies as the vehicles telescoped into each other. Colonel Mount suggested that there would have been significantly less serious injuries had carriages with rigid, all-steel, bodies been in use – thereby bringing into question the whole Southern Railway policy of constructing 'new' electric trains using ancient wooden bodies placed upon modern underframes. The chief mechanical engineer, Richard Maunsell, when giving his evidence to Colonel Mount, could only limply opine that as the accident had occurred at high level on a viaduct, 'rigid' bodies may have remained intact but possibly would have toppled down to ground level below.[41]

A few weeks later the Southern Railway, still reeling from this horror and the resultant bad publicity that ensued, suffered a second tragedy at Swanley on 28 June. The steam-hauled 8.17pm from Margate on its way to London via Ashford and Maidstone had joined the former LCDR Sevenoaks branch just south of Otford to proceed north along the recently electrified railway to join the LCDR main line at Swanley. The train of seven SECR 'birdcage' coaches was hauled by L class 4-4-0 No. 1768. This train was not scheduled to stop at Swanley but, due to the late running of another train, about 80 passengers were standing at Swanley station having missed the last connection of the day to London. On his own initiative, the station master decided to stop the train at Swanley to pick up the stranded passengers, but his decision could not be passed on to the engine crew. The semaphore 'starter' signal at the London end of the 'branch' platform was duly set to danger, but at 11.18pm the train ran past the signal – which was not easily 'sighted' by the driver – at considerable speed. Not anticipating that he would have to stop at Swanley or seeing that the junction 'distant' signal was at caution (despite the warning call from his fireman across the noisy cab), the driver had inadvertently ignored the signal with disastrous results.

The interlocking arrangement at the junction was such that with the starter signal at danger, a set of points just past

Map of the Sevenoaks electrification scheme. (Author)

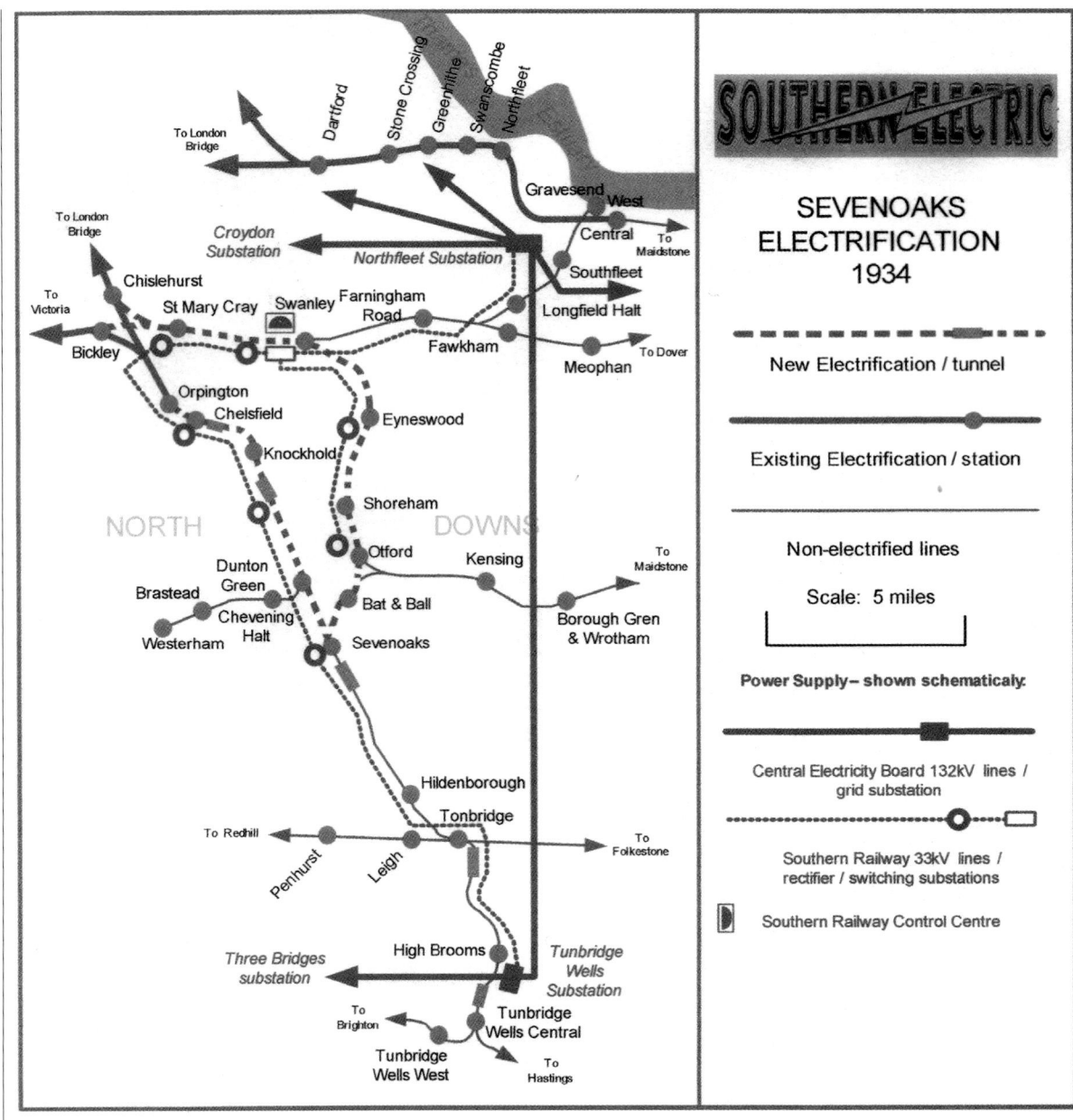

the signal was set to turn any overshooting train into a siding to avoid fouling the main line. That evening 'standing on the siding' was a loaded open goods wagon and one of the two-car unpowered trailer sets. Beyond the siding's buffer stop, and fully in the line of fire, was Raworth's Swanley switching station. The heavy steam locomotive, after destroying the wagon and the two-coach set, was propelled into the palisade fencing, wrecking the heavy reinforced concrete post and transom structure supporting the 33,000V cable feeding the two substations on the line towards Otford. Amazingly the driver and fireman escaped without injury but were severely shaken. Four of their passengers perished in the wreckage of the carriages behind the locomotive, which was left turned upon its side above the remains of the vehicles that had been standing in the siding.

The massive impact on the cable resulted in the immediate tripping of the circuit breaker. But the operation of this circuit breaker would not have disconnected the direct current feeds to the conductor rails at the scene of the accident. Nonetheless the Southern Railway did not contradict the press reports which gave the impression that all electrical equipment was disconnected when the crash occurred. The conductor rails remained alive, energised from the first substation in the other direction – that is towards St Mary Cray. A different revised account was given in Colonel Mount's report. He had been told that it was only due to an urgent telephone call to the nearby Swanley control room that by using the ASEA control equipment the electricity supply was fully disconnected at 11.31pm – 13 minutes after the accident.

As soon as it was safe to do so, the conductor rails were energised. Alfred Raworth had designed his electricity distribution system in such a way that one substation could be switched out without detriment to the running of trains – but not two. Consequently, due to the excessive voltage drop along the conductor rails, for the next three days the electric trains were forced to run at a much-reduced speed. There was a further complication as the track circuiting near Otford was put out of action due to the loss of the alternating current supply. Following the removal of all the wreckage from the substation, a temporary cabled connection was installed on 2 July.[42] The lack of an alternative supply for the two substations might have been a piece of bad design by Raworth but, as usual, this arrangement was only intended to be temporary. The completion of the Sevenoaks electrification was, as for all the schemes during this era, a jumping off point for the next scheme. In 1939 this 'dead-end' would be connected back to Northfleet Grid as part of the Maidstone electrification.

Until 1936 the Southern Railway, rather irrationally, had to purchase its electricity from the local electricity companies despite being supplied directly from the CEB National Grid. This was due to the protection given to the local supply authorities by the 1919 legislation that set up the Electricity Commissioners and the Joint Electricity Authorities. The Southern Electric received its electricity from the Brighton and Croydon Corporations and the West Kent Company finally was to bill the railway for a traction supply at Tunbridge Wells substation. With the passing of the Electricity Supply Act of 1935 the Southern Railway could negotiate a much better agreement directly with the CEB. After a temporary six-month agreement and following some very tough negotiations, a 17-year contract was agreed for the Southern Railway to take supply from seven Grid substations. The tariff was made up of a service charge for each site, other than for the first connection, of £2,500; a maximum demand charge and then a unit kilowatt-hour charge. With all these costs put together the railway expected to pay less than half an old penny per kilowatt hour of electricity consumed (about 0.18 pence). A further advantage of buying directly from the CEB was that it had powers to lay cables in the public highway, thus it would be easier to obtain supplies from Grid substations not directly adjacent to the railway.[43]

The suburban electrification might be deemed complete by the mid-1930s but one addition to the Southern Electric network was the electrification of the 2-mile link between Lewisham and Nunhead. The original former LCDR Nunhead to Greenwich Park branch had been included in Stage One of Raworth's 1919 proposals despite the branch being closed in July 1917 as a wartime economy. The Greenwich Park branch was never reopened, but at the point where it crossed above the SER line close to St Johns station, in 1929, the Southern Railway had built a connecting line facing Lewisham allowing goods trains from Hither Green to leave the SER line at Lewisham and to run through Nunhead and Loughborough Junction and then make their way northwards via the LCDR's cross-city line to destinations in LMS or LNER territory. The line towards Greenwich Park was abandoned. In 1934, following electrification of this new link, some trains that had previously used London Bridge were switched to Holborn Viaduct.[44]

CHAPTER 18

TWO ELECTRIC ROUTES TO PORTSMOUTH

The Southern Electric continued to be a financial success. Following the electrification of the lines to Eastbourne and Ore the passenger numbers rose by 22 per cent and receipts rose by 41 per cent. There were even better returns after the Sevenoaks electrification when passenger numbers increased by 55 per cent and receipts by 41 per cent.[1] While the Southern Electric part of the railway was performing well, due to the general economic climate and the overall financial health of the railway, the raising of capital for further extensions became problematic. Nonetheless, Sir Herbert Walker was able to convince his directors to commit to electrification on the former LSWR Portsmouth Direct line.

If the Brighton electrification scheme was deemed to be brave by many commentators, the Portsmouth Direct scheme must have seemed heroic if not foolhardy. The line to Brighton with its express services had the look of a main line railway but its traffic pattern – as observed by Lord Weir – was that of a suburban commuter railway. The Portsmouth Direct was undoubtedly a main line and exhibited similar characteristics to many such lines in Great Britain. It linked the capital with an important provincial centre, with miles of open country along the route. There was no commuter traffic other than at the London end of the line and the traffic peaks were seasonal not daily. But by 1935, Sir Herbert had moved away from his original view, once shared with others such as Sir John Aspinall, that the economic driver for any electrification proposal was that passengers who had deserted the railways would be enticed back by improved electric services and the additional revenue thus generated would cover the capital cost of the schemes. Walker now believed that on many routes the potential operational savings alone justified electrification. This view went against contemporary opinion as most in the railway industry continued to view the capital cost of electrification as being too expensive. Nonetheless, the directors, taking Sir Herbert's advice, approved his scheme to extend the LSWR electrification from Hampton Court Junction, through Woking and Guildford to Portsmouth and, as a bonus, to electrify the line from Woking via Pirbright Junction to Farnham in Surrey.[2] No immediate public announcement was made of this proposal. When it was later made public a majority of those in the railway industry were to greet the Portsmouth electrification proposals with negative comments, noting the lack of local and commuter traffic along the line and the enormous difference between the holiday traffic to the Solent area compared to the off-season traffic. Sir Herbert's critics were once more to be proven wrong, for 10 years later, when peace returned in 1947, 7.5 million passenger journeys were recorded compared to 3 million in 1936.[3]

In response to the effects of the Great Depression, the 'Third National Ministry' was formed in 1935 with the Conservative Stanley Baldwin as prime minister. To stimulate economic activity and reduce unemployment the new government established the Railway Finance Corporation. Under this arrangement, the railway companies were to submit to Parliament details of their large capital expenditure schemes and subject to approval the railways could borrow the necessary capital from the Railway Finance Corporation which had a £26 million pot of money available for suitable schemes.[4]

Obviously, the Southern Railway wanted a slice of this cake for its electrification projects. But for what was undeniably a 'main line' scheme, it feared that the Ministry of Transport would insist on the use of its preferred 1,500V scheme, an electrification system denigrated by Alfred Raworth and Herbert Jones.

The original secrecy of the Southern Railway about the Portsmouth Direct electrification might seem strange, but Sir Herbert Walker and his directors were again taking the stance that it is best not to tell the government anything until it is a *fait accompli* – just as had been done in 1912 prior to the LSWR electrification, thereby avoiding the need to seek approval or scrutiny from the Board of Trade. With £3 million of expenditure approved for the Portsmouth electrification, contracts for equipment signed, the newspapers buzzing with rumours and no official notification to the Ministry of Transport, Alfred Raworth made a visit to what must have seemed like the lion's den for an uncomfortable two-hour confidential updating meeting with Ministry of Transport civil servants on 11 July 1935.

Raworth met the chief inspector of railways, Lieutenant Colonel Mount, who was accompanied by Lieutenant Colonel Trench at the Ministry of Transport's offices.[5] The colonels gave the impression that it had been Alfred Raworth on his own initiative who had made the appointment to unofficially meet with them. That Raworth would have done anything so politically dangerous without Sir Herbert's knowledge is inconceivable, but such a ruse would allow anything that he was to say, or anything said that was misconstrued, to be later denied. Possibly Raworth had suggested to his general manager that a face-to-face meeting with the Ministry might clarify many issues without recourse to a lengthy and convoluted exchange of correspondence. Nonetheless, at times during this meeting Raworth had to use all his 'stonewalling' powers, at which his master, Sir Herbert, was so adept.

Colonel Mount went immediately on the offensive by reminding Raworth that the Southern Railway had been urged to increase its conductor rail voltage to the standard 750V but had not so far done so. Raworth replied that he believed that there was no advantage in increasing the voltage above the existing 660V for each of the small incremental extensions. The question that most concerned the colonels however was the rumoured Portsmouth electrification. Raworth 'confidentially' informed them that the Southern Railway directors had approved the spending for this scheme and that various contracts had been signed.[6] Raworth said that following the board's approval he had promised his general manager that he would complete the scheme by July 1937 – only two years hence. This immediately raised the question as to the system of electrification to be used – was it to be the 1,500V overhead electrification in accordance with government policy? Raworth divulged that the 1,500V system was not going to be used and added that when the Brighton scheme was in the early planning stages, consideration had been given to doubling the third-rail voltage or using the 1,500V overhead electrification, but these had been discarded for cost and technical reasons. He affirmed that the Southern Railway was confident that the third-rail system then in use was the best system for its requirements, as costs were low and failures few.

Colonel Mount wanted to know how much a 1,500V overhead scheme to Portsmouth would cost. Raworth said that he did not know – it would take a year to carry out a design and feasibility study to produce an estimate for a 1,500V scheme and the railway needed the electrification in two years' time. This caused the colonels to become slightly annoyed: how could the railway discard a nationally approved scheme without even knowing how much it would cost? Alfred Raworth of course had some insight into what the unit cost per mile of both systems would be. In 1933 the chief electrical engineer of the LMS had read a paper to the Institution of Electrical Engineers about the electrification of the Manchester, South Junction and Altrincham Railway, when he divulged that the cost to electrify using 1,500V overhead wires was £3,370 per single track mile – about twice Raworth's costs.[7] These LMS costs – for what was effectively a trial installation – would later be shown to be excessive, but they were the best that were available at the time. Without a feasibility study Raworth could not quantify what the additional costs would be to modify bridges, platform canopies and signals to accommodate the overhead structures. Furthermore there was the nightmarish worst-case scenario of having to increase the size of the tunnels through the South Downs at Buriton or through the North Downs at Guildford where there was chalk tunnel and sand tunnel. Also, considerably more 1,500V electrification would be required to complete the scheme from Waterloo to Portsmouth – not just Hampton Court Junction to Portsmouth.

Raworth was asked about potential savings from having fewer substations if the 1,500V system was to be used. The view of Raworth and Jones was that at *only* 1,500V the smaller section overhead wire, while only needing to carry slightly less than half the current for the same power delivery compared to a 660V conductor rail, was still too small to carry sufficient current for a busy railway. Without offering a complete explanation Raworth opined that a reduction in substation costs ' … would be more than counterbalanced by increased costs of other equipment'. When quizzed about the possibility of any future electrification schemes such as to Basingstoke and then on to Southampton, Raworth stated that such an extension was probable. Less likely however, in his opinion, was the extension into minor lines in Kent. Interestingly he made no comment about infilling schemes within the London–Portsmouth–Hastings 'triangle'.

Raworth was asked if the Southern Railway had considered dual operation of some of its routes, using both its direct current third-rail and the 1,500V overhead system. This Raworth objected to, insisting that it was too expensive to be seriously considered, but this was at odds with Herbert Jones' evidence given to Sir John Pringle's 1923 technical conference.

The colonels now turned to the question of electric locomotives. If electrification was to be spread across the country they argued, locomotives, not multiple unit trains, would be required to haul the freight and parcels traffic as well as the long-distance passenger trains. The current view of some railway professionals was that goods traffic was so important that any proposal to electrify main lines should not be solely for passenger trains. Therefore, they argued, electric locomotives must be used and the idea that only multiple unit trains be used on electrified routes was condemned, and furthermore while the Southern Railway operated the greatest suburban electrification in the world, it had so far been unable to build a single electric locomotive.[8] Many, including Lieutenant Colonel Mount, believed that unlike the 1,500V overhead system it was not possible to build a practical electric locomotive for use on the third-rail electrification.

Raworth argued that he was preparing, with others, a practical design for a third-rail electric locomotive. He reminded the colonels that it would cost more than any steam locomotive, thereby giving the clear hint that there was no reason at the present time why the cost conscious Southern Railway should not use steam locomotives to haul the relatively infrequent goods trains over its mainly passenger network. He revealed that when available the new electric locomotives would be used on the Newhaven boat trains running up to 65mph and for the longer distance freight trains. To protect staff at the goods yards the Southern would build short sections of 660V overhead lines for his proposed locomotives.

Colonel Trench asked why the Southern Railway was out of step with the rest of the engineering community. How could it be that the 1,500V system was at the time being adopted in so many overseas countries? Trench cited successful schemes in France, Holland, South Africa and Australia. Raworth refused to offer any opinion on the technical merits of these foreign schemes for which he had no intimate knowledge but he did suggest that in these countries where there was more open space it was safer to use overhead lines as the railways were not fenced off. This led on to the issue of safety and maintenance. There had been many injuries and fatalities resulting from the live conductor rails, mainly to railway staff, but occasionally to members of the public. Raworth conceded that track maintenance was indeed more convenient with an overhead system but not less expensive. At this remark, Colonel Mount interjected by reminding him that the Southern Railway had to pay track workers extra 'danger money' payments, which would not have to be paid if the live equipment were safely up above the tracks.

Raworth was then reminded about the 3,000/1,500V three-wire system that he had proposed for the SECR. Following the company line, he stated that the Southern Railway had dutifully followed the recommendations of George Gibbs' report and not used 'his' system – but fortunately the colonels did not quiz him about Gibbs' other recommendations concerning the LBSCR 6,700V system and the use of 750V for the third-rail which were *not* followed.

Despite their interrogation the two colonels seem to have understood that the Southern Railway could only build on what it had, so Raworth was asked what he would propose if the Southern Railway board were to give him a completely clean sheet. Loyally he stated that the third-rail system was best for the Southern Railway's present needs but imagining a completely different railway he probably surprised the colonels by advocating a single-phase alternating current system running at a much higher voltage than 1,500V. He was possibly contemplating a 20,000V system as was then being trialled in the German Black Forest on the Hollantalbahn line from Freiburg to Seebrugg via Titisee.

The meeting closed cordially with Colonel Mount thanking Raworth for his assistance but firing a parting shot by rebuking him again for not being willing or able to produce any cost estimate for electrification to Portsmouth using the 1,500V overhead system.

Mount had to advise the Minister of Transport on three aspects of the proposed Portsmouth electrification. First, should the Ministry interfere with the railway's engineering decision as to how to carry out its Portsmouth electrification; second, should the Ministry take a view on the safety of the third-rail system; and finally, if the scheme was not to the national standard or deemed unsafe should it, or any further extensions, receive financial support through the Railway Finance Corporation? Mount was to base his advice on his informal interview with Raworth and with other more formal written correspondence, covering much the same issues, with Sir Herbert Walker.

Mount was not in favour of the third-rail system. Others at the Ministry of Transport concurred, arguing that while the minister had no powers to veto the scheme (due to the rather limp words in the Standardisation Order), on grounds of general safety and setting an example, he should be reluctant to recommend a system which would inevitably result in an increase in casualties as the third-rail system was extended.[9] The Southern Railway had some friends at the Ministry of Transport however, as another civil servant advised the minister that it would be unwise, uneconomic and unjustified for the railway to depart from the third-rail system, using multiple unit trains for its incremental extensions. He identified the railway's methodology as a 'creeping process'. The same civil servant admired the company's courage and noted that the Portsmouth extension was to be economically difficult, thereby making government assistance necessary; however he hoped that electrification of the West of England line would be at the higher voltage. This memo had a hand-written postscript advising that any approval or government blessing of the third-rail extension to Portsmouth might expose the Ministry of Transport to criticism in the engineering world.[10]

The transport minister in the latest coalition government was Leslie Hore-Belisha, a National Liberal who had achieved immortality for the Belisha beacons he introduced on pedestrian crossings. He was passionate about safety and was appalled at the road casualty figures. During the previous year, 1934, there had been a record number of casualties on the public roads: 7,343 deaths and 231,603 injuries.[11] (This may be compared with the 2014 figures of 1,760 deaths and 191,530 injuries with considerably more traffic on the roads.) Casualties due to accidental contact with railway live conductor rails in the previous 10 years had been a total of 33 fatalities and 343 injuries,[12] most being railway workers, but also some trespassers. These statistics, awful as they appear to our modern sensibilities, must be seen in the context of the contemporary national railway workers' casualty rates, which were typically between two and three hundred fatalities and several thousand injuries per year.[13] One civil servant when offering his advice to Hore-Belisha expressed the view that the casualty rate attributed to accidents due to the live third-rail might be insignificant when compared to road traffic casualties, but it was a bad argument to compare the countrywide road traffic statistics with the limited application of electrified railways.[14] However, Hore-Belisha reasoned that railways were overall a safer mode of transport and anything that encouraged the public to travel by train should be promoted.

The permanent under-secretary to the minister of transport, Cyril Hurcomb, was instructed to write to the Southern Railway in September 1935 to give the Ministry of Transport's official response. In his letter he stated that even in the absence of any comparative approximate estimate of the additional cost of using the 1,500V overhead scheme, Hore-Belisha was satisfied that there would be no advantage in using the higher voltage system. He appreciated the Southern Railway's arguments in favour of a third-rail system, but also appreciated that there were others who disagreed. He was concerned, and would continue to worry, about the safety of the third-rail system, particularly in rural areas. He also noted the Southern Railway's continued use of multiple unit trains and hoped that it would soon be successful in developing a practical electric locomotive for the third-rail system. Finally, Hore-Belisha inferred that the Southern Railway's 'creeping process' must not in future be used to justify further third-rail electrification outside the London–Portsmouth–Hastings 'triangle'. Beyond this zone, he argued, the higher voltage system must be used.[15]

This was good news for the Southern Railway: it could now proceed with the Portsmouth electrification and apply for government financial assistance. Officials, despite misgivings, must have understood that the Southern Railway's choice was not between its third-rail and government approved electrification – it was a choice between third-rail electrification or no electrification at all. Furthermore, in the difficult political climate, the Portsmouth electrification was a welcome economic activity when the nation was still recovering from the Great Depression.

In November 1935, a formal agreement was made between the Railway Finance Corporation, the Treasury and the Big Four railway companies. The four railways were represented by their chairmen and general managers at talks with the Chancellor of the Exchequer, Neville Chamberlain, and Leslie Hore-Belisha, when it was agreed that lists be prepared of projects that could be carried out within the next five years. The Southern Railway's share of the loan money was a generous portion of between £7 and £8 million.[16]

The entire loan to the Southern Railway was to be spent on electrification. The list of electrification schemes sent to the Railway Finance Corporation included the Portsmouth Direct scheme, to which, with the new finance available, was added an extension from Farnham to Alton and the line between Weybridge and Staines. The other proposed schemes on the list were the LBSCR route to Portsmouth via Horsham and Three Bridges (completing the electrification between Brighton and Portsmouth along the south coast), due for completion by July 1938; the former SECR main line from Sevenoaks to Hastings and the Bexhill branch, due to be completed by July 1939; electrification to Reading and the branch via Camberley to Aldershot to be completed by January 1939; and electrification to Maidstone by both routes and to Gillingham, due for completion in 1940.[17] Also proposed was a completely new electrified railway line – a loop on the former LSWR New Line – from Motspur Park to Leatherhead. This was a bold and impressive plan that was mostly – but not all – to be completed. Alfred Raworth kept his promise to Sir Herbert Walker to complete the scheme by July 1937, for within 20 months, on 8 March 1937, the trial running of electric trains to Portsmouth began after 42 miles of railway had been electrified. (See plate 15)

The Portsmouth Direct electrification followed the same electricity distribution design as the Brighton and Eastbourne schemes. Grid intake points were established at West Byfleet substation at the northern end of the route and at Wymering substation at Portsmouth. Both of these CEB substations have a common boundary with the railway. West Byfleet substation is adjacent to the LSWR main line and Wymering substation is by the Portsmouth to Fareham line which had been built in 1848 and was not to be electrified as part of this scheme – it had to wait until the 1980s. Between December 1935 and November 1936, 309 miles of single-core 33,000V cable (three separate cables for each three-phase feeder) and 218 miles of pilot (control cable) were installed[18] to electrify 242 miles of track. These cables were laid from wagon-mounted drums – a total of 188 special trains were run during the construction period. The standard Raworth type of substation was used – the 'Portsmouth Direct' substations can still be identified today for the buildings have the brickwork covered with a light grey rendering. The 33,000V cable routes mostly had an alternative Grid supply, but the cable route to Alton was left as a 'loose end' (and was to remain so until the 1980s). Also temporarily without an alternative supply were the two substations on the loop between Weybridge and Staines – but here there was the intention to interconnect with Reading Grid when the Southern Electric reached Reading. At the Portsmouth end of the scheme, the substation at Fratton – a site where two transformer-rectifier units were installed – was provided with its own private ring from Wymering.[19]

Two more new control rooms were built, one at Woking and the other at Havant. Both buildings, while no longer required for 'control', still exist and are visible from the trains if you know where to look. Although of a general art deco design, these buildings lacked the stylish curved tower design of the Three Bridges building. Both buildings had additional space available for the future planned electrification programme. Using the ASEA system the Woking room controlled the substations down the line to Portsmouth as far as Shottermill substation between Haslemere and Liphook, and the Havant room controlled the substations up the line as far as Wheatsheath substation between Liphook and Liss. The Portsmouth Direct scheme was, like the other main line electrifications, an extension of the original suburban electrification using the 11,000V distribution cables from the railway power station at Durnsford Road, which had one of its 5,000kW generators replaced with a 12,500kW unit for the anticipated additional demand between Waterloo and Hampton Court Junction.[20]

Siemens Desiro unit passes the Grid substation at West Byfleet. The close proximity of the substation to the railway chimes with Lord Weir's desire for the railways to power their trains by electricity. This substation has had various names over the years, first Woking, then West Byfleet, and when the author was employed in the electricity supply industry it was known as West Weybridge. The Southern Electric substation is some way down the line, behind the train. (Author)

A Portsmouth Direct substation at Pirbright Junction (where the Alton line leaves the Main Line). The date is 22nd April 1967, and work is in hand in preparation for the extension of the electrification to Bournemouth. (Photo LRT 552)

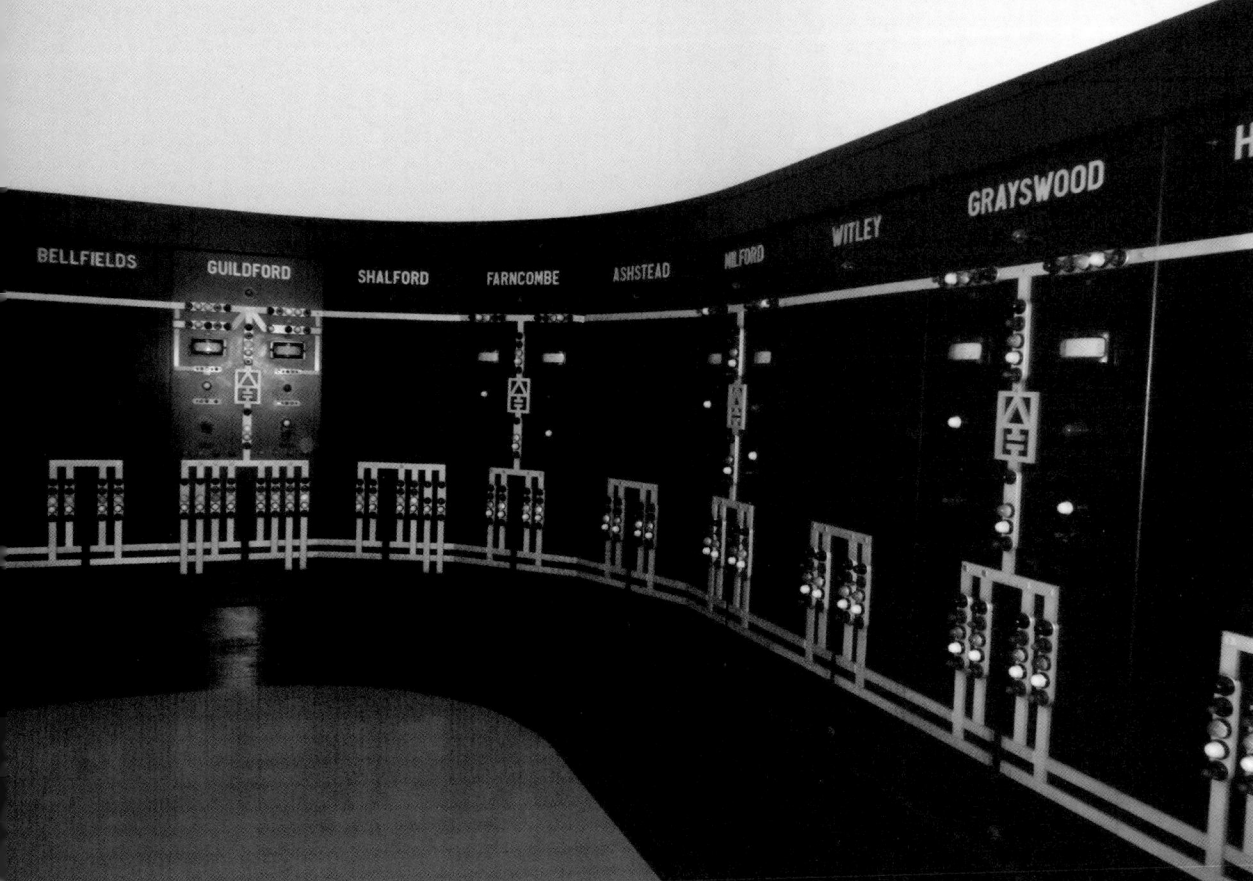

Front (west) side of the Woking control building. (Author)

A view inside the Woking control room showing some of the panels for the Portsmouth Direct electrification. (Author)

Woking control building. (Author)

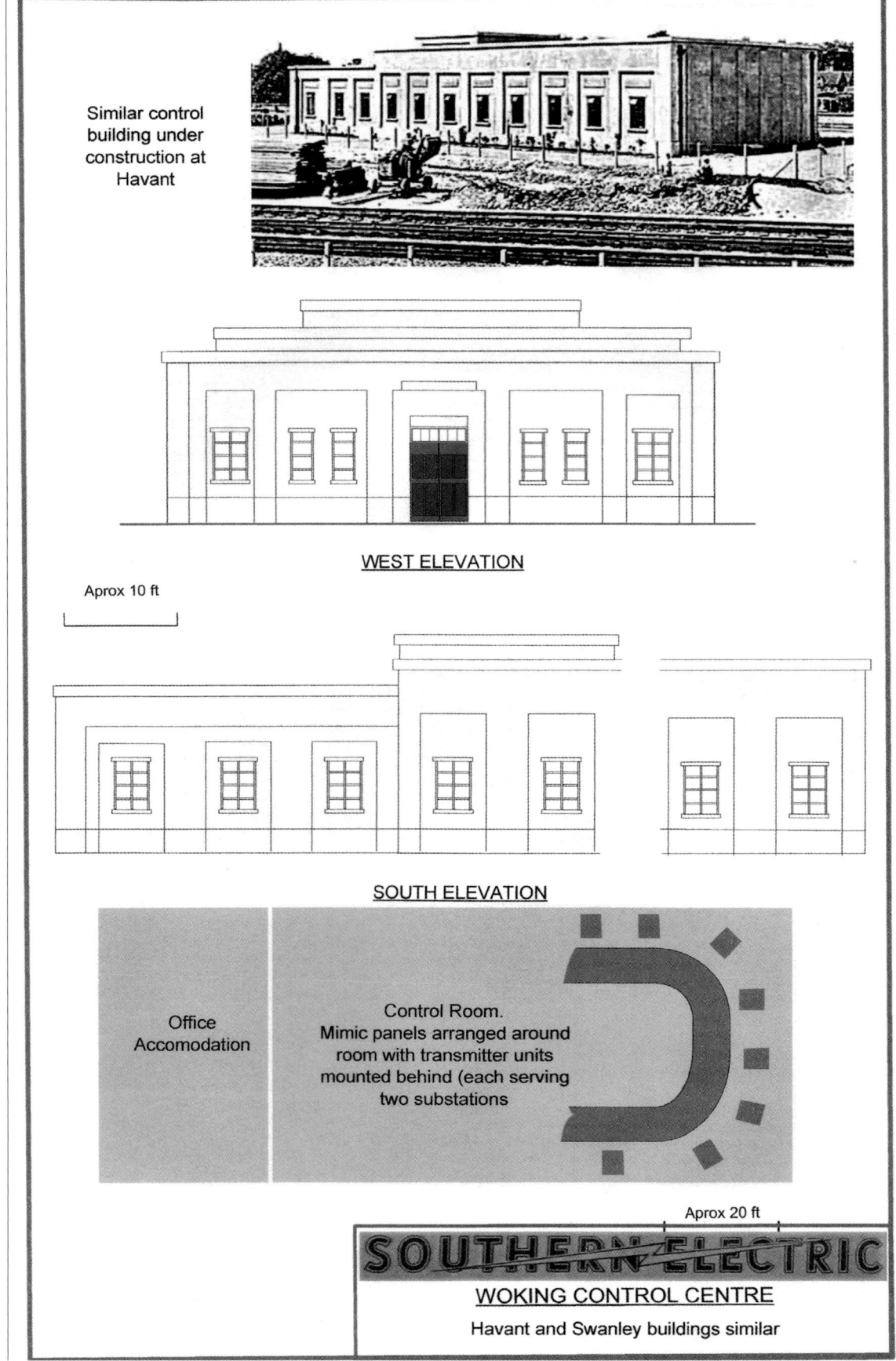

At Guildford, where the distribution system met the old 25 cycles per second 11,000V network, the existing rotary converters and the 11,000V switchgear were removed from the large substation building and a new standard substation installed slightly further down the line towards Guildford station. With the usual Southern Railway frugality, the rotary converters were re-used to reinforce substations in the suburban area and the pair of 11,000V cables from Clanford to Guildford which had formed the last leg of the original power distribution from Durnford Road were sold to the electricity department of Guildford Borough Council for incorporation into its high-voltage network. Since the cables remained on railway land the Southern Railway was able to charge the council an annual rent of £22.[21]

Every month, at the peak of the installation programme, the heavy electrical equipment for three complete substations was unloaded from special trains. Each train consisted of 18 wagons plus a 36-ton steam crane and a brake van. Typically, these trains were hauled to site by a former LBSCR E4 0-6-2 tank engine. The unloading was usually carried out between 11.00pm on a Sunday evening and 4.30am on Monday morning.

Tough procurement negotiations enabled significant savings to be made. For example, a modest price reduction from £5,140 to £5,000 for each of the 24 transformer-rectifiers purchased from Edinburgh-based Bruce Peebles, which was building the mercury arc rectifiers under licence from Brown Boveri. British Thomson-Houston agreed to provide the 217 direct current circuit breakers for installation in the substations and track paralleling huts for a 2 per cent reduction on its previous price.[22] Such were the benefits to the Southern Railway of a depressed economic climate when suppliers were desperate for business.

Despite the proven performance of the Raworth power supply arrangements at the time of the formal inspection, Colonel Trench, besides noting the expected and easily fixed minor points such as fencing and bonding issues, questioned the suitability of the tried and tested mercury arc rectifiers. The rectifiers were a Swiss design, but the colonel, perhaps exhibiting the government's continuing bias against the third-rail system, suggested that they should comply with a new *British* standard which was to be introduced later in the year. His main concern was the lack of adequate 'smoothing' equipment on the rectifiers. By employing six alternating current phases to form the direct current, the output was less than completely 'smooth'. In theory the 300-cycle 'ripple' on the direct current output – effectively an alternating current component – might pass through the blocking circuitry to cause faulty operation of the signalling equipment. This issue had not been highlighted previously and there was no history of problems with the signals. To resolve this, the Ministry of Transport agreed that the Southern Railway must provide written reassurance that the rectifiers would not cause a fault in the signalling and track circuiting equipment. This Sir Herbert was happy to provide, obviously with advice from both Alfred Raworth and Colonel Hall, the railway's signalling engineer.[23]

As was to be expected, concurrent with the Portsmouth Direct electrification, station and signalling improvements were put in place and the most notable was at Woking station which was given the full J. Rob Scott treatment with a new main station building situated on the 'down' side – that is the Portsmouth direction. The new buildings on the 'up' side were smaller with reduced facilities compared to the main buildings, which was unfortunate because the main shopping and commercial area for the town is on the up, London, side. All the platforms were extended to accommodate the proposed 12-coach trains.[24] There had been three signal boxes at Woking – East, West and Junction – and these were all replaced by a single modernist glasshouse signal box.

Despite being a busy main line, the Portsmouth Direct lacked any sections of quadruple track beyond Woking to provide 'fast' and 'slow' lines, allowing express trains to overtake stopping trains. To alleviate this difficulty, steps were taken to increase line capacity by installing multi-aspect colour light signals to provide more signalling blocks. Semaphore signalling at Woking was replaced with colour light signals, and these were continued with automatic signals further down the line through Worplesdon station to Guildford. The signal box at Worplesdon was retained but was normally switched out (probably only used when the local steam-hauled 'pick-up' goods train was in the goods yard).[25] To further increase line capacity, additional signalling blocks were provided by using two-aspect intermediate block signals. These were between Buriton Siding and Idsworth Crossing boxes south of Petersfield, and between Liss and Liphook. Another intermediate block signal was installed in the London direction only between Whitley and Greyswood. Further track and signalling upgrading (all with colour light signals) was carried out between Portsmouth and Fratton.[26]

Another station to get the full J. Rob Scott modernist treatment was Havant, where the original LSWR Portsmouth Direct meets the LBSCR coastal route. Other stations along the line – Esher, West Weybridge (since 1962 renamed Byfleet and New Haw), Godalming, Haslemere, Rowlands Castle, Fratton, Portsmouth and Southsea and Portsmouth Harbour – had platform lengths and canopies extended to accommodate the new 12-car trains. At Haslemere, where there is an extra platform line, the centre line was signalled to allow a train in either direction to overtake trains held in either of the other two platforms. New carriage sheds with carriage washing facilities were installed at Farnham and Fratton and a new repair shed was built at Wimbledon.[27] All this work amounted to considerable investment by the Southern Railway to improve the line that the LSWR had been forced to lease and then buy off Thomas Brassey.

Sir Herbert Walker gave Alfred Raworth strict instructions to keep both the capital and running costs to an absolute minimum, and this included the cost of the electrical equipment on the new trains. The overall design of the electric multiple units must be attributed to Richard Maunsell – the chief mechanical engineer – but on the critical issue of the rating and number of the electric motors it was Raworth, as authorised by Sir Herbert, who made the decisions. The Portsmouth Direct was a route quite different to the Brighton main line, which is a straight and mostly level track, perfect for fast running. Thomas Brassey had bequeathed his Portsmouth Direct route with not only steep inclines but also sinuous curves which absorb power due to the friction imposed on the wheel flanges. It might be expected that the new Portsmouth trains would have to be at least as powerful as the Brighton trains to cope with this more difficult task – but no, these new trains were to be provided with less horsepower than the Brighton sets because fast services would not be possible due to the difficult route and also the express services would have to be integrated with the slower stopping trains. Places where the fast trains would be able to overtake the stopping trains were limited, south of Guildford, to Petersfield and Haslemere where platform loops permitted overtaking. The new 12-car trains were to consist of three four-car sets, each of these to have its two motor coaches equipped with only one power bogie. Compared to the Brighton stock, this reduced the overall horsepower from 3,600 (2,700kW) to only 2,700 (2,000kW).

The new 4-COR sets built for the Portsmouth electrification scheme did not have any catering facilities as these were to be provided within the similar 4-RES (restaurant) sets. A 12-coach train consisted of a 4-COR at either end with a 4-RES in the centre. This necessitated corridor connections between the sets – hence the 4-COR designation. The 29 4-CORs built for the Portsmouth Direct electrification had a motor coach at either end with the intermediate trailers being a full third class coach and a first and third class composite coach. The 19 4-RES units comprised of two motor coaches with a first class carriage with five compartments and a small open saloon for the well-heeled diners, and a third class kitchen car with an open saloon for the third class diners.

It was intended that every fast train to Portsmouth was to have a 4-RES set included in the formation, but because the builders of the restaurant cars had not been able to obtain the necessary equipment in time, on the first few Saturdays after the new electric services began (the busiest day during the summer months) several trains had to run with hungry passengers.[28] What they were missing was food cooked fresh from a cooking range with four boiling plates and two grills, four roasting ovens and a steaming oven. Also in the kitchens was a coffee-making machine, a milk boiler, a soup boiler and an egg boiler. There was also a refrigerator and a wine cooling cabinet. The larger items such as the ovens took their supply direct from the nominal 660V traction supply, while smaller items used a 200V supply obtained from the onboard motor-generator set.[29] Breakfasts, lunches, afternoon teas and dinners were served in the restaurant cars depending on the time of day – a far cry from the trolley service on the South Western Railway trains on the Portsmouth service today.

As a further economy, the motor coaches on the 4-CORs and 4-RES units were, unlike the all-steel Brighton motor coaches, built using the traditional wooden framed method in line with the trailer vehicles. But inside the trains the passengers found the same high standard as the Brighton stock – comfortable seats and lots of natural wood features. The first class compartments had four decorative schemes employed using contrasting veneers of Indian silver greywood, African mahogany and American black walnut.[30] At the front of the new trains the single motorman's window was balanced by the route indicator panel on the other side of the corridor connection; this gave the vehicle a 'one-eyed' appearance and – given that their destination was Portsmouth – the nickname of 'Nelson' stock was very soon to be applied.

These new trains' apparent lack of power must have raised a few eyebrows both from

4-COR No. 3101 near Guildford in 1965 (probably between Sand and Chalk tunnels). (Rail online)

2-BIL No. 2098 has arrived at the end of the electrified lines at Alton in the 1950s. The train is standing in the platform now used by the Mid Hants Railway. Note the 'steam railway' atmosphere with the traditional oil lamp about to be placed on the rear of the train ready for its journey back to Waterloo. (Rail online)

within the Southern Railway and from any technically astute member of the public. The 4-CORs compared unfavourably with the then new 2-BILs which had two 275 horsepower motors per power bogie compared with the 225 horsepower motors on the 4-CORs – and for both the 4-CORs and 2-BILs there was a power bogie for every two coaches. Following the official inauguration ceremonies on 1 July 1937 there followed what amounted to a demonstration run for the benefit of the distinguished guests and the press from Portsmouth to Waterloo. The then 'Southern Electric' correspondent of *Modern Transport* magazine was Charles Klapper. Alfred Raworth invited him into the tiny motorman's cab for the journey back to London.[31] Charles Klapper seems to have had a good relationship with Raworth, and described the electrical engineer as a ' … remarkable character … always genial and stimulating company'.[32] The impression that Charles Klapper has given is that Raworth seemed to be concerned that the

4-CORs had only 75 per cent of the power of the Brighton trains and needed to find out for himself how the trains performed. It is unlikely that after the necessary test running Raworth needed to wait for the press run to find out how good, or bad, the 4-CORs were at tackling the difficult route. The test run proved however that concerns were unfounded, for the train easily sped up for the 2-mile-long climb to Buriton tunnel on the summit of the South Downs and Raworth's comment to Charles Klapper was: 'these trains are just right, just right'. This view was reinforced by the ease that the train coped with the 1 in 80 climbs to Liss and Liphook and then the 1 in 100 climb up to Haslemere. On the downhill section at Whitley, 78mph was achieved. As they raced through the Hampshire countryside, Raworth confided to Charles Klapper that he believed that in different circumstances, on another railway, with another general manager, he could build trains capable of 120mph.[33] The ever-loyal Raworth was not criticising

2-NOL No. 1859 at Chertsey on 2 March 1957. (Meredith/Transport Treasury)

his general manager, for he understood that Sir Herbert was a commercial realist and could only commit to projects that the Southern Railway could afford, and it was only by the incremental extensions to the third-rail network that the Southern Railway could continue its electrification ambitions at all.

After proving that the train could hill-climb, Raworth, outwardly at least for the benefit of the journalist present in the cab, continued to urge on the motorman to run as fast as possible on the flatter, faster sections beyond Woking to test the 4-COR's high-speed running capability. The train performed well even if the important passengers returning from the inauguration ceremony were thrown about somewhat over certain sets of points on the approaches to Waterloo, to such an extent that a representative of the Southern who was on the train burst into the cab demanding that the speed be reduced immediately. Despite this stirring performance Raworth had to concede that the train did not respond as well as the Brighton stock, but he was able to reassure the young Charles Klapper that they would be able to perform well enough to cope with the new timetable requirements.[34] The comment made by Raworth that the trains were 'just right' can be construed as being analogous to the modern logistics term 'just in time' – that is, engineered to do precisely what is required, no more and no less. Incidentally, Klapper's contemporary *Modern Transport* account of the Portsmouth Direct inauguration does not mention his exciting ride in the 4-COR cab.

The new electric railway ran to a fresh, new recast timetable. Following the time honoured 'clock-face' principles, at 50 minutes past each hour, a fast train left Waterloo for Portsmouth Harbour calling at Guildford, Haslemere and Portsmouth and Southsea (High Level) stations, with the corresponding London-bound train departing from Portsmouth Harbour at 20 minutes past the hour. Additional semi-fast trains using the 12-coach express sets were run at peak hours during weekdays to cater for commuters. Most workers now had paid annual holiday and the Isle of Wight had become a popular destination with families seeking the sun at the beach resorts on the sheltered south-eastern flank of the island. This popularity had made summer Saturdays an operational nightmare for the Southern Railway when using steam trains, so during the summer months additional electric trains could now be run on Saturdays between Waterloo and Portsmouth Harbour to connect with the Southern Railway steamers to Ryde Pier. At these times three fast electric trains were provided every hour to Portsmouth: one ran non-stop between Waterloo and Portsmouth Harbour; another made stops at Guildford and at the Portsmouth and Southsea (High Level) station while the third called only at Havant. This latter Saturday extra made connections with trains for Hayling Island, another popular holiday destination, for those wishing to take the former LBSCR branch to the slightly down-market island resort which then consisted of holiday camps, caravan parks and an untidy collection of holiday accommodation resulting in what might fairly be described as a shanty town.

For most of the day local services left Waterloo at 27 and 57 minutes past the hour and ran non-stop to Surbiton from where they continued stopping at all stations. At Woking, the train – usually eight coaches made up from four 2-BIL units – was divided, with the front two 2-BILs continuing as a stopping service to Portsmouth and Southsea, while the rear pair became a stopping service to Alton. After 10pm the Portsmouth line trains ran only to Guildford with an hourly interval. There were, of course, corresponding London-bound local trains – these left Portsmouth and Southsea precisely on the hour and on the half hour.

This timetable allowed a Portsmouth bound stopping train to arrive at Guildford around six minutes before a Portsmouth-bound fast train, giving sufficient time for passengers to change to the express service and also allowed for those who had boarded at Waterloo and wished to travel to stations beyond Guildford to transfer to the local train which then followed the fast train down the line. Fast trains – particularly the Saturday extras – could overtake the local trains using the third platform lines at Haslemere and Petersfield and, when arriving at what had been LBSCR territory at Havant, use the additional pair of fast through lines between the platforms.

Additional trains to suit the business traffic were provided on the Alton line and on the Chertsey loop. To suit the new timetable the timings of steam trains on the Midhurst branch (from Petersfield), the Meon Valley line to Gosport and the Mid-Hants line to Winchester (both from Alton) were rearranged. Following the electrification to Alton the passenger service which had been run through Tongham on the LSWR's original Guildford to Farnham branch was withdrawn.[35] The pull-push steam services from Ascot to Aldershot now ran twice an hour and were timed to connect with the Waterloo to Alton service at Ash Vale. The long-established service from Woking to Ascot using Frimley East curve was discontinued except for a few business-time trains.[36] It was not possible to add vans for the conveyance of market produce, milk or parcels traffic to the electric trains as had been possible with the replaced steam trains, so additional steam-hauled van trains had to be run on the Portsmouth and Alton routes. Some 203 passenger coaches were released for other services and 25 less steam locomotives were required. Seventy less pairs of footplate men were needed but 32 more guards were to be rostered (12 more on summer Saturdays) plus 100 extra motormen.[37]

A rather charming feature of the Portsmouth 4-CORs was the roof-mounted destination boards fitted in

British Railways, Southern Region days and possibly earlier. These informed prospective travellers that the train was a service from Waterloo to Portsmouth and Southsea, Portsmouth Harbour and Ryde Isle of Wight. A train that went overseas! But the Southern Railway owned the ferry service which was part of an extended single transport link that continued on throughout the island with a network of steam-operated railway lines. To cope with the anticipated extra travellers between Portsmouth Harbour and Ryde Pier following electrification, the Southern Railway augmented its fleet of ferries with a new vessel.[38] One might assume that the ultra-modern Southern Railway would order a state-of-the-art motor vessel but no, it ordered a paddle steamer from William Denny and Brothers of Dumbarton. The new ship, a proper steam ship with a three-cylinder, triple expansion engine fed by steam from a coal fired boiler, was to be named *PS Ryde* and was a sister vessel to *PS Sandown* bought earlier in 1934. Both ships survived in service long enough to receive British Rail's new corporate double-arrow livery in 1965 but neither remained in service with British Rail for much longer than a year after the repaint.[39]

With the completion of their work on the Portsmouth Direct, Raworth's team immediately set to work on the equally ambitious 'Portsmouth Number Two' scheme – a further 75 route miles of electrification. This was the former LBSCR's route to Portsmouth, Littlehampton and Bognor Regis – the Mid Sussex Line. At the dawn of the railway age part of this route had been proposed as the London to Brighton main line, which would have secured prosperity for Horsham and Arundel, but the early 1836 schemes came to nought and the main line took the more direct route via Haywards Heath. Later in 1863 the LBSCR completed the through route from Victoria via Three Bridges, Horsham and Arundel to join the coastal route at Ford. The LBSCR also built the Littlehampton branch. As part of this new route the railway upgraded its Three Bridges to Horsham single track branch to double track. The LBSCR needed to have a shorter and more competitive route to Portsmouth following the LSWR's somewhat fortuitous acquisition of the more direct route from London to Portsmouth in 1859. At Littlehampton, the LBSCR developed the quayside attempting to operate a cross-Channel steamer service to France, and for a short while there were sailings to St Malo and Honfleur. Later the LBSCR completed the full Mid Sussex route by linking Horsham with Victoria via Dorking and Leatherhead, after taking over from a local company that had met with financial difficulties.[40]

The suburban electrification at Dorking was to be extended southwards through Horsham and Arundel to meet the LBSCR coastal route. Also included in the scheme were lines linking existing main line electrification from Three Bridges to Horsham and from West Worthing to Havant and the important branches to the coastal resorts of Bognor Regis and Littlehampton. This whole scheme was estimated to cost £2,775,847, financed by the loan from the Railway Finance Corporation supplemented by £1,250,000 from the Southern Railway's locomotive, carriage and premises renewal accounts. Sir Herbert had informed the directors that a 9.75 per cent increase in receipts would be needed to financially break even when the scheme was complete and the existing 2,083,898 annual train miles hauled by steam locomotives would become 3,812,544 annual train miles using electric multiple unit trains.[41]

No more control rooms were required for this scheme, as the existing buildings at Three Bridges and Havant were to be used. A total of 77 miles of 33,000V cable was installed, connecting 20 new substations spread out along the new electrified routes. The 33,000V cable route from West Worthing was extended to Havant; a new cable route was established between Three Bridges Grid and Horsham; and from an extra Grid supply point at Leatherhead a cable route to Horsham. The first substation south of Leatherhead was at

Dorking where space was reserved for plant to serve the Reigate to Guildford line as Raworth anticipated that this too would be electrified at some time in the future.[42]

Colour light signals were installed at Horsham station and at some additional intermediate locations to break up the block sections. New modernist signal boxes – now being described as the *standard type* – were built at Horsham and Bognor Regis. Chichester, Horsham and Littlehampton stations were rebuilt and there were extensive improvements at Arundel and Bognor Regis stations. At Dorking, track alterations to allow permitted fast trains in either direction to overtake stopping trains were installed. Colour light signals were also put in place, all controlled from another new signal box.

Due to the ground conditions, piling had to be carried out to build some of the new substations, for example at South Stoke, situated on the marshes just north of Arundel. Constructing the foundations involved driving 50-foot (15 metre) piles into the ground and to protect the substation from flooding it was elevated so as to appear to be standing on stilts. The lifting bridge across the River Arun at Ford was removed and replaced with a more convenient fixed structure – this was made possible because the Admiralty no longer required that the River Arun be navigable by their tall ships.[43]

New electric trains were built totalling 292 carriages. There were a further 26 'Portsmouth' 4-CORs and 68 2-BIL units. To run with the 4-CORs, 13 4-BUF units were built, each comprising of the usual two motor thirds, a trailer with compartments for first and second class accommodation and a new design of buffet car. To service the increased fleet of electric trains a new depot was built at Streatham Hill to accommodate up to four 12-coach trains and four eight-coach trains. For repairs, the carriage repair shop at Slade Green was further extended.

On Sunday, 3 July 1938, regular electric train services commenced. The existing hourly service from Victoria to West Worthing was extended to Littlehampton. An hourly semi-fast service from Victoria was divided at Barnham to serve Bognor Regis and Portsmouth Harbour. An hourly all stations Three Bridges to Horsham and Littlehampton train connected with the London to Brighton trains at Three Bridges. There was an hourly service calling at all stations between Three Bridges and Horsham and Bognor Regis – this train ran via Littlehampton thus reversing at the terminus and connecting with the London to Brighton trains at Three Bridges. There was a half hourly semi-fast service from Brighton to Portsmouth Harbour (stopping at different smaller stations giving them an hourly service). Additional trains were run during peak times and along the coast route during the summer months.[44]

In July 1938, the provision of multiple electricity supplies from the Grid did not prevent the complete loss of power to most of the newly electrified Brighton and Eastbourne lines. It had been a hot and sunny summer's day during the morning of 5 July, but later a violent thunderstorm crossed London, Surrey and Kent. The CEB's 132,000V overhead lines were subjected to multiple lightning strikes resulting in Southern Railway passengers experiencing the abrupt and unexpected stopping of their trains.[45] It is usually possible to reclose circuit breakers immediately after a lightning strike as there will not have been any permanent damage to the lines. On this occasion however, several concurrent lightning strikes resulted in the Grid sites at Three Bridges, Brighton and Eastbourne being isolated from the rest of the Grid for a short period before there was time to take any restorative action. This not only affected the electric trains but also caused the failure of all track circuiting and colour light signalling. Restoration of supply was severely hampered as telephone circuits had been damaged making communication between the various electricity control centres and the signal boxes difficult, causing even more delay to the restoration of train services.[46]

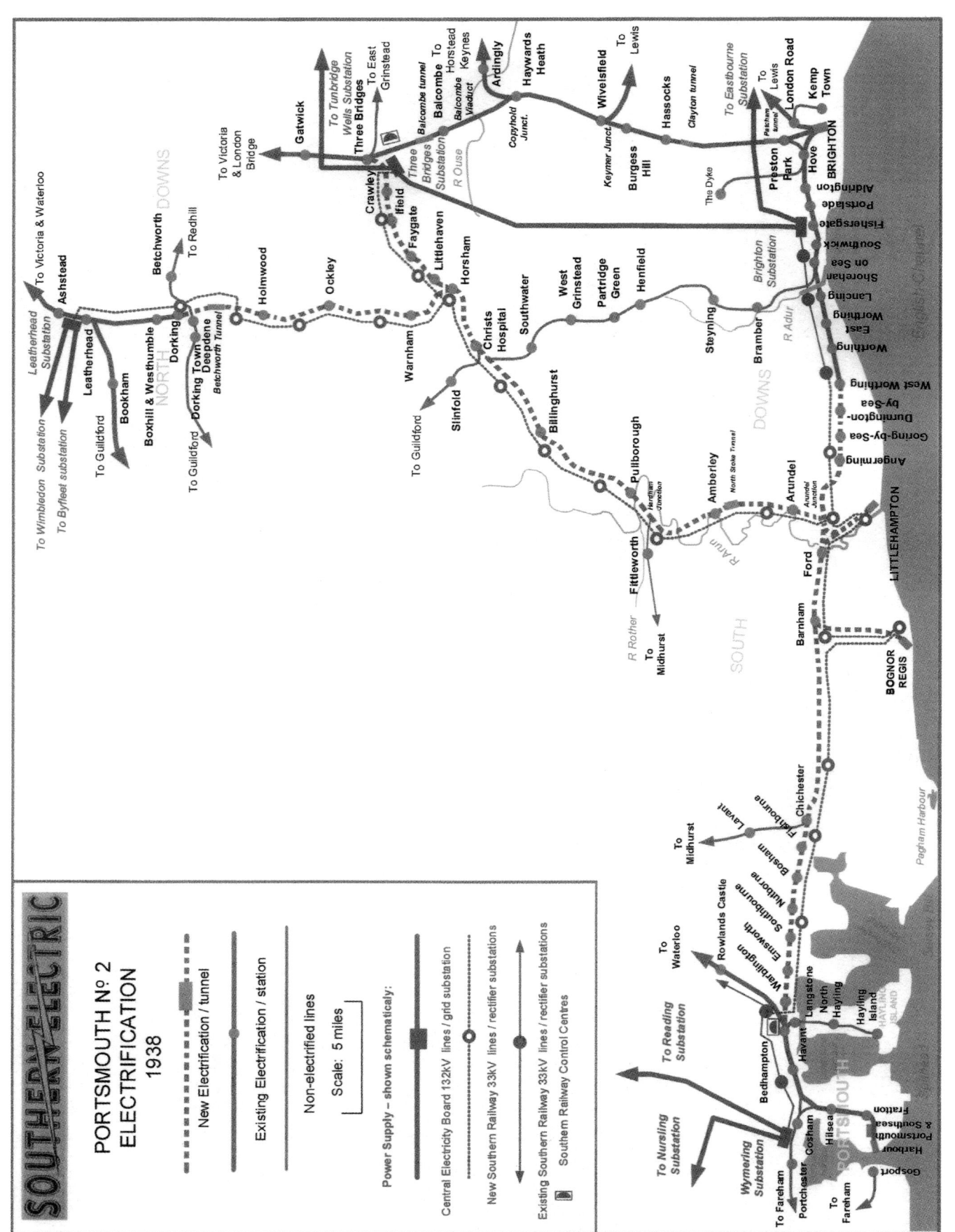

Map of Portsmouth No. 2 electrification scheme. (Author)

CHAPTER 19

THE SOUTHERN ELECTRIC IN THE COMMUNITY

In 1930 Sir Herbert Walker moved John Elliot from his publicity role to the traffic department to learn more about the operational and practical aspects of running a railway, working directly under Edwin Cox. During his five-year tenure in the publicity and advertising section of the Southern Railway, Elliot had set a high standard for promoting the railway and the Southern Electric. His innovations had included an excellent monthly magazine for season ticket holders, *Over the Points*, which was intended to highlight improved train services and modernisation. It was written by E.P. Leigh-Bennet, who on Elliot's instructions, produced a magazine to represent the concerns of the passengers and was unafraid of telling the truth about the railway – good or bad. Leigh-Bennet's flowery prose eulogised over such matters as the beautiful countryside which could be enjoyed by city dwellers using the Southern Electric, however he tended to focus disproportionally on the railway's steamer services to the Continent rather than on the trains.

Elliot had also ventured into poster art, being unhappy with the Southern Railway's 'plain ordinary' offerings. He was impressed by the work of Frank Pick for London Transport and W.M. Teasdale for the LNER. Both advised the Southern on poster format and which artists to commission and Elliot easily persuaded Sir Herbert to authorise a budget of £10,000 a year for posters.[1] This resulted in a succession of excellent (and very collectable) posters organised by his successors, depicting Southern Railway destinations with stylistic representations of the 'new' electric trains – often with a lightning flash over the train to emphasize not only the power of the trains but also the aura of mystery and magic attributed to electrical power. Besides posters, the Southern Railway produced guidebooks to encourage travel on the Southern Electric. Titles included *Winter Resorts in Southern England* published in 1929 and *Hints for Holidays* published in 1934. The potential of the South Downs as a hikers' paradise was recognised with publications such as *The South Downs* and *Hills of the South*, the latter famed for its excellent artwork by Audrey Weber. Other Southern Railway publications such as *Country Homes at London's Door* publicised the new housing developments that were appearing across the Southern Electric area and enthused over the joys of moving out of the city to live in clean air in spacious modern developments – while still being able to travel to your place of work with ease by electric train.[2]

The massive increase in passenger journeys is testament to the popularity of the Southern Electric. Faster and more frequent electric trains were replacing steam trains and people were choosing to live in the new outer suburbs. A letter was printed in the *Daily Express* in 1937 from a lady who had travelled to Lewes on three consecutive days and described the trains as 'perfect' and 'clean and prompt' in both directions.[3] Not all travellers however were fans of the Southern Electric. One such was the prolific writer, orator, satirist and politician Hilaire Belloc who is best remembered for his *Cautionary Tales for Children*. Writing in his column in the *Sunday Dispatch* on 15 November 1936, he complained that his one-hour journey to Brighton in a new electric train was more fatiguing than a four-hour journey from London to York by steam train. He probably travelled in one of the rough riding motor coaches but, after complaining about the noise and vibration he had experienced on his electric journey, he continued his article with a polemic against large monopolistic organisations – such as the Southern Railway – who he claimed were restricting the individual's range of choice, for this was the main thrust of his article.

There was however serious public concern about the safety aspects of the unprotected 'live' conductor rails. Sir Herbert's minimal cost strategy so enthusiastically and loyally followed by Alfred Raworth had, with our modern hindsight and safety sensibilities, arguably given insufficient attention to public safety. At the level crossings the conductor rail was discontinued, and jumper cables buried under the track maintained the electrical connection. Slats of timber were fixed either side of the public access to act as a cattle grid to prevent stray animals endangering themselves. Such crossings were not only on public metalled roads but there were also many pedestrian crossings and private 'occupation' crossings for farmers to move cattle or vehicles across the rails where the railway cut through a single farm or estate. At the time of electrification, where a reasonable alternative access was available, many of these private crossings were closed by agreement and compensation paid to the landowners. Many in the rural community did not seem to understand that provided they remained on the designated crossing, either road or footpath, there was no danger whatsoever from the live rail and there was never any need for any member of the public to have to step over the conductor rail. The fact that the railway embankments on the electrified railway could easily be accessed directly from these crossings was sooner or later to incite local opposition at best, or end in tragedy at worst. Prominent white signs with red letters were installed at platform ends and at crossings to warn the public not to touch the conductor rail but, of course, small children and domesticated or wild animals cannot read.

An early encounter with negative public opinion had occurred when the former LBSCR Wimbledon to West Croydon line was being electrified. At Merton Park station there was a public foot crossing giving children illegal access to railway land which they habitually used as a playground. A local resident wrote to the Home Secretary insisting that the crossing be replaced by a bridge to provide 'the protection for children that any but an insane criminal would make first charge on any scheme'.[4] In this instance the footpath was diverted to the nearest bridge, and at other locations the situation was amicably resolved by the provision of warning signs, extra fencing or the closure of the crossing as appropriate. Often it was the Railway Inspectorate that spotted a dangerous situation and instructed the railway to make the necessary improvements.

On 27 June 1934 a five-year-old boy who had been playing in Eltham Park – a large public open space – wriggled between the thin wire strands forming the railway boundary fence and wandered onto the electrified railway between Welling and Eltham Park stations on the former SECR line between Blackheath and Dartford, a line that had been electrified eight years previously. The young boy was electrocuted upon touching the live conductor rail and, horrifically, his body was subsequently run over by a train. At the inquest, the coroner remarked that the fencing had been passed as suitable by the Ministry of Transport, and this point was noted by the local authority, Bexley Heath and Dartford Council, the owners of the parkland.[5] Subsequently on 3 August 1934 the council wrote to the Ministry of Transport demanding that the railway be fenced off at all points, to prevent children getting onto the railway lines.[6]

This tragic incident brought into question the whole issue of the extent of the Southern Railway's duties and liabilities concerning the fencing of railways and of the extent that the Ministry of Transport was responsible for inspecting and enforcing the safe fencing-off of electrified railways. The law concerning the fencing of railways was inadequate for twentieth-century Britain. The 1845 Railway Clauses Act – passed long before electric railways were ever contemplated – required that the then new railways provide 'sufficient' boundary fencing to delineate their land and to protect the tracks from straying cattle, but there was no obligation to fence to a standard to keep out trespassers.[7] In line with what was described as 'government regulations' (probably unofficial custom and practice following inspections by the Railway Inspectorate), railway fencing usually had eight strands of wire and the Southern Railway was applying an extra two strands when fencing off an electrified line.[8]

The issue was discussed within the Ministry of Transport and Colonel Trench opined that it was not necessary that the Southern Railway's electrified lines should be fenced off at all points so as to prevent children gaining access to the tracks as demanded by the Bexley Heath and Dartford Council. He conceded that in the interests of safety, it was reasonable that at certain locations where there was a likelihood of trespassing by children improved fencing might be required – but not necessarily at all other places.[9] It was noted by the civil servants that the fence bordering Eltham Park consisted of concrete posts with wire strands and they speculated that if a wire mesh was fitted over the wires then by this simple modification Bexley Heath and Dartford Council's demands would be met. As to the Ministry of Transport's duty to enforce the fencing off of electric railways, it was agreed that the department had no statutory power to deal with this matter, either at the time of any formal inspection of a line or at any other time. It was conceded that the railway companies in general and the Southern Railway in particular had always shown themselves ready to provide any additional protection the inspecting officers might suggest as reasonable.[10]

A senior Ministry of Transport civil servant, Mr E.W. Rowntree, wrote to Sir Herbert Walker requesting his views on the Eltham Park tragedy. Sir Herbert replied insisting that the fencing at Eltham Park fully met the Southern Railway's 'statutory obligations'.[11] Sir Herbert continued by stating that experience had shown that no fencing will prevent a trespasser: even the 'unclimbable' iron fencing had been surmounted, steps had been cut into wooden sleeper-built fences and any additional barbed wire or extra strands that had been provided following complaints of trespass had been torn aside. In some cases, where local authorities had not been satisfied with the standard of the fencing, additional protection had been provided but at the expense of the local authority – thereby Sir Herbert pushed the onus back to Bexley Heath and Dartford. But he did state that the Southern Railway was concerned about the dangers from trespassing children. He asserted that it had been pointed out to local authorities and other bodies on more than one occasion that parental control would be an effective way of preventing such distressing incidents as Eltham Park. This did not impress those at the Ministry of Transport as the 'statutory obligations' were virtually non-existent and one civil servant described Sir Herbert's reply as 'superficially plausible nonsense'.[12]

Elsewhere other concerns were being expressed as to the safety of the third rail. While the Southern Electric had been expanding across the urban area of South London and straying outwards only into the immediate suburbs it had not had to face the rural elite. As early as 1932 while the Brighton main line was being electrified, East Sussex County Council began a campaign against rural railway electrification based on public safety grounds and was supported in its cause by the East Sussex Federation of Ratepayers' Associations.[13] The County General Purposes Committee had written to the Ministry of Transport outlining the council's concerns about the risks to people and animals in rural areas. The committee noted that the Pringle Report had recommended the adoption of either overhead electrification at the higher voltage or third-rail at the lower voltage, but this had been purely engineering advice – Pringle had not sought the views of organisations such as the Central Landowners Association or the National Farmers' Union. The incensed county councillors saw the implementation of the Pringle Committee recommendations as an unsatisfactory example of 'legislation by departmental order'. The Ministry of Transport confirmed that the Southern Railway was to use the lower voltage third-rail system, but that the intention was to insist on additional safety measures.[14]

With the announcement of the Portsmouth Direct scheme a more widespread protest movement against the third-rail electrification brought together local landowners, farmers and others who might loosely be described as 'hunting folk'.[15] The Hampshire Federation of Women's Institutes also formally entered the fray, together with the Central Landowners Association and the National Farmers' Union.[16] The protestors contained many influential people, for example Lieutenant Colonel Clarke MP who lived within two and a half miles of the London to Brighton main line and had suffered the trauma of having four of his hounds plus another of his dogs electrocuted on the main line; and now that the Horsted Keynes branch, which actually crossed his land, was also electrified, he claimed that he and others were suffering a great deal of anxiety both for the safety of their children and the local wildlife. He claimed that on one occasion the bodies of seven badgers had been discovered along a two-mile length of the branch.[17]

Following several protest meetings, a delegation of country gentry met the latest Southern Railway chairman, Robert Holland-Martin – who had recently succeeded Lord Wakehurst – and Sir Herbert Walker. Despite being treated with courtesy and being plied with sherry, the delegation came away feeling very dissatisfied, complaining that the company had not listened to their genuine concerns. They had been told that it would be too expensive to carry out what they considered to be a relatively minor improvement to the electrified lines' security by covering the existing fences with what they termed 'rabbit wire' – that is a fine mesh made from galvanised steel wire.[18] Perhaps learning that the Southern Railway was considering using overhead wires in goods yards, the delegation suggested that at safety critical sections along the railway the conductor rail should be substituted for overhead wires, but of course a thin overhead wire might be acceptable for slow shunting manoeuvres around a goods yard, but an overhead conductor on the main line would need to be more akin to a suspended steel girder than a wire, to be able to carry enough current to match the conductor rails at a voltage as low as 660V. The delegation was rather patronisingly assured that there was little danger to human life from a shock of only 660V since a healthy person would experience no more than a sharp shock. The delegation was told that instances of fatalities had been confined only to those persons who were trespassing. The delegation however was as much concerned about animal welfare as risks to human life. The Southern Railway representatives had to concede that some stray cats and dogs had been electrocuted on their railway as well as some hunting hounds. Otters and badgers had been killed, but it was stated these animals (otters) were particularly prone to electric shock due to their damp coats and the ease with which they could squeeze under the conductor rail.[19]

The delegation came away from the meeting feeling that, while being treated courteously, all reasonable pleas had fallen upon deaf ears. These privileged elite were accustomed to getting their own way and were not prepared to give up without a fight. In March 1937, the latest Southern Railway Bill was due to have its second reading. This was the usual technical 'housekeeping' bill and should not have been contentious, but the 'country' lobby included many MPs, who saw this as a chance for their concerns to be publicly aired in Parliament. An amendment was proposed that the Bill be delayed for six months, and this instigated a debate allowing the detractors to pour out their scorn on the Southern Railway under the protection of parliamentary privilege. The amendment was proposed by Brigadier General Clifton Brown, the MP for Newbury – whose constituency was far away from any electrified railway.[20]

The brigadier general opened the debate by claiming that people in rural areas were deeply concerned about the dangers from the electrified railways and that they were dissatisfied with the Southern Railway's response. He reminded the House of how many children had been killed over the last three years when touching the live rails – seven in 1934, six in 1935 and 11 in 1936. He argued that the Southern Railway was proposing to spend millions of pounds on further electrification, yet could not afford to pay for adequate fencing. He then attacked the third-rail concept by citing the 1928 Pringle Report and the 1931 Weir Report, which had both recommended that the overhead system should be used as was then proposed by the LNER and the LMS. Clifton Brown reminded the House that West Sussex County Council had recently passed a resolution urging the Ministry of Transport to hold an inquiry into alternative and safer methods of electrification, and that this inquiry should examine the possibility of the Southern using the overhead system and furthermore, if the Southern would not adopt both systems at least it should be made to protect the public from these dangerous lines.

The amendment was seconded by Rear Admiral Beamish, who after first claiming that there was hardly a live-rail system in use anywhere on the Continent, restated that the 1928 and 1931 committees had never seen fit to consult any of the rural public or private interests. He referred to a recent speech by the Southern Railway chairman who claimed that the Southern took 'every and ample care' to prevent unauthorised access to its lines, but the rear admiral contradicted this by insisting that the company displayed neither 'every' or 'ample' care to safeguard animals and people. He argued parents had serious concerns about the safety of their children and that thousands of people who lived within yards of the electrified lines thought it was unfair that they could not let their dogs or cats out for fear of them wandering onto the line.

The Southern Railway did have some support in the House. Lesley Boyce, MP for Gloucester, spoke up for the Southern, insisting that it had shown a readiness to take reasonable steps whenever representations had been made to avoid dangers to small children. He argued that a railway track which everyone knows to be electrified is a much safer place than say, any public road, any canal, or the cliffs at Dover.

The two-and-a-half-hour debate had given the detractors the opportunity to let off steam. All sorts of other issues relating to the Southern had been aired – such as complaints that the railway did not allow taxis sufficient access to Victoria station, resulting in them congregating in residential areas to the annoyance of the local populace in nearby up-market Eccleston Square. At the end of the debate the parliamentary secretary to the Minister of Transport, Captain Austin Hudson MP, reminded the house that the Bill in question had absolutely nothing to do with the electrification of railways in Sussex or elsewhere in southern England, and it made little sense to block this particular legislation. This was understood of course, but the debate had been an opportunity for the rural community to express its feelings about the third-rail electrification and it had caused embarrassment to the Southern– and this deeply concerned Sir Herbert Walker. When the vote was called the amendment motion was defeated.

Three months after the debate, the Southern, still smarting from the attacks in Parliament, had to endure the month of June 1937 – an appalling time for child fatalities on the electrified railway. On the 2 June, a young boy was killed on the Brighton to Lewes line between Falmer and London Road after passing through a fence described as being 'in perfect condition'. On 11 June, another boy gained access to the railway via some private land in Tooting, through a hole recently made in an advertising hoarding, and then from this private land gained access onto the railway through the fence which, while generally deemed to be in good condition, had one broken section which had made climbing over very easy. He then climbed up the railway embankment and stood with his back to the track. He slipped on the loose stone ballast and unintentionally sat down on the live conductor rail. The coroner returned a verdict of accidental death. On 24 June, a three-year-old boy was killed after he had passed through the fence at the end of the garden of his parents' house in Orpington. At his inquest, the coroner was told that at this location, children were frequently seen playing on the railway embankment and on the electrified railway. The coroner's verdict was misadventure but with the rider that the post and wire fence should be replaced with more efficient fencing to divide the gardens of private houses from railway land.

Following these incidents the Ministry of Transport again wrote to Sir Herbert Walker who reiterated that no fence would prevent a determined child from getting onto the railway if they had a mind to do so. He explained that the fence at the site of the Orpington tragedy was to the railway's standard design which was 4.5 feet (1.4 metres) high. It comprised of 10 strands of plain galvanised steel wire supported by concrete posts spaced at 18-foot (5.5 metre) intervals. The government's view was that this would not deter a not particularly determined child, who could easily squeeze through the resulting 10-inch (250 millimetre) wide gap. It was proposed that a 2-inch (50 millimetre) chain-link mesh be affixed to the standard fence at locations where trespass was likely. This recommendation was communicated to all the railway companies who used conductor rail electrification.[21]

Accordingly the Southern Railway took immediate steps to improve the fencing in its electrified areas. To protect children, two improved designs of fence were to be installed, both incorporating the Ministry of Transport's chain-link recommendation. To improve the safety of foot crossings in the electrified areas more 'cattle grids' were to be provided and improved fencing was to be erected on either side of public footpaths over railway land. At locations where there was a history of child trespass and the existing wire strand fence was in good condition, it would be covered with the 4-foot 6-inch (1.4 metres) high chain-link mesh and extra concrete posts erected to support the additional weight. For all new construction in the electrified areas, or when renewing existing fencing, the mesh was to be carried by three horizontal wires on the railway side of the concrete posts spaced at 9 feet (2.75 metres) apart. Above the meshing was fitted what might be described as 'outriggers', or as the Southern Railway preferred to call them, 'cornices', projecting outwards from the railway, linked by further strands of wire. The railway also announced that it was considering modifying the fencing in hunting country due to the special risk of hounds getting onto the line.[22]

The country and hunting lobby did therefore achieve a small victory. There was however one group even more influential than the country folk: the British military establishment. With rumours rife that the Southern Railway was to electrify to Portsmouth, a retired rear admiral wrote

Improved fencing and signing at a public access point. A 4-SUB train is about to cross the foot crossing at a location near Cheam. (Southern Railways Group/John Sutton Collection)

to the Ministry of Transport concerned that an electric railway would be vulnerable to enemy attack, and that an important naval establishment such as Portsmouth must have a secure railway link to all other parts of the country. This view was taken seriously enough to warrant consultation with the War Office. The reply was loosely in the Southern Railway's favour. It would be better, according to the War Department, if the Southern Railway did not adopt the third-rail system but joined the other railway companies in what was rather optimistically expected to be a national interconnected 1,500V overhead electrified network to facilitate the movement of troops and military materials across the country. The admiral was told that the Ministry understood that the Southern Railway would still own many steam locomotives for work on other routes and for goods trains on the Portsmouth line which would be available to haul through trains from other parts of the country.

This rather sanguine view was not held by all military men, and during the debate in Parliament concern had been voiced about the use of the third rail on national security grounds, in particular that in electrification terms it was not compatible with the proposed national 1,500V overhead standard. The House was assured by Brigadier Clifton Brown MP that, apparently on the advice of the War Office, the Southern Railway was to retain steam locomotives to be available in a national emergency.[23] The reality was that in the upcoming conflict, the Southern Electric was to prove to be surprisingly resilient.

CHAPTER 20
THE END OF AN ERA

Despite Sir Herbert's wish that he and Edwin Cox – a crucial member of the Southern Electric project team – should retire together, the Southern Railway's traffic manager retired in October 1936. One of his first tasks in retirement was to give a presentation to the Institution of Transport in which he proudly recounted what the Southern Electric had achieved to date. Cox gave details of the service provided by Southern Electric – London Bridge alone received 94 trains an hour during the busiest period of the morning inward commute. Examining the performance of 122,000 trains the average late arrival was only 0.63 minutes; train defects cost 453 minutes (0.4 per cent of the time) and delays due to electricity supply failures amounted to only 59 minutes (0.06 per cent of the time).[1]

Now without Cox, the last significant cooperation between Raworth and Sir Herbert was to submit to the board details of the proposed next phase of the Southern Electric expansion – the main line electrification to Hastings. This seaside town set among beautiful countryside was served by both the former SER line from London Bridge via Tonbridge and the LBSCR coast route which was now electrified from Portsmouth to Ore. Given the shorter distance to London prior to the grouping it was the SER, later of course the SECR, line that was Hastings' principal route to the capital. The SER line had an unfortunate start to its operations because the contractors who had built the railway had used insufficient bricks to line the tunnels and subsequently (after a tunnel collapse) the railway had to remedy this defect by lining the tunnels with an additional course of bricks. The consequence of this was a reduction in the size of the bores, inhibiting the use of full width carriages along the line. Relations had not been good between the SECR and the good folk who lived in Hastings. In 1903, when remarking on the reduction in receipts from traffic to Hastings, H. Cosmo O. Bonsor caused anger in the town when he was unwise enough at a shareholder's meeting to opine that the loss of trade was not the railway's fault but due entirely to the town not giving sufficient attention to the provision of amenities for its seaside visitors. This caused a storm of retaliation, unleashing a series of complaints about the pitifully slow and unreliable service to London Bridge and the cramped facilities at the station where passengers had to negotiate piles of fish boxes stacked on the platform awaiting dispatch for London. One correspondent to the local newspaper condemned the SECR for its arrogance, insisting that the railway was there to serve the town, not the other way round.[2]

In the years leading up to the Great War relations improved and following the grouping the Southern Railway further improved the train service to the town. Due to the restricted tunnels, special narrow coaches had to be built, designated 'restriction 0'. These had a maximum width of just over 8 feet (2.5 metres) which resulted in them having straight sides, cramped interiors and a very boxy appearance. In 1925 the Pullman Company provided six restriction 0 Pullman cars for the Hastings line services which provided the catering facilities on the express trains. The trains had an apparent high provision of first class accommodation, due partly to the narrowness of the vehicles which resulted in only two-a-side seating being possible in the first class compartments of these corridor coaches, but also to the importance of the Hastings line's first class commuter patronage. A three-coach portion from each express train was usually detached at Crowhurst to continue down the branch to Bexhill West.[3] This former SECR branch had been temporarily closed during the Great War as a wartime economy measure. After the grouping, the Southern had concentrated the through traffic to Bexhill on the SER route through Tonbridge, in preference to the Brighton route. This policy was reversed following the electrification of the LBSCR service to Hastings.[4] In 1931 and 1932 the Southern Railway had built 69 new narrow coaches for use on the Hastings route only, and the express trains were exclusively hauled by what many consider to be Richard Maunsell's finest creation, the Schools class 4-4-0s introduced in 1930. Thirty-nine of these modern three-cylinder express locomotives had been built and given the 'Elliot' treatment by being named after British public schools – including one named *Dulwich*, Raworth's old school – and were the most powerful British 4-4-0s ever built, with only slightly less pulling power than the King Arthur class 4-6-0s. While not designed for working exclusively on the Hastings line, part of the Schools' specification was that they were narrow enough to work through the offending tunnels and they worked all the express trains on the SER Hastings route with many of these locomotives based at St Leonards motive power depot near Hastings.

Meanwhile, the new Eastbourne electrification services were not popular with the business community in Hastings. They wanted a fast service to London Bridge for the City of London and the new electric services usually took them to Victoria in the West End. The lengthy reversal at Eastbourne and the additional stops reduced any time benefit expected from the introduction of new electric trains. A delegation of local politicians from the town led by the mayor had visited Edwin Cox at Waterloo to complain about these issues. Cox was unrepentant, and he appealed to the local dignitaries to support the improved train services and not to decry them at public meetings.[5] He did however tell them 'confidentially' – for the proposal required the approval of the Southern Railway directors – that there were plans to electrify the 'direct' former SER route to London via Tonbridge. Not such a wild promise – the Hastings line electrification was on the list of schemes that had been presented to the Ministry of Transport for loans from the Railway Finance Corporation and, due to Raworth's foresightedness on earlier schemes, almost a third of the required 33,000V cable had been laid for the Sevenoaks scheme to make connection with Tunbridge Wells Grid substation and there was the underemployed Southern Electric control room at Ore awaiting further business.

Alfred Raworth's report, submitted early in 1937, called for the building of 192 new passenger vehicles all to the narrow restriction 0 width.[6] Drawing on the success of the Brighton electrification, the hourly fast train service would use the proposed 10 four-car sets – which would include a single Pullman car – and six four-car 'City' sets. These sets would be augmented as required with two-car 'express' units of which 34 were proposed for construction. These two-car sets – a 'narrow' version of the 2-BIL design – would also provide the hourly stopping service. Raworth proposed using the existing 33,000V cable from Sevenoaks to Tunbridge Wells to feed three rectifier substations along the route and to establish a new 33,000V cable route from Tunbridge Wells to Ore with six new rectifier substations.[7] It was also proposed to electrify the Bexhill West branch. No high voltage cable would be laid, or substations built on the branch as the supply would be through the conductor rails as was done on the Horsted Keynes branch. The total cost was to have been £1,585,231.[8]

At one of the last board meetings Sir Herbert attended as general manager, the Hastings electrification proposal was discussed for the first time, but the decision was deferred. The Southern

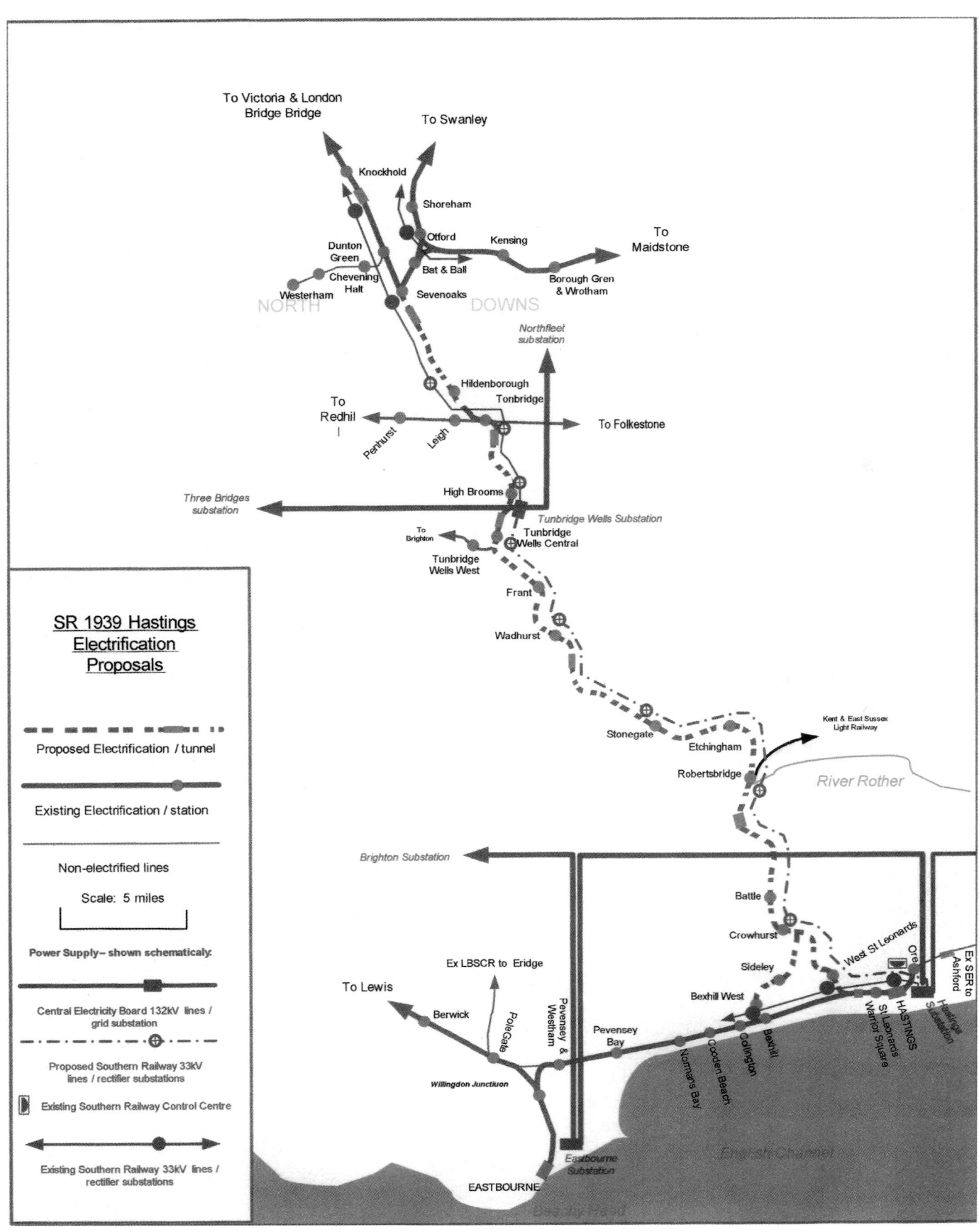

Map of the proposed Hastings electrification scheme. (Author)

On August Bank Holiday Sunday – 31 July 1938 – Schools 4-4-0 No. 907 *Dulwich* hauls an excursion train to Hastings. The train consists of the special restriction 0 stock – including one of the 'narrow' Pullmans – all of the train including the locomotive was relatively new at the time. (Mike Morant Collection)

Railway directors' ethos was to only make unanimous decisions. When there was any disagreement – such as for the Hastings scheme – the matter would be deferred so that any dissenting director could, away from the boardroom, be subjected to the pressure of argument, logic, or even sheer willpower until a collective agreement was reached.[9]

The directors were apparently unhappy about the expense of providing more new restriction 0 rolling stock given that so much money had recently been expended on the Hastings line steam stock which, due to its cramped interiors, would not be popular if transferred to other lines. Furthermore they doubted that the Pullman Car Company could be once again persuaded to make any financial contribution towards the cost of the proposed Pullman cars. Every scheme to date had been expected to cover its costs from day one of operation from increased receipts – the now expected 'sparks effect' – but there was commercial uncertainty due to the alarming international situation as many were concerned about the warlike noises coming from Nazi Germany. Prior to the First World War most of the British public had enjoyed decades of peace and were untroubled by fears of an imminent conflict. The rapid way that the political crisis in 1914 had developed into all-out war took most people completely by surprise, so in the late 1930s the public was much more wary of the political situation in continental Europe. A year later in October 1938 the now notorious Munich Agreement between the British prime minister, Neville Chamberlain, and Adolf Hitler was signed, but a sceptical government began preparing for war with the mobilisation of the nation's armed forces and plans were made for the government to again take control of the railways through the REC. The 'sparks effect' would be stifled if the summer trade to seaside Hastings collapsed due to war. Robert Holland Martin seemed to be uncomfortable about creating demand on the south coast. At the formal opening celebrations for the Portsmouth Number 2 electrification held at Bognor, while lauding the latest extension to the Southern Electric, he warned the populace about excessive new housing that would result and hoped that it would not spoil the natural beauty of the area.[10] Not an attitude that would encourage more commuter trade.

Meanwhile there was to be an important new arrival at Waterloo. In 1937 the 69-year-old chief mechanical engineer Richard Maunsell had retired. Maunsell, a man who was thought of by many as being irascible and wild tempered, had been ill for some time. His replacement was to be the 56-year-old Oliver Vaughan Snell Bulleid.[11]

Oliver Bulleid was born in Invercargill, New Zealand, in September 1882.[12] Tragically his father, William, died of pleurisy in August 1889 and his mother Marian brought Oliver and his younger brother and sister 'home' to Wales. One of his cousins, Reverend Edgar Lee, a High Anglican and vicar of Christ Church, Doncaster, decided that he should oversee Oliver's moral welfare. Among Reverend Lee's flock was a good friend, H.A. Ivatt, the locomotive superintendent of the Great Northern Railway (GNR), and a premium apprenticeship at Doncaster works was arranged for Oliver.[13] Socially, young Oliver Bulleid developed a warm friendship with the Ivatt family. He shared their taste for classical music and would regularly escort Ivatt's daughters to church on Sundays. His friendship with the youngest, Marjorie, blossomed into romance and led to a successful marriage. At the end of his four-year apprenticeship he continued to work at Doncaster as the personal assistant to the locomotive running superintendent but in 1908 he journeyed to Paris to work for the French subsidiary of the giant Westinghouse Company at its factory near Paris and was soon promoted to assistant works manager and chief draughtsman. Later Bulleid was employed by the Board of Trade overseeing the exhibiting of British industry in Brussels, Paris and Turin.[14] In 1912 he wrote directly to Nigel Gresley, now GNR chief mechanical engineer, requesting to be re-employed and, to his delight, Gresley offered him a post as his own personal assistant.

In 1914 Bulleid enlisted in the army as a lieutenant and was posted to the Railway Transport Office at St Omer in northern France. Despite being in a reserved occupation he had patriotically volunteered to serve his country. While not being directly in a front-line unit, the carnage of the war affected Bulleid, such that after witnessing how the Roman Catholics had coped with the fear and misery of the battlefields, he converted to full Catholicism. He retained his fervent faith for the rest of his life. He was promoted to major and posted to the Royal Engineers' establishment at Richborough in Kent as works manager. This was not to Bulleid's liking, being far away from hostilities, and he viewed his staff at the establishment as 'war evaders', working in a safe environment which ironically was akin to what he had willingly left behind at Doncaster.[15]

On cessation of hostilities, Bulleid returned to the GNR as its carriage and wagon superintendent and then, after the grouping, he became assistant to Nigel Gresley who was appointed as the LNER chief mechanical engineer. Due to Oliver Bulleid's nomadic career before returning to Doncaster he had become fluent in technical French. This French connection brought Bulleid into contact with the latest steam locomotive technology. The French were using a rigorous scientific approach and not just trial and error, which was more often the case for most contemporary steam locomotive development. Foremost among the French locomotive engineers was André Chapelon, arguably the greatest locomotive engineer of the twentieth century. Gresley and Bulleid were a strong partnership and both admired Chapelon's work and consequently French ideas not only influenced improvements to Gresley's designs but also Bulleid's later creations.

Due to the imminent retirement of Richard Maunsell, it has been suggested that it was the prospective general manager, Gilbert Szlumper, who arranged Oliver Bulleid's invitation to visit Sir Herbert Walker's wood-panelled office at Waterloo on 11 May 1937 to discuss the possibility of his appointment with the Southern Railway.[16] What they discussed has not been recorded, but Bulleid was soon to make it clear to others that he believed that the Southern had already taken electrification too far – not a view that would have endeared him to Sir Herbert. The outcome was that Bulleid was advised to apply for the position of chief mechanical engineer – not a ringing endorsement for Bulleid, but Sir Herbert was due to retire so it was for others to make the appointment. Bulleid was cultured and dignified and his easy charm and social graces immediately won over the new chairman, Robert Holland Martin. He was held in high regard within the railway industry, partly due to his association with Gresley, and was at the time vice-president of the Institution of Locomotive Engineers. Many queried Bulleid's decision to leave the LNER and move to the Southern Railway which, with its focus on electric multiple unit trains, was being derided by many at the time as a 'tramway'. But Bulleid was ambitious – he wanted to be a 'chief' and progress his own ideas on steam locomotive development, but Nigel Gresley was only six years older than himself and at 55 years old he saw that his time was running out. Three weeks after the meeting at Waterloo, Bulleid was confirmed as the new chief mechanical engineer. At first, Oliver Bulleid was to be paid only £3,000 a year[17] which was £1,500 less than had been paid to the retiring Richard Maunsell and £1,000 less than Alfred Raworth was to be paid when he was promoted to chief electrical engineer.[18]

Sir Herbert Walker had been hinting for some time that he would soon retire. This roused mixed feelings, as so great was his influence that to many he *was* the Southern Railway.

The board members were not professional railwaymen themselves and had become over-dependent on Sir Herbert's vast experience and knowledge. Rumours were rife that following Sir Herbert's retirement, dramatic changes were to be introduced, allowing the directors more day-to-day insight into the workings of the organisation, resulting in Eric Gore-Brown, the deputy chairman, letting slip in conversation that he viewed Walker as a 'benevolent despot'.[19]

In May 1937 Walker, after months of prevarication and speculation, finally announced his wish to retire the following October.[20] Sir Herbert's last board meeting as general manager was on 14 October 1937. Holland Martin, who had always given unflinching support to Walker,[21] was wholesome with his praise for Sir Herbert, admiring his 'courage, inspiration, energy and soundness of judgement'.[22] After all the speculation a change in leader did not mean a change in management structure. Sir Herbert was succeeded by a relieved Gilbert Szlumper, who had almost given up any hope of being promoted to the top job.

Szlumper was the son of the former chief engineer Alfred W. Szlumper who had been in overall charge of the original LSWR electrification when Alfred Raworth was Herbert Jones' assistant. The long-retired Alfred Szlumper had died in 1934, and Alfred Raworth had duly attended the funeral of his former boss at St John's church Richmond.[23] The new assistant general manager was to be the aspiring John Elliot who was appointed with a salary of £3,500 a year – £300 more than Alfred Raworth was being paid at the time.[24] It had also been an anxious time for John Elliot; he too had his eyes on promotion, but Sir Francis Dent, the former general manager of the SECR and still a Southern Railway director, tried to block his appointment on the basis that it was not necessary to have an assistant general manager. It was to take the persuasive powers of Sir Herbert to convince the board that an assistant was required, and that John Elliot was the man for the job.[25]

Curiously, there is a note in the margin in the Southern Railway board minutes, next to the entry appointing John Elliot, which reads: 'see note in Chairman's private memorandum kept in Secretary's safe'. Were there conditions on, or further reservations about, Elliot's appointment?

Sir Herbert was not to remain away from the board room for long for he was later to return to the Southern Railway as a director in February 1938[26] and he served in this capacity until the end of the Southern Railway's existence on the last day of 1947.

To the outside world little appeared to have changed. In November 1937 the new traffic manager, Eustace Missenden, spoke publicly on the 'wisdom of electrification' and derided those who had previously expressed doubts about the policy (and this included some of the Southern directors). He produced impressive traffic statistics that showed that since 1932 passenger journeys had increased by 22 per cent which was in stark contrast to the general reduction in business being suffered by the railways throughout the rest of Britain. He claimed that the receipts from electrification would continue to increase as the population of Britain continued to grow, arguing that it was likely to increase to a peak in 1952 and then level off. This was in the area then served by the Southern Electric, and London was drawing in people from the North and this migration would continue beyond 1952 to a level 'beyond the calculation of anyone'.[27] His praise for the Southern Electric was based on what had been achieved so far – there was no hint of significant further electrification as the emphasis was on the consolidation of the gains that had been made to date.

With the change in management Bulleid was to lobby for change. He wrote a report for John Elliot calling for an urgent steam locomotive building programme maintaining that, despite what the directors had previously been led to believe, the time was fast approaching when further extensions to the Southern Electric network would not be economical. His main argument was that on the lines operated by steam locomotives it was essential that the locomotives must be modern and efficient. He advocated that new steam locomotives were urgently required and that the Southern should desist paying for electrification by raiding the locomotive renewal fund.

The fact that from the earliest days the relationship between Bulleid and Raworth was fractious and, as time went on, deteriorated even further[28] has been so often reported that it is impossible to convincingly offer any contradiction. It was Raworth who made the first move after they (probably) met for the first time at the official opening ceremony for the Portsmouth Direct electrification on 1 July 1937. The festivities took place in the Portsmouth Guild Hall with the mayors of Portsmouth and Chichester sharing the banqueting table with two important dignitaries from the Isle of Wight, the mayors of Ryde and Newport. Sir Herbert Walker and Robert Holland Martin presided with several of the directors who all mixed convivially with representatives from the equipment manufacturers such as ASEA, English Electric, Bruce Peebles and Metropolitan Cammell and a representative from the CEB; plus, an unexpected guest from the LNER: Oliver Bulleid, still with his title 'assistant chief mechanical engineer LNER'[29] He may have been invited to attend the opening celebrations so that he could socially meet all his new colleagues at the celebrations.

After being introduced to Alfred Raworth Bulleid would have been unlikely to hold back on explaining his views on electrification – principally that the railway he was about to join had already done too much. So, as well as impressing the young Charles Klapper for his potential report in the *Modern Transport* magazine, Raworth, by urging on the motorman driving the 4-COR train, attempted to impress the 'assistant chief mechanical engineer LNER' travelling behind that it was not only the steam hauled LNER streamlined express trains that could spill the coffee in the buffet cars racing over the points, but so too could the new Southern Electric express trains. The early part of the journey back to Waterloo demonstrating the 4-COR's hill-climbing ability was a worthwhile demonstration, but the gratuitous speeding over the pointwork at the London end of the route was bad behaviour.

With the passing of time and the impossibility of testing any of the evidence by speaking directly with those who worked with Raworth and Bulleid, the precise depth of the animosity is difficult to quantify. In his book *Bulleid, Last Giant of Steam*, Sean Day-Lewis maintains that Raworth was a 'dry, witty, testy and irritable man'.[30] Raworth had worked harmoniously with the fiery Irishman Richard Maunsell; and the Walker, Cox and Raworth collaboration had been highly successful. John Elliot described Raworth's attitude to his work as professional, and it is not very professional to engage in petty point scoring too often. Other senior officers spoke of Raworth's cooperation and willingness to give advice on technical matters.[31]

Sean Day-Lewis also asserts that when Bulleid advocated more development of steam locomotives and Raworth fought hard for the cause of electrification, the electrical engineer 'was the more efficient fighter' but 'Bulleid the more winning'.[32] Seen from this distance in time the in-fighting might only have been the all-too-common rivalry between two powerful men with opposing views, who, despite outward behaviour, in fact respected each other's honest opinion but entertained themselves by mischievously indulging in harmless banter and antics such as 4-CORs being driven too fast. It did not help that the two men were also of a different personal character. Raworth, being both 'dry' and 'witty', probably made remarks that were tongue-in-cheek, for example his comment that Bulleid and his ilk were all 'steam dogs',[33] might have been taken too seriously by any lesser mortal overhearing. The urbane and educated Bulleid would probably have enjoyed 'clever' repartee and banter with Raworth. They both had very different managerial styles. Raworth was focused, precise and positive; Bulleid more of a dreamer, always looking to go a bit further with an idea, when what was needed was an urgent decision. Posed photographs of Alfred Raworth from this period show a balding middle-aged man with a stern countenance, which might be because he felt uncomfortable being photographed, but they do give the impression of a terrier eyeing up the postman; while the mild-mannered, tall and elegant Bulleid, a devout Roman Catholic, exudes a calmer appearance – looking as if he is about to step up into the pulpit in a cathedral to deliver a sermon on Christian forgiveness.

In truth it was unlikely that these two chiefs could ever have worked particularly well together, because despite differences in outward appearance and demeanour, in many respects they were just *too alike*. Both were to spend much of their career carrying out work which was at total variance with the current conventional thinking within their respective disciplines and both were passionate innovators. They both had a vision of the future of railway traction that they wished to develop, but unfortunately these two visions were diametrically opposed. Raworth saw in his version of the future most of the Southern Railway network becoming a complete electric railway, while Bulleid wished to develop steam locomotives much further. Raworth and Bulleid have both been described at different times as having the quality 'genius' – and two geniuses intent on pulling in different directions in the same organisation was a recipe for continuing confrontation.

Bulleid began to persuade senior management into seeing things the 'Bulleid' way. This would have irked Raworth even more than any genuine honest difference in opinion and looking back it is hard to understand how Bulleid, apparently using only charm and gentle persuasion, was able to influence the new general manager, the chairman and senior officers around to his way of thinking. During the Walker regime Raworth had enjoyed the complete trust of his general manager and consequently had returned complete loyalty. Following Bulleid's arrival, a poisonous environment developed when those who had found Raworth 'difficult' seem to have sided with the new chief mechanical engineer. Matters got worse and by the time of his retirement an embittered Raworth had lost the trust of his general manager and the man who had once been loyal to his patron Sir Herbert Walker was perceived as being loyal to 'few other people'.[34]

Bulleid was right however about the state of the Southern steam locomotive fleet. Many former LSWR, LBSCR and SECR locomotives remained in service and were likely to be so for some years to come. Little steam locomotive development had taken place since the grouping. The construction of the excellent SECR 'mogul' 2-6-0s (N and U classes) designed by Maunsell continued until 1934 and many more of the LSWR 'King Arthur' family of 4-6-0s – the named express locomotives (N15), the mixed traffic locomotive (H15) and the freight locomotive (S15) – were, with many improvements instigated by Maunsell, to be built between 1925 and 1936. There was some new build in production when Bulleid arrived, in the form of 20 'Q' class goods locomotives – a conventional and uninspiring typically British 0-6-0 tender engine. The only other completely new Southern Railway designed locomotives were the 16 impressive 'Lord Nelson' 4-6-0s and the 39 successful 'Schools' class 4-4-0s. Bulleid was right – the lack of new, modern steam locomotives was largely due to lack of finance after Walker, Cox and Raworth had taken money from the railway's locomotive renewal reserve fund to finance further electrification. Bulleid perceived a need for a large locomotive to haul the continental boat trains to the Channel ports, and so avoid the necessity of providing more than one locomotive on these very heavy trains. The first fruit of his lobbying came in March 1938 when the Locomotive and Electrical subcommittee approved construction of 10 large main line steam locomotives of an undetermined design.[35]

CHAPTER 21

CHIEF ELECTRICAL ENGINEER

In July 1938 Herbert Jones announced that he wished to retire. The Southern Railway directors warmly thanked their chief electrical engineer for his valuable service and complimented him on his contribution towards the railway's electrification schemes.[1] Jones left the Southern Railway on 30 September 1938. The new general manager and chairman had decided that a single electrical department was now more appropriate for the company's present needs and henceforth the new chief electrical engineer would take full charge of all the company's electrical activities. Arguably this single-headed structure was an indication that the new men at the top had lost some of the Walker enthusiasm for further new extensions to Southern Electric. There was, of course, only one likely choice to fill the new combined role and, predictably, Alfred Raworth was duly appointed to the post from 1 October 1938. Perversely, despite his promotion, other changes in the Southern Railway management structure resulted in Raworth – who would have willingly accepted the work of amalgamating the two sections – having less real power as chief electrical engineer than he had previously enjoyed as merely electrical engineer new works.

Raworth was to receive a salary of £4,000 a year. Herbert Jones' long-time technical assistant, F.G. Cole, followed his master into retirement in November 1938. Given the tensions between the two halves of the Southern Railway's electrical department and Raworth's reputation for being 'difficult' to work with, it is probable that Mr Cole did not relish the prospect of working for this new chief. It was Raworth's principle assistant, W.C. Moore, who had worked loyally with Raworth since SECR days, who was to take Cole's place with a £400 a year pay rise elevating his annual salary to £1,500.[2]

Any delight that Raworth had after receiving his promotion was tempered with sadness because only one month later his mother died. She had been widowed for 21 years. *Modern Transport* reported her death and offered condolences to Alfred Raworth. The main 'news' value of Mrs J.S. Raworth's death was that she was the mother of the Southern Railway's

'Official' Southern Railway photograph of Alfred Raworth taken at the time of his appointment as chief electrical engineer. Some 'touching-up' of the picture has been carried out to remove his facial scars. (Southern Railway)

new chief electrical engineer, indicating that Alfred had now achieved status in his own right. The article did however give a brief account of John Smith Raworth's achievements and this was probably the last public recognition Alfred's father was ever to receive from the engineering community.[3]

Despite the apparent diminishing of enthusiasm for electrification Raworth had plenty of work already authorised. Following completion of the Portsmouth schemes, the Southern Railway next embarked on two significant extensions into outer suburbia. The first of these was to extend the LSWR suburban electrification from Virginia Water to Reading, including the significant branch from Ascot, which runs due south through Bagshot and Camberley to pass beneath the LSWR main line to a junction with the Alton line at Ash Vale. At Frimley, where the short section of single track passes under the quadruple track main line, there was then a connection to the main line facing back towards London – the Frimley East curve. This was also electrified together with the short section of main line back to Pirbright Junction – the point of departure for the Aldershot and Alton line. Only the slow lines received the conductor rail between Frimley Junction and Pirbright Junction. This curve was little used by electric trains and in later years was abandoned, and the track lifted. Within the folklore recounted by Southern Electric enthusiasts is the anecdote that, to maintain the integrity of the direct current interconnection, the Southern Region left the conductor rail in situ – live and obscured by vegetation. Also included in the Reading scheme was the link between Aldershot and Guildford. The conductor rails were installed on the curve between Aldershot North Junction on the former LSWR line to Ash and the former SECR Reading to Redhill line, and then continued on to Guildford. Alfred Raworth had prepared his usual report for submission to the Southern Railway board in October 1936. The proposal was approved and the necessary application for funding duly forwarded to the Ministry of Transport.[4] (See plate 16)

The design of the power supply faithfully followed the established 'Raworth' principles. A new Grid supply point was installed at Reading. Reading Grid substation is adjacent to the Great Western main line between Reading and Sonning and the Southern Railway negotiated a 'wayleave agreement' with the GWR for the 33,000V cable to be laid alongside its line, to reach the first new Southern Electric substation just outside the former SECR terminus, Reading Southern. Three new cable routes were installed, the first from Reading through Wokingham and Ascot to Virginia Water where the cable paralleled the electrified line to make connection at the existing Chertsey substation. At Ascot, the second cable route ran to Pirbright substation. A third route ran from the existing substation at Ash Vale to the substation at Guildford with a single rectifier substation at Wanborough on the route. Still to be clearly seen at Wokingham substation is space in the outdoor switching compound for another circuit breaker to potentially connect a further 33,000V cable, which would have been needed for the electrification of the former SECR line between Wokingham and Ash – an obvious filling-in scheme never undertaken, but obviously planned by Raworth in the 1930s. Furthermore, Sturt Lane substation was built about a mile away from the electrified Ascot to Ash Vale line, besides the former SECR line, at the point where it passes under the then non-electrified LSWR line. This required a long diversion of both the 33,000V cables and the direct current cables. These were laid on railway land beside the main line, but this extra route length could only be justified if it were proposed ultimately to electrify the Ash to Wokingham line.

The new Grid connection provided extra resilience to the overall Southern Electric electricity distribution network, as there was now an alternative supply route from Reading Grid to many of the Portsmouth Direct scheme's substations.

The Southern Railway's ASEA cabinet that was used to operate the circuit braker at Reading Grid. The cabinet has been 'rescued' after the equipment was modernised. These cabinets were designed to be able to operate two circuit brakers, but only one was installed at Reading. (Author)

The design of the rectifier substations was unchanged from the earlier installations, but it is possible to distinguish a 'Reading' substation from the earlier ones, such as those for the Portsmouth Direct electrification, as the newer buildings have non-rendered brickwork consisting mainly of small bricks, often arranged in an intricate pattern and are testament to the skill of the railway's own bricklayers and a joy to behold if not recently daubed with thick brown 'protection' paint, or worse, tasteless graffiti.

This scheme did not involve much engineering work other than that necessitated by the electrification. The charming country stations at Bagshot and Frimley were unaltered apart from the conductor rails on the main running lines and, at Bagshot, the intrusion of the substation tucked discreetly away in the corner of the goods yard. Ascot however did receive some extensive alterations. Prior to electrification Ascot station was effectively two stations side by side, one for the Reading line, the other for the Camberley branch, without any track connections at the west end of the station.[5]

A double track interconnection was provided at the west end with additional signalling controlled by a new glasshouse signal box. This arrangement allowed trains from Waterloo to be divided into two portions. The front stopped at all stations to Reading, and the rear portion, also stopping at all stations, was then able to proceed to Guildford – a journey that involved reversing the train at Aldershot.

The electrification divested the Camberley line of its lowly status as a branch since now, unlike prior to electrification when local pull-push steam trains had worked the traffic, a full through service to Waterloo was provided. Usually the combined train ran non-stop from Waterloo to Staines, then stopped at all stations beyond and this pattern was held for several decades. The new electric services commenced on 1 January 1939 with trains generally running every half an hour – every 20 minutes at peak times – and this amounted to an 85 per cent increase in the number of trains.[6] To operate the train services several more 2-BILs had been built. Production of these two-car units finished in 1938 after 140 sets

Bagshot substation building in 2020. The original intricate brickwork – a credit to the Southern Railway's own bricklayers – has been spoilt by first the application of 'protection' paint and then the efforts of a graffiti artist. (Author)

had been constructed by the Southern Railway.⁷

With the commencement of the Reading area electric train services, another major extension in Kent around the River Medway was being planned which amounted to a further 53.25 route miles of electrification. In October 1937 Herbert Walker had presented to the directors Alfred Raworth's proposals for the electrification of the former SECR lines to the Medway towns: from Gravesend Central to Maidstone West; from Swanley Junction to Gillingham; from Strood to Rochester, Chatham and Gillingham; and from Otford to Maidstone East. The original plan was that this scheme would follow the main line electrification to Hastings and would be completed in 1940. The business case assumed an increase in train miles per annum from the present 1,208,452 using steam locomotives to 1,866,592 using electric trains. The total cost was to be £1,689,296, of which £847,181 would be capital expenditure backed by the Railway Finance Corporation, £152,469 from a suspense account and £689,646 from yet another raid on the locomotive, carriage and buildings renewal budgets.⁸

The power supply arrangements followed the usual pattern with control panels for 16 more substations added to the Swanley control room. There was extensive 33,000V cabling installed for connection to the Grid at Northfleet substation. The existing 33,000V cable from Northfleet to Swanley supplied two new substations at Fawkham and Tweed Hill

Southern Railway publicity material for new 'Reading' services. (Author's Collection)

(near Farningham Road station) which were both built to the standard design. The other 14 new substations spaced along the new cable routes were built to a new compact design. The substation buildings were smaller and did not have the distinctive tower at one end. The use of minimum oil circuit breakers permitted reduction in the footprint of the 33,000V outdoor switchgear. In this design of circuit breaker, fault current arcing is quenched by oil, driven from an internal piston, rather than the rapid natural circulation of oil in a bulk oil circuit breaker.

As trains built with the new standard 62-foot (19 metre) underframe were to be used, there were many places on the former SECR routes out to Maidstone where the track clearances had previously prohibited the use of the wider carriages.

Additional civil engineering work was required not only on the newly electrified lines, but also back to Holborn Viaduct, Victoria and Charing Cross. At the London termini, platform 1 at Holborn Viaduct was lengthened to accept eight-coach trains and at Cannon Street two more platform lines were electrified. Between the metropolis and the Medway towns, minor improvements to obtain safe clearance for the new trains was required, particularly on the lines through Greenwich and Bexley Heath on the way to Gravesend. A new bridge had to be constructed near Gravesend and extra refuges for permanent way staff working in the tunnels had to be provided. Station improvements were required at Swanley, Strood, Rochester, Gillingham and Maidstone Barracks.[9] There were also improvements to the signalling,

2-BIL No. 2053 on a service from Waterloo enters Camberley station on 5 November 1961. The train is hiding the substation, but the building can just be seen. (Rail online)

with some new two-aspect colour light signals, together with one of the new style glasshouse signal boxes built at Strood.[10] Additional signals were installed at Borough Market Junction to squeeze in an additional 'section' by shortening the existing block lengths.[11] It was intended to divide the combined Gillingham and Maidstone East trains at Swanley but the station layout with platforms on the 'main' and 'branch' lines made this impractical. The station was therefore rebuilt a quarter of a mile further west where two 820-foot (250 metre) long island platforms were provided.[12] Money was only spent when necessary; at many locations the 'steam railway' atmosphere was retained with semaphore signalling and mechanical signal boxes and it was still possible to see at Cuxton station – between Strood and Maidstone East on the former SER line – traditional level crossing gates operated by hand 75 years after completion of the Maidstone electrification.

Meanwhile Bulleid was making his mark on the Southern Railway with the construction and look of the new carriages. The olive green livery used by the railway was rather drab and had caused many adverse comments from the public. After Bulleid arrived a much brighter green with bright yellow lettering was applied and this was to be officially referred to as 'Bulleid Green'.[13] Legend has it that bright malachite green was chosen by Sir Herbert Walker on one of his last inspection trips to the Isle of Wight, when he spotted a green cord in a draper's shop window. Buying a piece of the cord he declared that this was to be the new colour of the Southern Railway's carriages. But Bulleid, disliking the colour originally chosen by Sir Herbert, persuaded the Southern Railway's new general manager and the directors to use 'his' version of bright green.[14] The overturning of a general manager's decision as to the precise shade of green with which to paint the railway's carriages might not seem a significant victory for the persuasive Oliver Bulleid, but it did seem to set the scene for what was to come. Bulleid could do what Raworth had not

CHIEF ELECTRICAL ENGINEER • 255

Map of Maidstone electrification scheme. (Author)

been able to do – that is, to persuade the senior managers and directors that his way was the best way. Raworth had accepted company policy and by good management and innovation provided the directors with what they needed on time and to budget. After nationalisation Bulleid – by then reduced in status to being the chief mechanical and electrical engineer for only the Southern Region of British Railways – hosted a lunch at the Old Ship Hotel in Brighton for senior engineers from the new central British Railways mechanical and electrical engineer's department, when

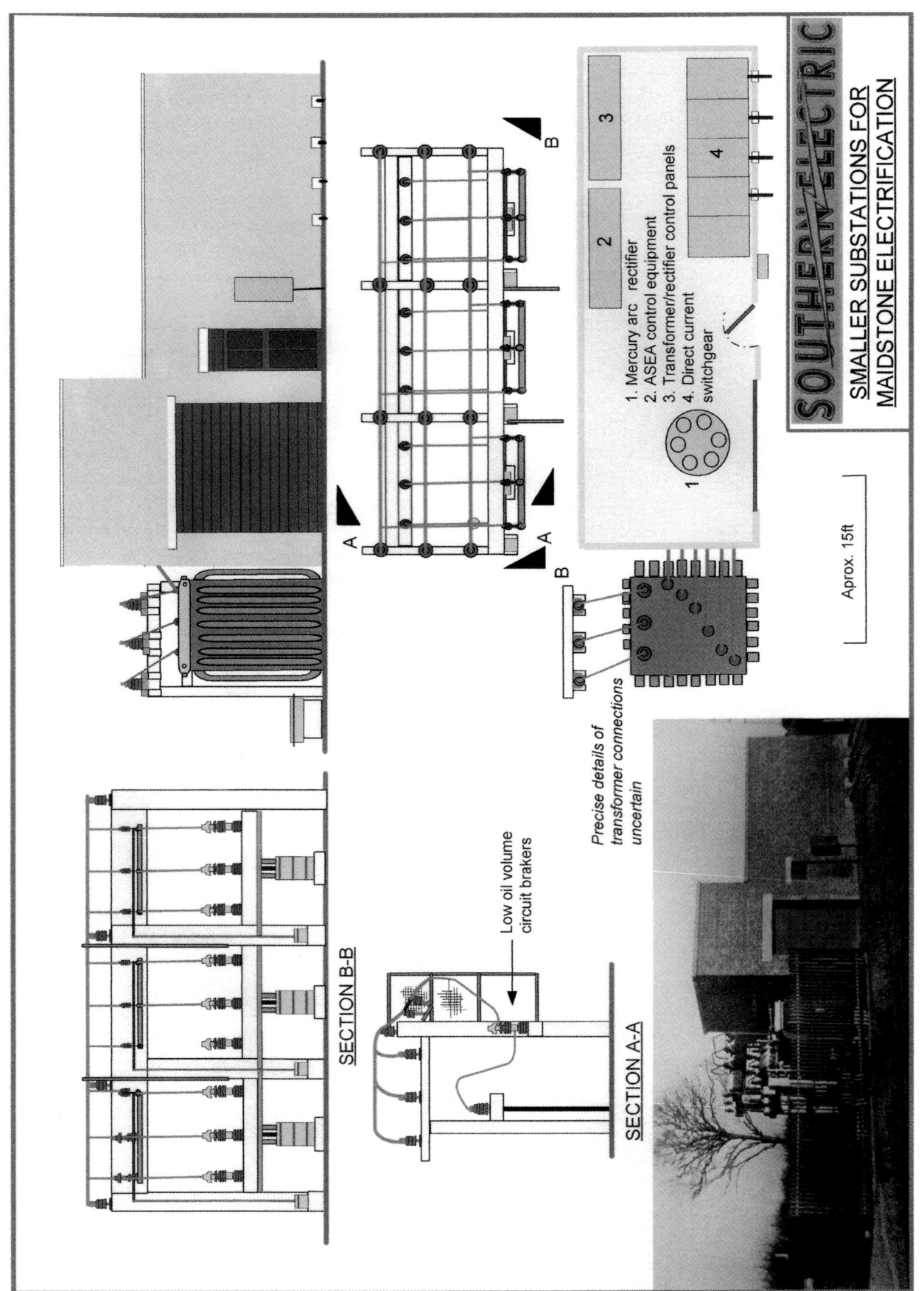

The 'compact' substations installed as part of the Maidstone electrification scheme. (Author)

according to E.S. Cox – the new assistant chief mechanical engineer responsible for the design of new locomotives[15] – his guests assessed him to be a 'supreme autocrat', who was able to impose his will upon management to an extent not seen on the other railways. The gathering also concluded that Bulleid was an individualist, who would not accept a painstaking step-by-step approach to improving steam locomotive efficiency and performance but wished 'with brilliant and dramatic improvisation' to solve all remaining issues by new and innovative methods. Furthermore, Bulleid had politely made clear that if the new order could not learn from his work with the Southern Railway – so much the worse for them.[16]

Before Bulleid's steam locomotives were to draw gasps of surprise he was to make radical changes to the passenger rolling stock. The earliest examples of his thoughts on passenger accommodation were the buffet cars for inclusion in the four-car sets built for the Portsmouth Number Two electrification scheme. The cultured Bulleid broke away from the stuffy Victorian and Edwardian style of railway carriage interior design using natural woods and provided these new coaches with interiors offering a fresh modernist look, befitting a modern form of transport and complimenting the new station architecture and highly acclaimed artwork on the publicity posters. The passengers were presented with bright art deco interiors with the plain walls contrasting with a royal blue Wilton carpet with its bold diagonal gold stripes.[17] Seating was either at the tables on moveable chairs or at the bar on stools. The stark interiors were broken up by scallop shaped ornamental curves. The only other ornaments were decorative plaques designed by Mr G. Kruger Gray CBE FSA, representing the food served on-board: a pig, a hen, some vegetables, a fish, a bullock and a leaping lamb. These were arranged in two groups of three, placed horizontally, and each group of three positioned either side of the exit door.[18] The buffet cars were painted in 'Bulleid Green', at first mismatching with the olive green on the rest of the set.

For the Maidstone electrification the staid 2-BILs were to be replaced with Bulleid's 2-HAL sets of which a total of 92 were to be built between 1939 and 1940.[19] Electrically they had the same equipment as the 2-BILs. These new two-car sets had only one of the carriages with a side corridor giving access to a lavatory, hence 2–HAL (half lavatory). Externally, Bulleid designed driving ends which were built from welded rolled steel sections and contrasting with the 2-BILs which had windows with square corners and glass held in a wooden frame, the 2-HALs' windows had stylish rounded corners flush with the coach body giving a generally more modern and uncluttered appearance. The two vehicles were a motor brake third with a guard's compartment, seven full width third class compartments, and a driving trailer with four first class and four third class compartments plus a lavatory linked by a side corridor. Internally the 2-HALs could only have been seen by the passengers as a retrograde step for the comfortable hammock-slung seats with lift out cushions were superseded by cheaper bench type seating while the level of comfort in the first class compartments was only about as good as the previous third class seating.[20] Gone were the natural hardwoods, which were replaced by man-made Rexine, in both first and third class compartments. A leather-grained nut-brown colour was used on the interior surfaces up to a level with the bottom edge of the windows. Pastel shades were used for the upper panels, a stone colour in the third class compartments and lichen green in the first class compartments. The seats in third class were a dark jade green material with a fleur-de-lys pattern and in first class the material was brown with a darker 'renaissance' design. The floors were covered with a granite design of linoleum for third class and rugs in the first class compartments.[21] The public were not happy with these new, rather spartan, coaches and many third class passengers chose to sit illegally in the first class compartments. One of the first miscreants to be challenged by a travelling ticket inspector was John Elliot's mother.[22]

The usual civic celebrations were held on 28 June 1939, this time involving the latest minister of transport, the Conservative politician Euan Wallace MP, who gave a speech applauding the benefits of fast and frequent train services to country districts, stressing how important this was to City workers. The new services commenced on Sunday, 2 July. The usual regular interval timetable was in place, providing 49 trains from London to Chatham via Strood – an additional nine trains with a journey time reduced by seven minutes; 33 trains from London to Chatham via Swanley – an additional five trains with a journey time reduced by 10 minutes; 43 trains from London to Maidstone West – an additional 19 trains with a journey time reduced by 13 minutes; and 22 trains from London to Maidstone East with a journey time reduced by 16 minutes.[23] During July and August 1939 the Southern Railway passengers using the new electric trains made an additional 54,000 journeys and in so doing paid an additional £4,000 into the ticket office.[24] (This shows how 'cheap' rail travel seems to have been to our modern eyes, as the average ticket price was about 1/6d – 7 pence.)

There was one other important Southern Electric project initiated by Sir Herbert. This was the proposal to construct a new railway from Motspur Park to Leatherhead. Construction of this line duly commenced, and it was opened to traffic in two stages – from Motspur Park to Tolworth in May 1938 and to Chessington South in May 1939. Unfortunately, the project was halted in September 1939 with the outbreak of the Second World War. The new railway had been built only as far as a point just beyond Chessington South station. Unlike

At a location near Chislehurst, four 2-HALs form a train on 13 September 1956. (Rail online)

the other Southern Railway new line, the nearby Wimbledon to Sutton railway, the more conventional design of stations with separate platforms for each direction was used, but the stations were still built in the distinctive Southern Electric modernist style. This was a purely suburban line anticipating further new housing being built along the route. To provide electrical power for the opening of this new line, a new substation was built at Chessington using rotary converters transferred from the Southern Electric substation at Leatherhead.[25] Unfortunately, after the war the project was not resumed – the post-war 'Green Belt' planning legislation curtailed the expansion of housing on the edge of London thus taking away the main reason to build the railway.

With the departure of Sir Herbert as general manager there was a major change in policy relating to the formation of the suburban trains, which would eliminate the use of the ungainly unpowered two-car trailer sets. This was sensibly achieved by the addition of a second trailer coach to the 3-SUBs, making them 4-SUBs. An uncontroversial decision satisfying everybody including Bulleid and Raworth.

The Southern Electric was now suffering from its success by proving just too popular with passengers, resulting in overcrowding on many of the suburban area lines. The number of passengers travelling through Charing Cross and Cannon Street at peak times had increased from 80,000 when electric trains first began operating in 1925 to 140,000 in 1937.[26] Woolwich Borough Council, which in 1922 had so fervently urged the SECR to electrify the lines through its borough to fulfil promises made many years before, by 1938 was complaining of overcrowding on the electric trains and suggesting a reduction in the relatively unused first class accommodation as a part solution to alleviate the problem.

The original intention was to use pairs of the reformed 4-SUB trains augmented by a further powered two-car unit to create 10-car suburban trains. But Oliver Bulleid was more forward thinking and designed an alternative in the form of new 9-foot (2.7 metre) wide high capacity all-steel carriages, these to be made up into four-car sets running in pairs to form eight-car trains. These carriages were to have the distinctive 'Bulleid' curved sides to maximise the body width within the railway loading gauge. Originally there was to be some first class accommodation provided, but this feature was soon dropped, producing the quintessential Southern Electric high capacity 'people mover' that was to roam the suburban lines south of London for over three decades. At the Southern Railway's annual general meeting held in February 1939 the shareholders had been informed that at last the Southern Railway was to increase its fleet of suburban electric trains[27] and that some of these would be the completely new Bulleid 4-SUB suburban units of all-steel construction, or alternatively a new all-steel trailer incorporated into an old 3-SUB – this resulted in a very odd-looking set with one coach markedly wider than the other three.

Meanwhile the issue of the Hastings electrification remained 'deferred'. The directors were unconvinced that the 16 per cent rise in revenue required to make the scheme viable was attainable. At Sir Herbert's bidding Raworth made attempts to reduce costs, including converting the existing carriages into electric multiple units, a process in keeping with the early Southern Railway 'recycling' ethos, but the directors remained unconvinced. Conspiracy theorists will imagine Bulleid lobbying Gilbert Szlumper and Cox's successor as traffic manager, Eustace Missenden, discreetly persuading them that the scheme was unnecessary. Certainly, the cancellation of the Hastings electrification would have helped with Bulleid's desire to build more steam locomotives, since had the scheme gone ahead, the modern Schools class locomotives would have been transferred to other steam lines reducing the need for his proposed locomotive building programme.

In December 1937 at a meeting not attended by Sir Herbert Walker – it was in the short period between his retirement and appointment as a director – the indecisive directors finally lost their nerve, and the Hastings proposal was rejected.[28] This was the only occasion that one of Sir Herbert Walker's major proposals was rejected by the board and the disappointed former general manager never forgave them for what he considered to be a bad decision.[29]

The scheme was subsequently deleted from the Ministry of Transport's list for financial assistance from the Railway Finance Corporation.[30] The news of the cancellation was greeted with dismay in Hastings. The town's mayor, Councillor E.M. Ford, hurriedly met with some of the Southern Railway directors. He accused them of ignoring their own statistics and reminded them that earlier Southern Electric schemes had resulted in an increase in traffic receipts of 148 per cent, accompanied by a reduction in the operating costs. The mayor could not believe that following such success the company was now predicting a financial loss if the Hastings electrification was to be carried out.[31] The Southern Railway directors may have thought however that as both Bexhill and Hastings had an electric service to London via the old LBSCR route, any increase in revenue on the SER route would be at a cost to the existing electric services, a case of 'robbing Peter to pay Paul'. This did not take into account that the existing service was not popular with those in the business community who commuted to the City of London.[32] The impression given to the populace of Hastings was that their town continued to be the Cinderella of the south coast, receiving an inferior train service compared to its close rivals, Brighton and Eastbourne.[33]

Having made an undertaking to Parliament to borrow money from the Railway Finance Corporation for further electrification and then reneging on the promised Hastings scheme, the directors debated the possibility of a 'compromise' scheme that would enable them to borrow the allocated money to spend on what they saw as a less risky proposal.[34] What was to emerge from their deliberations was a proposal that would have added a further 42 route miles to the Southern Electric by electrifying the 'Oxted' route from South Croydon on the electrified Brighton main line to Tunbridge Wells West, including the line from Hurst Green to Horsted Keynes through East Grinstead.[35] Following the abandonment of the Hastings scheme some of the directors might have believed that an opportunity had been lost to tap some of the existing and potential commuter traffic of a more 'suburban' nature from in and around Tunbridge Wells.

The proposal to electrify the Hurst Green to Horsted Keynes route belatedly threw some light on the earlier rather strange decision to electrify the Horsted Keynes branch from Haywards Heath. Sir Herbert Walker's opinion of Raworth's father has never been recorded, but part of Alfred's appeal might have been that he would share the same vision as John Smith Raworth, who had been a passionate believer in developing ways to economically provide viable modern transport systems to areas not deemed suitable for such investment. It is reasonable to suspect that it was Alfred, backed by Sir Herbert, who in 1932 wanted to provide 'rural' electrification to Horsted Keynes as the first part of a wider electrification programme. The Southern Railway had demonstrated that both suburban and main line electrification could be a huge success, so why not continue the third rail onto minor secondary lines and the short branch lines? The earlier decision to electrify the Horsted Keynes branch neatly fitted into the grand strategic plan, serving as a pilot for future rural schemes. The electricity distribution could be intelligently designed to minimise additional cabling and substations. Modern electric trains with a crew of two (motorman and guard) could do the work of a short steam train made up of old and worn-out equipment with a crew of three (driver, fireman and guard) at a similar cost, after allowing for capital charges. Electric trains would, as had been proved so many times before, coax back passengers from competitors – in this case the rural bus and the private car. Such a scheme would also stimulate new housing in these greenfield areas. Furthermore, when these trains reached the main line junction, they could either be attached to other electric trains or proceed on their own to other destinations as did the Horsted Keynes trains that continued to Seaford, something which was not practical with the branch steam trains. This argument to extend the Southern Electric into rural areas would have been the subject of informal discussions between Walker, Cox and Raworth many years before – even if this vision were to be kept from the directors for the immediate future.

Raworth prepared a report for the directors, advising them that on the minor rural lines the track restrictions limited the speed and train load that could be hauled by steam locomotives. Alternatively, if modern electric trains were used on these routes overall speed could be raised and more passenger seating provided. The directors may, or may not, have been impressed by this argument as no decision was made.

The board did seem to be keener on another aspect of a South Croydon to Horsted Keynes electrification, which was that it would reduce congestion on the recently electrified line from Croydon to Haywards Heath which was only double track – not fast and slow lines in each direction. It can be assumed that all the East Grinstead to Haywards Heath via Horsted Keynes line would have been made double track – not likely to be a problem as most single-track railway was built with bridges and tunnels having sufficient space to accommodate a second track if this was later required. Also, this addition to the Southern Electric network could be used in an emergency when the main line was blocked for any reason.

But what sort of 'emergency' were the directors contemplating? The Balcombe Viaduct situated three miles to the North of Haywards Heath carried the LBSCR main line across the Ouse Valley and is 1,500 feet (450m) long. This elegant brick-built viaduct has 37 arches and ornamental towers at either end, and the directors, now fearing a war, may have believed that on a moonlit night the viaduct, at such a strategic location on the Southern Railway network, would be an easy target for an enemy bomber.

Raworth's plan for this proposed electrification was to install a 33,000V cable from Tunbridge Wells Grid substation – situated in the village of High Brooms – through Tunbridge Wells West station to Hurst Green and provide five rectifier substations along this route. At Hurst Green this would connect to another 33,000V cable route laid from South Croydon to Haywards Heath via Hurst Green, East Grinstead and Horsted Keynes. Seven rectifier substations would have been built along this cable route.[36]

The Southern Railway board, which now included Sir Herbert Walker as a director, approved the South Croydon to Horsted Keynes electrification, agreeing to spend over **£1,250,000** on the scheme. But they rejected the Hurst Green to Tunbridge Wells West proposal. However, mindful of the new electrified routes' potential as an alternative route to the coast, the general manager, Gilbert Szlumper, was requested to produce a detailed report on the possibility of electrifying the link between Horsted Keynes and Lewes – a task that Alfred Raworth was pleased to accept.

Despite the directors' approval the war intervened, and the scheme was abandoned. The Southern Electric had hit the proverbial buffers. In 1939 the network comprised a total of 702 route miles of electrified railway; 402 miles had been added since 1930. This amounted to 1,751 miles of electrified track, 951 miles added since 1930. To provide the electricity for these lines there were 158 substations, 112 having been built since

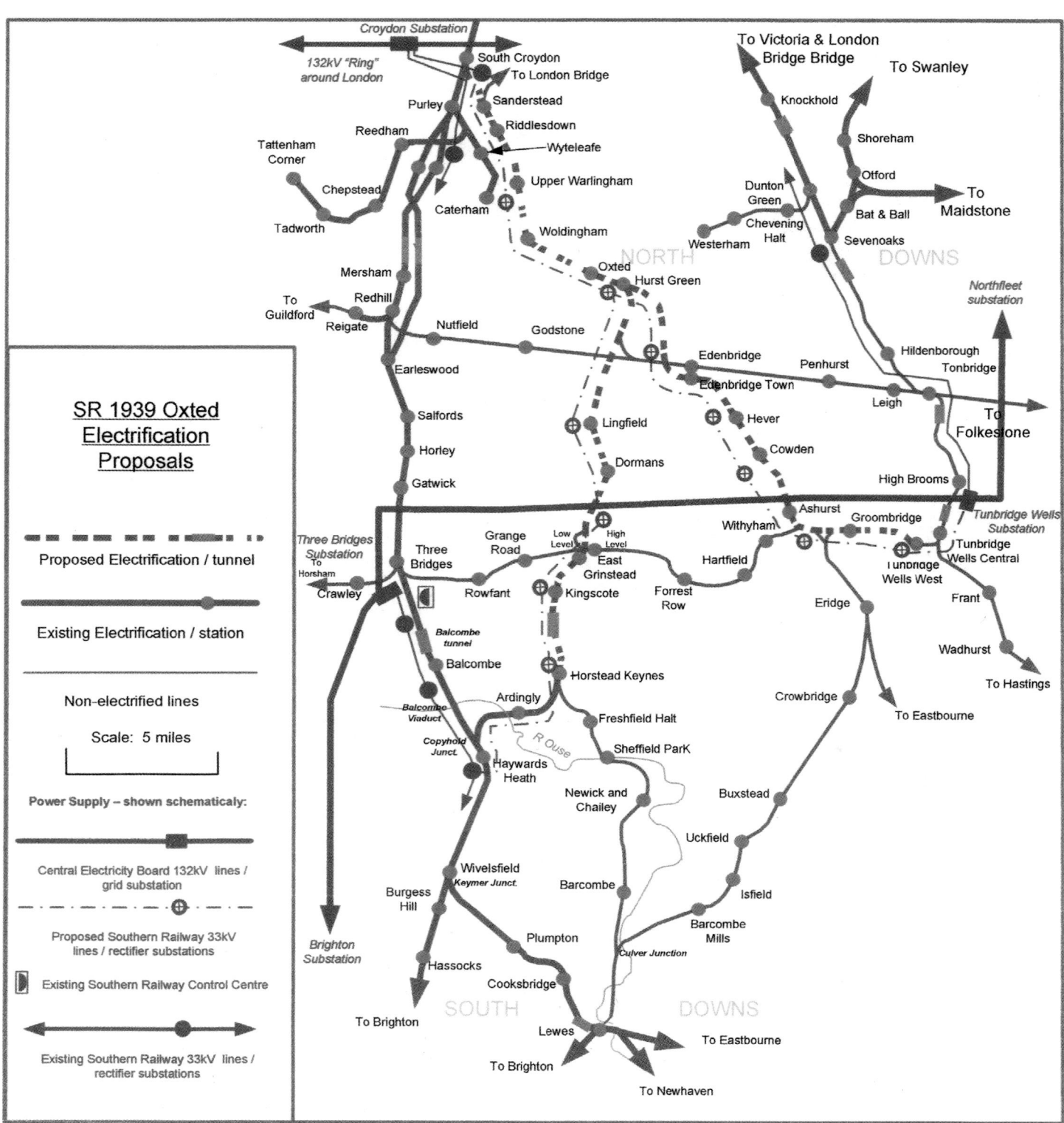

Map of the proposed Oxted electrification scheme. (Author)

1930. The Southern Electric operated a total of 3,189 electric train carriages. In 1932, 20,651,000 electric train miles had replaced 8,152,820 steam train miles and these figures had risen by 1939 to 41,241,682 electric train miles replacing 18,796,806 steam train miles.[37] It had been an unprecedented effort to carry out such a massive programme without the support of the engineering establishment or the wholehearted approval of the government. The concept of the low voltage direct

Class 207 diesel-electric multiple unit with a train from Uckfield to London Bridge via Hever at Hurst Green – the proposed junction of two electrified lines. Without electrification as desired by Raworth and Walker it was these diesel trains that were to replace steam. In 1987 however, the line to East Grinstead, which can be seen diverting to the right, was electrified – a partial completion of Raworth's plan. (Peter Winchester/Southern Notebook (Southern Railways Group))

current third rail may have been deemed inadequate for modern use, but Alfred Raworth, backed by Sir Herbert, made the very best of the system by introducing new equipment such as the ASEA control system, modern rectification and the 33kV distribution network interconnecting with the new National Grid. The aim had been to electrify the railway, improve services for passengers, increase revenue and pay dividends, and not to impress the electrical engineering purists or comply with the policies suggested by the Whitehall mandarins. By using in-house management and labour as far as possible, it was not necessary to spend a large portion of the budget on consultants or subcontractors. Furthermore, what made the Southern Railway's efforts even more commendable was that the work was carried out during one of the worst economic depressions in modern history.

CHAPTER 22

THE SOUTHERN ELECTRIC AT WAR

With the declaration of war, following a well-prepared plan, the Southern Railway moved its headquarters from the vulnerable Waterloo station in central London to Dorking in rural Surrey. The railway's wartime home was to be the rather dilapidated Deepdene Hotel which was a short walk from both of Dorking's railway stations and insulated from the town by a pleasant park. It had been well prepared for its new role with the provision of a telephone exchange and air raid shelters. The building had once been a stately home frequented by Churchill and Disraeli,[1] but it had latterly gained a rather unsavoury reputation after it was bought by Maundy Gregory. Gregory earned notoriety after the Great War by receiving commission for his part

The Southern Railway's wartime headquarters at the former Deepdene House – later Deepdene Hotel – in Dorking. The building was demolished by British Rail in 1967. Some of the gardens remain as part of Dorking Golf Club and form part of the Deepdene Trail. (Author's Collection)

in a clandestine scheme to sell baronetcies and knighthoods to the wealthy. Allegedly Gregory passed between £1 and £2 million into the coffers of the ruling Liberal and Conservative parties and with his cut he bought the *Whitechapel Gazette*, the Ambassador Club in London's Soho and the Deepdene Hotel at Dorking. Apparently, Gregory blackmailed many of his clients frequenting his establishments after gathering gossip about their sex lives, and the Deepdene Hotel subsequently acquired a reputation as the biggest brothel in south-east England.[2] Despite its unwholesome reputation, it was the respectable local resident Bulleid who recommended the hotel to Szlumper.[3] Most of the railway's senior officers, including Alfred Raworth and John Elliot, set up what they hoped would only be temporary offices at Deepdene. Sadly, the Dorking licensing justices refused to transfer the drinks licence from the hotel to the railway company.[4]

No sooner had the Dorking office been established than Gilbert Szlumper was unexpectedly seconded into the army. He was based at Southampton taking charge of all military traffic movements, and his replacement as general manager was not to be his assistant John Elliot, but the traffic manager, Eustace Missenden.[5] Missenden had risen through the ranks after being selected for promotion by Edwin Cox.[6] He did not have a warm personality and, in part because he had received only an elementary education, he developed a mistrust of anybody whom he deemed to be an intellectual.[7] It is unsurprising then that there was to be friction between himself and the public school educated technocrat Alfred Raworth, but paradoxically he too was won over by the highly intellectual, but nonetheless disarmingly charming, Oliver Bulleid. Szlumper did not return to railway service as he had hoped but was later to be appointed as the Ministry of Transport's railway control officer.

As during the Great War, the railways were now brought together under 'government control' through the REC, which was officially an advisory body to the Ministry of Transport with the remit to coordinate the wartime operations of the London Passenger Transport Board (LPTB) and the four main line railways.[8] In reality this resulted in control passing from the railway company directors, not to the government as such, but to the senior railway managers making up the committee – although there were to be many tensions and conflicts between the REC and the Ministry of Transport. The chairman of the REC in 1939 was the chief general manager of the LNER, Sir Ralph Wedgwood. Following Gilbert Szlumper's departure Eustace Missenden served on the committee representing the Southern Railway and occasionally John Elliot deputised for him. The Mechanical and Electrical Engineering (M&EE) subcommittee had been established early in 1939 in anticipation of hostilities and its first meeting took place in the LNER boardroom at King's Cross station on 3 March 1939. This subcommittee, consisting of the chiefs from the main line railways and the LPTB, elected Sir Nigel Gresley to be chairman.[9] During the war period Oliver Bulleid missed very few meetings. Alfred Raworth also attended the meetings but not as often as Bulleid. On occasions when a reduced committee discussed purely electrical engineering matters Raworth took the chair, making him the de facto chief electrical engineer for the whole of British railways. During the war period the M&EE committee dealt with a wide variety of issues such as the pooling of all private owner wagons, construction of new wagons, locomotive design issues, locomotive availability, the locomotive building programmes and railway staff wages issues. On 1 May 1941, the REC became part of the Ministry of War Transport with Frederick Leathers, 1st Viscount Leathers, being appointed as the first minister in charge.

During the first few months of the war the military authorities arranged for armed guards to be posted at many strategic railway locations, including Raworth's control rooms at Three Bridges, Woking, Swanley and Havant – but the underused Ore control room was of less strategic importance and was not to be guarded. By 1940, due to stretched resources, these guards were withdrawn and the Southern Railway itself was instructed to police access to these vital establishments by permitting admittance only to railway staff with proof of identity.[10]

Being situated in the south-eastern corner of England the Southern was in the front line for all the 'Big Events' such as the evacuation of children and mothers, the embarkation of the BEF; the retreat of the BEF from Dunkirk and later the mobilisation for the invasion of Europe on D-Day. To free up the railways for essential wartime traffic – which included the transport of coal which could no longer be safely carried by the coastal steamers – draconian reductions in passenger services were ordered by the Ministry of Transport – much to the distaste of the managers on the REC who put up some resistance to these unwelcome service cuts. At the outbreak of war the REC consultative operating subcommittee instigated a severely curtailed passenger service with trains running at reduced speeds to accommodate the anticipated military traffic and the immediate evacuation of cities – all taking place amidst air raids.[11] Despite the gloomy predictions there followed a period of relative calm – the 'phoney war' – and within days it became obvious that the original hastily produced timetables could be relaxed – although the term 'emergency timetable' would continue to be used.[12] The REC agreed that each company could reinstate cancelled trains subject to a speed restriction of 45mph between 'booked points'.[13]

The Southern Railway introduced a provisional timetable on Monday, 11 September which did not show any reduction in the number of trains running but indicated that services were slower with more intermediate stops.[14] When the first wartime winter timetable was published in October 1939, it showed many reductions in services compared to the September timetable. While the Southern Electric suburban services were to run at the same speed, there was a 50 to 70 per cent reduction in the service,[15] for example where there had been a 20-minute interval service from Charing Cross to Caterham this had been reduced to an hourly service. It was claimed however that the 'business trains' serving London Bridge and Waterloo for the City were providing a 'practically normal' service.[16] The main line Southern Electric services from London to Portsmouth were reduced to 40 trains a day with a 20 to 45 per cent deceleration. The electrified Brighton route had a more substantial reduction in trains running – from 100 to 46, but with only a 3 per cent speed retardation.[17]

The REC had proposed on 8 September the elimination of most restaurant and sleeping cars, reasoning that additional carriages providing more passenger accommodation could be added to the decelerated passenger trains without 'exceeding the locomotives' power capabilities'.[18] Nonetheless the Southern did not withdraw its catering vehicles, as most of these were part of the fixed formation electric trains making their removal and replacement by ordinary carriages problematic, and it would be difficult to explain to passengers why they could not have meals on the long-distance steam trains to the West Country when they could on the shorter electrified routes to south-east England. This made the position of the other companies invidious, so on 29 September the REC agreed in principle that restaurant and buffet cars could be reintroduced on 'limited lines'.[19] In 1942 however this concession was withdrawn and all the catering vehicles within the multiple units were taken out of service and stored awaiting the cessation of hostilities.[20]

The coastal areas served by the Southern Electric were deprived of custom as there were to be no more holidays on the south coast of England. While the nation was under threat of imminent invasion after the Dunkirk retreat, the minister of home security, Sir John Anderson, issued an Order creating a 'defence area' along the coast from the Wash to Rye in Sussex which extended 20 miles inland. Only residents were permitted to travel to the area unless there was a genuine business need, and police were expected to enforce this Order which specifically barred anybody making a journey for holiday or pleasure purposes.[21] Later, in July, the defence area was extended as far as Portsmouth and included the Isle of Wight.[22] Travellers would be stopped and questioned about their reasons for entering the area.

For the Southern Railway such passenger service reductions might have been a commercial disaster, but the government wished to ensure that the pre-war railway profits of £40 million should not be reduced.[23] Achieving this was possible due to the massive increase in wartime freight traffic and, as a consequence of the extraordinary wartime conditions, some classes of passenger traffic were on the increase, despite the government's 'Is your journey really necessary?' campaign. Much to the government's dismay, passenger traffic did not dramatically reduce. The reasons for this were complex.

The REC, when challenged by the Ministry of Transport, explained that much of the additional traffic was on government business or was government subsidised. This included: large concentrations of HM forces travelling to various parts of the country; the extra traffic due to evacuated workers following air raids; the evacuation of government departments and private businesses to 'safe' areas remote from cities such as London; the wide dispersal of munition factories and the need for workers to reach them; and extra traffic due to the dispersed evacuated civilian population. Meanwhile local and suburban traffic had continued, with passengers travelling on season, workmen's and cheap day tickets. After the initial organised evacuation, special reduced fares were being offered to relatives who were visiting evacuees in the 'safe' areas. Civilians also travelled to locations close to where their relatives were stationed at military bases – there were reduced rates available for them too. Many large companies and organisations had evacuated themselves to the provinces creating new passenger traffic. Furthermore there remained a 'considerable amount' of first and third class business travel using the long-distance trains.[24]

The Southern was not so fortunate – the coming Blitz would affect London suburban traffic. Also – despite protestations from the government – the British public still craved their annual holiday, and the other railways' holiday traffic was sustained. The whole of the coast served by the Southern Electric however, was for most of the war period within the 25-mile defence area. To advertise the extent of the prohibited zone, a list of railway stations to which you were not permitted to buy tickets for was published.[25] The list included Hastings, Bognor, Brighton and the whole of the Isle of Wight.

Early in the war two events which are now deeply etched into the nation's consciousness took place in which the Southern Railway was a vital participant. Over a period of three days in early September 1939 the Southern Railway operated 261 steam-hauled trains and 456 electric trains from London main line and suburban stations to evacuate children to rural or seaside destinations across the country. This was achieved in-between the morning and evening peaks to avoid any interruption to the regular business traffic. For example, electric trains transported children and some of their mothers to Reading for dispersal to welcoming households in the surrounding Berkshire and Oxfordshire villages. The second major challenge for the railway was the conveyance of evacuated troops away from the Channel ports after the defeat of the BEF. For this task, the Southern Electric took no part: the troops were all moved from the non-electrified Kent

Evacuated children arriving at Reading station. They would have arrived in the then new Southern Electric 2-BIL units, one of which can just be seen standing in the station. (Reading Borough Council)

coast terminals through the electrified area with only Aldershot being a military destination that could accept electric trains. The enormity of the task confronting the Southern can best be illustrated by the operation of the troop trains through Redhill. Between Monday 27 May and Tuesday 4 June 1940, a total of 560 special trains passed through the station.[26] Following the Dunkirk evacuation there was a second mass evacuation of children due to the perceived threat of imminent German invasion. The Southern Railway's contribution to this second exercise was 84 special trains conveying 42,391 schoolchildren away from Waterloo and other suburban stations in south-west London.[27]

One significant event, completely forgotten by the public, took place between these historic events. In early 1940 it was not the effects of any Nazi bombing that brought the railways to a standstill but the exceptional weather during the winter of 1939–1940 which was the coldest for 45 years. The reason why this severe winter has been largely forgotten is because the effects of the weather were unreported in the press and on the BBC due to wartime censorship. Of course, the public knew the weather was exceptionally severe, but what they were not told was how widespread the appalling weather conditions were, or how debilitating were the effects on transport, communications and the availability of essential supplies. The wintry weather started in the latter half of December with freezing fog and severe frosts. Heavy snowfalls accompanied by frost resulted in January 1940 being the coldest month since February 1895. On 23 February, a temperature of minus

20 degrees Celsius was recorded in Canterbury. The River Thames froze between Sunbury and Teddington and even the sea froze over at Bognor Regis, Folkestone and Southampton.[28]

The January storm resulted in precipitation on the lower lying areas across southern England falling not as snow but as freezing rain. This was due to a rare meteorological event that occurs when a warm front enters an area of very cold air, creating a 'sandwich' where a layer of warm air has cold air above it as well as below. Snow falling from the higher cold area melts as it falls through the warm air but is then supercooled, as raindrops freeze instantly upon reaching the surface. In 1940 this resulted in everything being covered by a sheet of ice as the air temperature at ground level never rose above minus 2 degrees Celsius. The weight of the ice that built up on the Southern Railway's telegraph lines broke and brought to the ground many wires, causing disruption to the telegraphic and signalling systems. Worse, the ice covered the conductor rails, often to a thickness of 1 inch (25 millimetres), bringing all electric trains to a near halt. Flashovers from the conductor rail to earth were also reported.

Any forward movement of the electric trains was accompanied by severe arcing as the trains could only advance at a slow walking pace at best. On the evening of Saturday 27 January many electric trains on the Western and Central Sections were stalled and stranded, although suburban trains on the Eastern Section fared better. On the following Monday, a few Western and Central Section suburban trains were running again, but for three days some country and main line electric trains were assisted by steam locomotives. This resulted in exhilarating performances when the collector shoes were able to make a good contact with the conductor rail and dramatic spectacles with blue-green arcs from the shoes and an eruption of exhaust steam from the locomotives when contact with the live rail became almost impossible due to the ice. The few electric services that were run were formed of either a single or a pair of the Brighton line six-car sets, or two of the four-car sets (4-CORs). The intention was that the trains would draw current whenever possible, with the steam locomotive assisting when ice was encountered. Where possible the electric current would be used to heat and light the trains and to drive the Westinghouse pumps for the brakes.

On the Western Section only the screw coupling connected the steam locomotive to the train, as the locomotives only had vacuum brakes, and it is reported that the steam locomotive driver took charge of the train communicating with the motorman (who had the brake lever) by hand signals. Steam drivers were keen to engage in rapid accelerations wherever possible to avoid the ignominy of being propelled by an electric train. On the Central Section 12-car trains were assisted by former LBSCR locomotives which were fitted with the compatible Westinghouse brakes.[29] The cold weather continued into February; when there was a short thaw followed by a return to the cold, but mild weather eventually prevailed on 20 February.

One of the perceived disadvantages of third-rail electrification is that it can be brought to a standstill by the build-up of ice on the conductor rail. The Southern Railway did learn from its winter 1940 experience, for in January 1941 as well as running extra trains at night to prevent the build-up of ice, trials with a special ice scraper were carried out.[30] One assumes there was some success, for in September 1942 Alfred Raworth obtained board approval for the fabrication of 200 pneumatic ice scrapers for use on electric trains, the cost of which was to be, with a stock of spare parts, £2,277.[31]

In August 1940, the Blitz began. Trains continued to be run during daylight raids with a speed restriction of 25mph in force for both steam and electric trains following the 'alert' warning.[32] During the raids staff would remain at their posts, taking what shelter they could, ready to attend to any incendiary devices. During the London Blitz between 24 August 1940 and 10 May 1941 there were only two days when the Southern Railway was not subjected to aerial attack and the railway staff had to bravely keep the trains moving and attend to the many fires caused by countless incendiary bombs damaging at one time or another all the principal stations.[33] When Waterloo was bombed, the boardroom was destroyed and subsequently meetings were held at the Charing Cross Hotel. Surprisingly, the public had little realisation of the railway's woes. This was mainly due to wartime censorship, but also the speed by which the dedicated staff managed to restore track and signalling, minimising the ensuing disruption and ensuring that services ran as advertised as soon as possible.[34]

Despite the fears of retired rear admirals, the Southern Electric proved to be very resilient during the hostilities. This was to be confirmed in a post-war Ministry of Transport report.[35] It was often possible to allow the lighter electric trains to pass at a low speed over damaged bridge structures when a heavy steam locomotive would have had to be barred. The interconnected network of electrified lines allowed local staff to organize ad hoc traffic rearrangements when the line ahead was blocked, by being able to reverse the electric trains over crossovers at stations then re-routing the service. This was particularly useful when due to bomb damage ahead the services had to be terminated at intermediate stations and reversed – often not possible with steam locomotives. The servicing and berthing depots for the electric trains were over a dispersed area and did not offer the Luftwaffe a single tempting target such as a large and conspicuous steam locomotive motive power depot. The target of the bombing was often the rolling stock, resulting in damage to the track and conductor rails upon which the targeted carriages or wagons had stood. The Southern Railway was able to repair the conductor rails and any local cabling as fast as it could remove the wrecked rolling stock. One of the worst periods for the Southern was 16 and 17 April 1941 when there was extensive damage to Waterloo station and the line to Clapham Junction – this did stop the electric trains when all lines were blocked, and priority was given to restoring lines for steam traffic before the restoration of the current to the conductor rails.[36]

No Southern Railway substations were damaged during the Blitz but had this happened and a single rectifier substation had been put out of action, the direct current for the conductor rails would still have been available from adjacent substations subject to any local repairs to the connections near the stricken substation.[37] Nonetheless there was concern about the possibility of air raid damage to the control centres and substations, so at some important locations a degree of camouflage was provided.[38] Raworth's 33,000V electricity distribution system which interconnected the Grid bulk supply points gave assurance that even the destruction of one of the Grid substations would not have disrupted the electricity supply to the trains. The National Grid proved to be reliable – a tribute to the 'gridiron' concept espoused two decades earlier by Lord Weir, for at no time during the Second World War did the Grid close in its entirety.

The most serious and potentially catastrophic incident to affect the Southern Electric was a Luftwaffe attack on Durnsford Road generating station on 14 October 1940, when a German bomb struck the west chimney and main flue causing serious damage to the boilers below.[39] Eight of the staff were injured.[40] The loss of the draught from the chimney resulted in flames billowing from the boilers into the boiler house. Wilfrid Smith, the charge engineer on the night, and junior charge engineer Frederick Sheldon organised a group of volunteers to draw the fires. They worked for over three hours in appalling conditions surrounded by showers of red-hot cinders. Bathed in light from the fires they were exposed to the enemy aircraft. Wilfrid Smith kept to his post despite suffering scalding injuries and was awarded the George Medal. Frederick Sheldon received the British Empire Medal.[41]

Fortunately, there was no damage to either the generators or the 11,000V switchgear but all supplies to the former LSWR electrified lines were cut. The Southern Railway understood that it had been lucky, and Raworth was praised for his 'remarkable good work'[42] in organising the improvised repairs so that 10,000kW – about a third of the station's maximum output – was made available[43] three days later. Raworth promised that 21,000kW would be available in two weeks' time.[44]

Due to the wartime timetable fewer trains were being run, but during this emergency those that were running had to run at a reduced speed[45] and without heating while full repairs were carried out. Fortunately, the original LSWR electricity supply network now had limited interconnection with the Stowage Wharf generating station at Deptford through Queens Road Battersea substation, which was situated where the SECR and LBSCR lines from Victoria cross the LSWR lines from Waterloo. A further 11,000V cable from Waterloo station to the SECR's Waterloo East station had also been provided.

The Durnsford Road 11,000V network was synchronised with and run in parallel with the Deptford supply during this emergency period. No part of the system was without power, but the voltage on the conductor rail at many locations was significantly reduced, slowing down the trains.[46] On 10 January 1941, Raworth reported to Eustace Missenden that he hoped to have full power restored in four weeks' time, but the general manager was not impressed. Missenden, not letting the war detract from the railway's public service ethos, was concerned that, as he perceived it, the irritatingly slow progress was causing serious inconvenience to the travelling public. He was not an engineer so may not have fully understood the issues, but he insisted that every effort be made to concentrate on work to complete the repairs as soon as possible. At a minuted meeting of all the senior officers of the Southern Railway, Missenden was openly critical of Raworth's efforts and insisted that he cease his independent actions and obtain assistance and any necessary extra staff from the chief engineer's department.[47] He believed that Raworth had become isolated from the management team as a whole and was continuing to behave in an independent manner as he had once been allowed – and encouraged – to do in the 'Walker' era.

Missenden offered to approach the REC for assistance in obtaining materials for the Durnsford Road repairs. On 17 January, Raworth was able to inform Missenden that the remedial work, which included a new chimney made from steel – a very precious commodity at the time – would be completed on 3 February and that he would duly notify the traffic department a few days in advance to allow it to resume a normal timetable, a timescale not significantly different from his earlier prediction despite the availability of 'assistance'.[48] This near disaster prompted Raworth to request the installation of a barrage balloon above the generating station.[49]

Following the repairs at Durnsford Road, Raworth oversaw the installation of three additional 11,000V cables to increase the interconnection between the former LSWR lines supplied from Durnsford Road and the former LBSCR and SECR lines supplied from Deptford. A new switch house was built at Wimbledon next to Durnsford Road generating station to connect these new cables which then ran east to Streatham.[50] The exigencies in carrying out such work in wartime were again evident, as Raworth experienced frustration due to the difficulty in obtaining switchgear for this project and Missenden had to again escalate the problem to the full REC.[51]

The Durnsford Road incident highlighted the vulnerability of the earlier electricity distribution arrangement

A post-war (September 1948) view of Durnsford Road generating station showing the replacement steel chimney. (Meredith/Transport Treasury)

with only one centralised power station which was vulnerable to enemy action. The concept of a pair of cables, securely located on railway-owned land to provide a duplication of supply was appropriate in peacetime but susceptible to enemy attack – an air strike on one cable would probably result in both cables being put out of use. This did happen on a few occasions. To expedite cable repairs as quickly as possible, Raworth's electrical department benefited from the full-time services of several cable jointing teams on loan from cable manufacturers to work as required on the direct current and the alternating current high voltage cables. Raworth had little regular work for them to do, but they were retained to be available for emergency repairs.[52] There was an incident due to an unexploded bomb adjacent to the 11,000V cables between Deptford and the main Lewisham intake substation. This bomb, described as a delayed action bomb (DAB), had to be dealt with by a bomb disposal team: had it exploded the total power available would have been reduced by more than 50 per cent.

On 11 January 1941, it was Portsmouth's turn to experience the full force of 153 Luftwaffe bombers. The city's two main railway stations suffered severe damage. A train of 4-COR stock standing at Portsmouth Harbour was marooned when part of the viaduct structure on which the station is built was destroyed. The damaged coaches were not to be removed by crane until September 1946.[53] The Southern Railway's woes on that dreadful night were slight compared to the trauma suffered by Portsmouth as a whole. Some 171 civilians were killed, and the major structural loss was the destruction of the Guild Hall, the venue for the celebrations when the Portsmouth Direct electrification had been officially opened. Also damaged were many other commercial buildings, a hospital and Clarence Pier, Southsea. The FA Cup, won by Portsmouth in 1939, was dug out unscathed from the vaults of a bank which had also been severely damaged.[54]

More serious than the damage to infrastructure was the tragic human cost. One morning in May 1940, the critic of railway electrification Josiah Stamp, now Lord Stamp and LMS president, failed to arrive for a regular meeting of the REC. After several telephone calls, the assembled railway managers were given the shocking news that Stamp had been killed, together with his wife and son Wilfred, when a German bomb had made a direct hit on their air-raid shelter at their home in Shortlands, Kent.[55] Equally shocking to the world of electrical engineering in 1940 was the death of Charles Merz, who, at the age of 66, was also killed during an air raid at his home in Kensington, together with his children and house servants.[56]

The war also directly affected Alfred Raworth's family. His brother Basil, who had been living in Hong Kong as the eastern director of the General Electric Company, managed to escape from the colony on the very last ship bound for Australia before the Japanese invasion. In Australia he re-joined his wife, Winifred, and his daughter, who had preceded him to safety in Sydney. Their eldest son, John Smith Raworth, who had been educated in England and spent his school holidays with Alfred and Ruby Raworth, was now Major Raworth in the Royal Artillery. Before leaving England to be posted in North Africa he had married Miss Marguerite Hone from Inverness.[57] In 1943 Major John Smith Raworth commanded 155 Battery Royal Artillery defending the remote Sidi Nsir railway station in Tunisia. Alongside 155 Battery were infantry from 5 Battalion Hampshire Regiment. On 26 February, the Germans tried to break through the British lines,[58] attempting to reach the town of Beja. Sidi Nsir station was attacked by mortar fire, Panzer tanks and later suffered a fierce aerial attack by Messerschmitt fighters. Virtually surrounded, the British soldiers doggedly defended their position enabling a defence line for Beja to be successfully established.[59] The enemy eventually overran Sidi Nsir station and only nine men from the 155 Battery, including Major Raworth, survived, and were captured and taken prisoner. Raworth was now a hero and when the news reached the British press, a proud Ruby Raworth was delighted to inform a reporter that her nephew was a fine 'all-round sportsman'. Alfred and Ruby Raworth shortly received a postcard via the Red Cross from their nephew in an Italian prisoner of war camp to say that all was well.[60]

Meanwhile, the work of the REC continued. In July 1941 the M&EE subcommittee dealing with purely electric matters and chaired by Alfred Raworth discussed the desirability of the railways joining with the CEB and the municipal electricity authorities to negotiate with the British Electrical and Allied Manufacturers Association a joint purchasing agreement based on a standard 'cost plus percentage' basis.[61] It was agreed that Raworth, with his proven ability to broker the best of deals for his own projects, was to represent the railway companies in the negotiations. Coincidently, on the following day at the full REC meeting a report was presented by Colonel Trench from the Ministry of War Transport recommending that all the electricity cables required by the railways should be ordered through the Ministry of Supply. This suggestion was to be forcibly rejected by the committee, which, led by Alfred Raworth in this instance, resisted such government interference and insisted that this proposal would only introduce delays, adding that there was a history of beneficial contracts between the railways and manufacturers such as that between the Southern Railway and Pirelli.[62]

One of Alfred Raworth's duties as a member of the M&EE subcommittee was to represent the REC on the Ministry of Supply's 'Standardisation of Electric Cables and Wires for Government Use' committee. The railways were requested to pass any technical queries relating to electric cables to Raworth for discussion at the meetings.[63] This for Raworth must have seemed like a return to his days during the First World War when he worked at the Ministry of Munitions. He told the Southern Railway senior officers' conference that any new cable standard would not affect the railway's electrical procedures but would mean an acceptance of modified cable specifications and revised testing procedures. The purpose of these changes was to speed up cable manufacturing.[64] One result of the committee's deliberations was the accelerated change from the use of rubber to PVC as an insulating material in cables and wires.

Just as in 1898 when John Smith Raworth had reached the pinnacle of his career, by 1941 his son Alfred probably thought that he had reached the peak of his. He had completed his most important work – the Southern Electric electrification – and was now engaged at the highest level in the governance of the nation's railways during the difficult war years. He had never proceeded from associate membership to full membership of either of the electrical or mechanical engineers institutions. Whatever his reasons for this oversight, his membership would have made little practical difference to his career to date, for unlike the academic institutions or the small consultancy firms, the larger corporations – such as the Southern Railway – cared little for such 'letters after the name' adornments as most senior managers had proved their abilities as they climbed the promotion ladder. But in the circles that Raworth was now moving most engineers had some form of professional membership, so perhaps it was now time for him to get himself 'elected' to an institution.

It must at first seem rather strange that given his education and training Raworth sought to join the Institution of Civil Engineers and be allowed to append 'MICE' to his name. But he was now a railwayman and a railway is a massive piece of civil engineering upon which certain electrical and mechanical contraptions just happen to run. He was also out of step with his colleagues in the electrical engineering world – they all appeared to believe in Charles Merz's 1,500V electrification design – and the mechanical engineers' views on railway motive power were most likely to be the same as those of his rival Bulleid. So, it is not so surprising that Alfred, like his father, felt more 'comfortable' in the 'civils'. Raworth may have been fully in charge of all the electrical equipment, but the electrification projects involved considerable civil engineering, not only the substation buildings but many relatively minor civil works involving modifications to stations, tunnels and bridges, etc. Furthermore, his toughest task during the war was probably the reconstruction of much of the Durnsford Road generating station. He was elected straight to membership of the Institution of Civil Engineers in 1942.[65] The proposer for his admission was his colleague, the Southern Railway's chief (civil) engineer, George Ellson. Those providing supporting evidence for his application included Gilbert Szlumper and C.E. Fairburn – the chief electrical engineer for the LMSR. Despite such solid recommendation, Alfred had to pass the harsh Institution examination – not a comfortable task for a 60-year-old man.[66]

The war dragged on and the Southern Railway increasingly played its part in moving military traffic in preparation for D-Day. After some respite from bombing the Southern along with the rest of south-east England began to suffer the effects of first the German V1 flying bombs – the 'doodlebugs' – and later the V2 missiles. The 'V' stands for *Vergeltungswaffen* which has been translated as 'retaliatory' or 'reprisal' weapon,[67] thereby relating to the German response to the massive RAF and USAAF bombings of the large civilian populations of Germany. Both V weapons held their individual terrors for the civilian population: the doodlebugs first announced their arrival by the throbbing noise from their primitive ram-jet engines and then, following cut-off, there would be incredible tension while those below waited to find out exactly where the bomb was to land. The V2s came with a different class of terror: travelling faster than the speed of sound they arrived without any noise warning whatsoever. The Southern Electric lines suffered considerable damage from the V1 doodlebugs at various locations, including structural damage to Victoria, Wimbledon and Falconwood stations and to bridges at Merstham and Peckham Rye.

The attack on Falconwood station on 31 August 1944 was particularly serious as the pair of 11,000V cables from Lewisham to Dartford was damaged, affecting the Southern Electric services to Dartford by all three routes. Raworth was requested to investigate the possibility of providing an additional cable route to reinforce the electricity supply[68] – this would have been achievable by extending a pair of cables either from Belvedere or Sidcup on to Dartford substation.

On the last day of the V1 attacks the bridge at Peckham Rye carrying the former LCDR Catford loop over the former LBSCR South London line was so badly damaged that it had to be taken down and replaced by a temporary military style trestle bridge. Raworth's 11,000V cables from Lewisham to Victoria were attached to this bridge and a single cable was laid on the temporary bridge to restore supply. Meanwhile, taking full advantage of the Southern Electric interconnectivity, trains were diverted to Holborn Viaduct.

Hampton Court Junction suffered the effects of a V2 rocket and in Bermondsey a bridge carrying the South London lines across Southwark Park Road was destroyed necessitating the construction of another temporary trestle bridge.[69]

After the war when speaking at Raworth's retirement ceremony, Missenden stated that he 'did not know where we should have been during the war had it not been for our electric services' and he then gave due credit for this to Raworth saying that his chief electrical engineer had 'relieved him of a tremendous amount of responsibility'.[70]

CHAPTER 23

THE ELECTRIC LOCOMOTIVES

The Southern Railway was to build three main line electric locomotives for use on its third-rail system. Railway enthusiasts know them as the 'Bulleid-Raworth booster locomotives'. Were it not for Alfred Raworth's name being attached to these locomotives – often otherwise referred to as 'Hornbys' – it is unlikely that he would ever have had any recognition from the amateur railway fraternity.

Raworth understood that to fulfil his dream of an 'all-electric' railway conveying all classes of traffic using the live third rail, it was a political necessity to design and build an electric locomotive to confound those critics who argued that only multiple unit electric trains could satisfactorily run on the Southern Electric third-rail system. At the time, the railway had only a minimal need for electric locomotives, but if the Southern Electric was to spread further east across Kent to the Channel ports, locomotives would be required for the boat trains and the heavy coal trains from the Kent coal field.

The evolution of Raworth's design for a main line electric locomotive had taken almost 10 years. There were those, including the Ministry of Transport's chief inspector of railways, Colonel Mount, who publicly affirmed in 1936 that on a third-rail main line electrification, multiple unit trains *had* to be used and therefore there could only be electric passenger trains.[1] The basis of this view was that a third-rail locomotive, unlike those drawing current from overhead wires, would become 'gapped' when passing over complex junctions and level crossings where the conductor rail was momentarily discontinued. Gapping was far less likely to occur on a multiple unit train where the pickup shoes at either end of the sets were sufficiently far enough apart to ensure continuous collection from one or other of the end bogies. It was not practical to follow the lead of the Metropolitan Railway which added additional pickups to carriages in the train to overcome this problem, since any Southern Railway electric locomotives would not be used to haul fixed sets of carriages on commuter routes – that was the multiple units' task. A second difficulty, relating to the gapping issue, was that goods trains at the time were mainly made up of unbraked four-wheeled wagons, loosely coupled together with three-link chain couplings. This meant that, unlike passenger vehicles which were closely drawn together by screw couplings, the goods wagons were free to bunch together and then be snatched apart if the train faltered in any way, such as would occur when a third-rail electric locomotive coasted across gaps in the conductor rail.[2] The resulting battering, it was claimed, would cause damage to the wagons and vans and there was a high probability of injury to the train guard travelling way behind in his wooden four-wheeled van. Furthermore, even the most sanguine of exponents for the third rail would never consider the use of a live rail in a goods yard where freight was loaded and unloaded by railway staff and local merchants.

To test out this generally held belief, a series of tests were carried out along the Brighton main line on two Sundays in October and November 1936 using one of the new express power cars as a third-rail 'electric locomotive'. As predicted, these tests not only confirmed that gapping was an issue but they also clearly showed that the standard multiple unit control system was not flexible enough for freight train working.[3]

At the time of the Brighton electrification, Raworth turned his attention to the possibility of building an electric locomotive suitable for hauling freight trains and for use in goods yards. Following discussions with Sir Herbert Walker, he began to actively consider a design for an electric locomotive which had on board a small auxiliary diesel engine powering a dynamo to facilitate movement at low speeds in sidings away from the electrified lines. Such a locomotive would be able to release itself when gapped and at complex junctions where gapping was inevitable it would be possible to temporarily switch over to diesel power. This type of locomotive has since been termed an *electro-diesel* and this Raworth 'innovation' did not materialise on British Railways until the Southern Region introduced in 1962 what were to become the Class 73 locomotives. In his 1919 report for the SECR Raworth had proposed the use of electric locomotives but did not provide any details. It is probable that he had in mind the electro-diesel concept and drew upon the research into diesel engines which he had carried out when considering the proposal to build dispersed diesel generating stations for the LSWR electrification. With his intention known, Raworth was not only actively canvassed for sales by British diesel manufacturers such as Armstrong Whitworth, but he also travelled abroad in his unsuccessful search for a suitable, reliable diesel engine that could be bought 'off the shelf' to avoid the Southern Railway incurring any development costs.[4]

Obviously, some form of auxiliary electrical supply arrangement such as a large battery was required, either to be continuously connected in parallel to the main traction supply – that is it would 'float' – or alternatively to switch in and out as required. In the 1930s however high speed heavy current switching devices were not available. Discarding the possibility of using heavy traction batteries, Raworth's thoughts then turned to the possible use of some form of kinetic energy device such as a large flywheel to overcome the gapping problem.[5] In 1936 the Southern Railway broke its long-standing relationship with Metropolitan Vickers and signed a new long-term contract for the supply of electric traction equipment with the English Electric Company. Raworth must have had some influence in this decision because even before the signing of the new agreement, he and the English Electric engineers had been working together on the design for an electric locomotive. This combined effort resulted in Raworth and the English Electric Company making joint applications for three patents, each titled *Improvements in Control Systems for Electric Motors*, in April 1936.[6]

These patents were duly published in October 1937. Two of them refer to relatively minor control details, but the main one – GB473137 (A) – described the control system for an electric locomotive 'drawing energy from an external source and controlled by a motor-generator set or booster set'. Put very simply, an on-board motor was to drive a generator – the motor-generator or booster set. The patent explained that these booster sets were to be 'designed to have a high moment of inertia, for example, by coupling it to a flywheel'. This would enable the motor-generator to keep running during any short break due to gapping and provide power to the traction motors. This is a bit of an over-simplification because the motor generator set formed an integral part of the traction motor control system. Raworth with his colleagues at English Electric had designed a 'buck-boost' control system – a development of what was at the time the well-known 'Ward-Leonard' method for finer speed control of electric motors, commonly used in mining applications. However, this came at the expense of extra weight and some loss of efficiency. Raworth and English Electric added a flywheel and some extra control contactors, and using this design the motor-generator was run continuously, and its output was used to control the speed of the locomotive by either assisting (boosting) or opposing (bucking) the voltage which was being supplied to the traction motors. The patent diagram shows four traction motors connected through the complex control system to a motor-generator set.

While the patents were being processed through the Patent Office, Raworth prepared a report in September 1936 for submission to the Southern Railway 'Locomotive and Electrical Subcommittee'.[7] In this report he recommended that initially two locomotives be built at an estimated cost of £31,000 for the pair. Herbert Jones, still the chief electrical engineer, and Alfred Raworth, together with Richard Maunsell, the chief mechanical engineer, had originally intended that much of the mechanical design and construction work should be outsourced. Initial enquiries had been made with Robert Stephenson and Company. The directors' final decision however was that the work must be done in house.[8] Speaking to the committee, Sir Herbert Walker explained that Raworth's estimated cost had been calculated on the basis that the English Electric Company would provide the electrical equipment as part of their long-term supply contract (much of the standard switchgear fitted in the latest multiple unit trains was to be used), and that the cost

A forlorn and discarded Raworth 'booster' set recovered from one of the electric locomotives when they were scrapped. (Nigel Calicott)

of the mechanical items, the body shell and bogies, would be charged to the chief mechanical engineer.

But when Oliver Bulleid arrived in 1937, he muscled in on the project and characteristically made it clear that he intended to take more interest in the design of the electric locomotives than had his predecessor, Richard Maunsell. So, for this project at least, there was a genuine and fruitful collaboration between Bulleid and Raworth. Significantly Bulleid did not wish to use the standard arrangement of two traction motors driving the two axles of a four-wheel bogie.[9] He designed a six-wheel bogie allowing the installation of three traction motors each rated at 245 horsepower (180kW).[10] This new design of bogie was pure Bulleid and described as 'bolsterless', for in place of a conventional centre pivot, four bearing pads were provided at each corner of the bogie frame. This novel arrangement allowed access to the centre traction motor, and this piece of Bulleid innovation enabled the six wheeled, three motor power bogies to be built.

With the onset of war and the lack of suitable design staff it is not surprising that a low priority was given to the construction of any electric locomotives. This contrasted with Bulleid's endeavours to proceed vigorously with his steam locomotive ambitions. When the REC was formulating its locomotive building programme, Oliver Bulleid had optimistically placed in his submission a request for 25 express passenger locomotives, and probably only Bulleid could be so totally self-assured and focused on steam locomotive development to make such an inappropriate proposition in wartime. These 25 locomotives were an increase from the original 10 that had been approved by the Southern Railway in peacetime 1938. The REC was not impressed with this proposal and in January 1940 the full committee was informed that the Southern Railway had 'substituted a high-speed passenger engine for one of mixed traffic type'.[11]

It has been generally accepted that the Second World War was a period when all technical and scientific development by the railways was put to one side to devote every effort to the winning of the war, even if this meant building up a backlog of maintenance and all plant and staff being worked to their limits.[12] But Oliver Bulleid, despite sitting at the heart of the wartime railway administration, superficially at least seems to have been totally oblivious to this reality. Such a controversial course of action as building a new design of large locomotive could not be carried out without argument and disharmony. The Southern Railway needed more powerful locomotives for the albeit less frequent but heavier wartime trains, and Bulleid's persuasive charms were used to the full when he endeavoured to obtain consent to build his new 'Merchant Navy' class locomotives. He believed that the thermal efficiency of the locomotive was not the most important element of the total cost per train mile, therefore it was not necessary to build locomotives for specific duties. His policy was to build fewer types of locomotive which could be used for a variety of duties, and this implied large locomotives. Bulleid's colleagues on the CM&E subcommittee appreciated his well-reasoned case and agreed on a 2-6-2 design. The very next day however, the 2-6-2 design had become a 4-6-2 which was approved at a full REC meeting. An error in the minutes or an overnight change of heart? Getting the proposal accepted by the tough Minister of Labour and member of the small wartime Cabinet, Ernest Bevin, was potentially a more demanding task. There was a meeting between Bulleid and Bevin to discuss construction of what the minister had understood to be an inappropriate high-speed locomotive. The two men were poles apart in their views: Bevin, a socialist, founder of the Transport & General Workers' Union, scourge of the pre-war London bus company managements and proponent of nationalisation, facing the autocratic Bulleid, a right-wing patriot opposed to nationalisation. Nonetheless Bulleid quietly convinced the minister that the Merchant Navies were to be 'mixed traffic' locomotives. But, also agreed at the meeting – probably as a quid pro quo – was that Bulleid was also to build 40 of his more appropriate Q1 'Austerity' locomotives.[13]

The first Merchant Navy locomotive emerged from Eastleigh works in February 1941. It was the first 4-6-2 'pacific' built for use on the Southern Railway. Bulleid's magnificent new creation was greeted with a mixture of awe and disbelief as its appearance was unlike any other contemporary steam locomotive. While the pre-war vogue had been for 'streamlining', Bulleid insisted that these locomotives were 'air-smoothed' having the round boiler shape hidden behind straight sides with a flat 'roof' above. The front end may have displayed less evidence of streamlining than Sir Nigel Gresley's speed record-breaking A4s, but any irregular features such as the outside cylinders were all neatly encased in a continuous wrap of sheet steel. Bulleid seems to have been influenced by many French semi-streamlined designs. The tender too was smooth sided to match the locomotive, with sides rising higher than was necessary with a neat fold inward at the top – these high side extensions being referred to as 'raves' at the time. Despite the current 'wartime black' policy the first of these new locomotives was painted in Bulleid's bright malachite green with three horizontal yellow stripes.[14] The entire valve actuating machinery for the three cylinders, most of which would on modern locomotives normally be on full display outside the driving wheels, was instead cleverly hidden away internally between the main frames. Here Bulleid took his lead from automotive technology by enclosing all the valve gear in a sealed oil-filled box and he used an ingenious chain drive to actuate the mechanism.

The locomotive's wheels were solid, with cut-away recesses and holes, resulting in a structure that was lighter but stronger than conventional spoked wheels. These were to be known as Bulleid-Firth-Brown (BFB) wheels due to Bulleid's collaboration with the engineering company Firth-Brown. Perhaps the most significant but least obvious feature was the magnificent Bulleid boiler hidden within the casing which, when first built, produced steam at 280 pounds per square inch – a higher pressure than the typical maximum of 250 pounds per square inch. The locomotive lacked any 'austerity' or 'utility' features. On the contrary it was equipped with a small turbo-generator to provide electric cab lighting and electric lamps on the front of the locomotive and on the tender, replacing the traditional oil lamps. The 'mixed traffic' description could – just about – be applied because the diameter of the driving wheels was only 6 feet 2 inches (1.9 metres), whereas 6 foot 6 inches (2.0 metres) was considered the minimum size of driving wheel for high-speed express passenger duty. Post-war experience was soon to show that these slightly smaller wheels were perfectly suitable for high-speed running; indeed, there are many accounts of Bulleid locomotives exceeding 100mph hauling heavy trains.

On 10 March 1941 there was a naming ceremony at Eastleigh where the first locomotive numbered in Bulleid's unorthodox notation, '21C1', was christened *Channel Packet*.[15] The choice of name honoured the Southern Railway's own fleet of cross-Channel ships and reflected that the new locomotives were destined to haul the heavy continental boat trains to the Channel ports, but it was a cheeky choice for these train services had been suspended for the duration of the war. At the naming ceremony the curtain was drawn back by the latest minister of transport, Lieutenant Colonel J.T.C. Moore-Brabazon, to display the elaborate new cast nameplate. Moore-Brabazon was a supporter of the Southern Railway, and he probably had some influence in allowing the locomotives to be built. The ceremony was witnessed by all the senior officers of the Southern Railway including the tetchy Alfred Raworth who would not have been impressed by such ostentation. Despite their obvious 'high speed express locomotive' pedigree, the Merchant Navies were put to good use hauling long and heavy wartime passenger trains to the West of England, as well as being used on lengthy goods trains. Twenty Merchant Navy class locomotives were to be built during the Second World War period, and then a further 10 post-war. Later naming reflected the shipping lines used by the Southern Railway's Southampton Docks, and naming ceremonies followed at regular intervals as the publicity-minded Southern Railway arranged for shipping line managers to 'name' 'their' locomotive. The second naming ceremony was 21C2 *Union Castle*. Raworth attended at least the first two ceremonies – probably under sufferance.

By comparison to Bulleid's Merchant Navy class locomotives, Raworth's two electric locomotives seem less of an achievement, but it was a more realistic wartime accomplishment. His third-rail locomotive had been largely developed before the war and had been 'promised' for several years. Nonetheless Raworth had to fight to get this project developed and built in an unsympathetic wartime climate. Despite his obvious desire to proceed with the project, realistically he knew that he had to be patient as there were other more pressing wartime priorities for both the English Electric Company and for Ashford works. In 1941 on the rather flimsy excuse that with the wartime shortage of footplate staff the use of an electric locomotive to haul the heavy goods trains in the Southern Electric area would require only one man in the cab, work on the first electric locomotives was resumed. Four and a half years after the scheme had been approved, Raworth was able to inform the English Electric Company that the first of the new locomotives had been towed from Ashford works and was now available at Brighton works for them to fit all the electrical equipment.[16]

Consequently, on 15 July 1941, the first of the 1,470-horsepower (1,080kW) electric locomotives emerged from Brighton works to be greeted without any of the fuss or ceremony that there had been for Bulleid's first steam creation. Unlike *Channel Packet*, to achieve a good monochrome photograph the new locomotive was painted in photographic grey with white lines to replicate the proposed yellow lining,[17] and it was to retain this rather drab garb until the end of the war when it was to be painted in the full bright malachite green livery with yellow lining, including the 'speed whiskers' on the cab fronts at both ends. Meanwhile the new Merchant Navies were being out-shopped in plain 'wartime' black.

Using Bulleid's unique numbering system which counted axles and allocated a letter of the alphabet to those powered, the new locomotives were designated 'CC' – two six-wheeled power bogies, no unpowered wheels, therefore the first locomotive was numbered CC1. It was almost 60 feet (17.5m) long, weighed 99 tons and had two booster sets on board, one for each power bogie. In an extreme situation when the locomotive was gapped and was not able to regain supply from a conductor rail and the motor-generator slowed down to a speed such that there was insufficient power to drive the traction motors, various contactors would open to safely shut down the system. For this rare event, a battery was provided to allow the locomotive to move off at a crawling speed. An electric boiler was provided to be used with the train's steam heating system. CC1 was designed to haul passenger trains weighing 500 tons at a maximum speed of 75mph or 1,000-ton freight trains.[18] To enable the locomotive to enter goods yards, overhead lines were to be provided just as Raworth had proposed for the Stage Three of his 1919 proposals for the SECR. Accordingly, a roof-mounted pantograph was fitted to CC1. The overhead conductors were to be of a lightweight construction with wires of a sufficient size to be able to provide enough current to gently start a 1,000-ton loose coupled train and run it at 15mph 'under the wires'.[19] The locomotive outwardly displayed

The first electric locomotive, CC1, in its 'wartime' livery. (Author)

THE ELECTRIC LOCOMOTIVES • 283

many Bulleid features: the 12 wheels of 3.5 foot (1.08 metre) diameter were of the BFB pattern and the driving cabs at either end closely resembled the contemporary 2-HAL units built for the Maidstone electrification.

After the appearance of CC1 the second locomotive, numbered CC2, was also completed but was to be stored until the end of the war. Trials of CC1 commenced but with little urgency since there were more important matters relating to running the railway in wartime. The trials began with local trips and running light to Three Bridges and back. Later, in October 1941, a 14-coach train was hauled from Brighton to Victoria, followed by a round trip to Selhurst depot, and then back to Brighton from Victoria. Further trial runs took place; an official Southern Railway photograph shows CC1 with another 14-coach train at Purley in November 1941.

Commencing on 24 November 1941[20] CC1 worked a 12-coach passenger train on the Waterloo to Portsmouth Harbour service for several weeks, and in doing so maintained the same schedule as the usual multiple unit 4-COR trains.[21] This was a significant achievement on this difficult route, as the locomotive had a 1,250 horsepower (900kW) deficit compared to a 12-car train made up of three 4-COR units. Only one man was in the driving cab, a practice that had been challenged by the footplatemen's union, the Association of Locomotive Engineers and Firemen (ASLEF), which initially took an uncompromising stand demanding that in no circumstances would it agree to the engine being worked by one man for either passenger or freight trains.[22] The Southern Railway's view was that, given the desirability of having two men when shunting goods wagons, it accepted ASLEF's demand in respect to freight trains, but for passenger trains insisted that there should be only one man in the cab as for the electric multiple unit trains, but this

CC1, now British Railways owned and numbered 20001, hauls a Victoria to Epsom Downs Derby Day special train. (Meredith/Transport Treasury)

CC2 as it first appeared in Southern Railway livery taking charge of a single parcels van at Bognor Regis station. After nationalisation it ran for a short period painted blue with 'British Railways' written in full on its side. (Southern Railways Group/John Sutton Collection)

issue with the union was not to be resolved during the Southern Railway's lifetime.

Meanwhile, in 1942, a year after CC1 appeared, Bulleid's promise to Bevin resulted in 40 'Austerity' class Q1 0-6-0s being built at Brighton and Ashford works. These locomotives, quintessentially 'Bulleid', can only be described as the ugliest locomotive ever to run on any British railway, but nonetheless their brutal modernist design was to be admired by many after the initial shock. The Bulleid Pacifics when originally built might be described as a rectangular box with the entire boiler neatly hidden out of sight, but on the Q1s each of the boiler components, the fire box, boiler barrel and smoke box was an individual box, each of which decreased in size from the fire box to the smoke box. These 'lumps' were not quite square for the sides curved slightly and the tops were a very shallow curve which, at the apex, on the smoke box at least, was not quite flat. A basic cab was attached to the fire box and the tender was like that coupled to the Merchant Navies but lacked the high ornamental 'raves' and the sides did not neatly match the cab side in height. The whole ensemble looked ungainly as there was no running plate above the wheels on which shed staff could stand to clean the beast or reach various boiler fittings such as the washout plugs. For essential tasks requiring access high up on the locomotive, the shed staff had to find ladders and there was no handrail along the boiler to hold on to. Without a footplate or splashers the six relatively small BFB driving wheels were starkly exposed to view.

During 1942, CC1 was out of service for several months, its return to service delayed by the ever-present wartime difficulty in obtaining material for modifications.[23] The problems included excessive wear on the flywheel bearings (for which special permission had to be obtained from the wartime Ministry of Supply to obtain spares from Sweden) and wear issues on the many contactors in the control system that resulted in excessive routine maintenance.[24] In October it was agreed that on its return to service the locomotive would haul freight trains between Norwood and Fratton, but it was still unavailable the following January.

Overhead contact wire was installed by British Insulated Callender Cables

THE ELECTRIC LOCOMOTIVES • 285

Limited at the Brighton inspection sheds and in the sidings at Balcombe station on the Brighton main line.[25] The installation at Balcombe was experimental to test both the locomotive and the overhead equipment.[26] In 1942 further tests were carried out along the Brighton main line with freight trains running between Hove and East Croydon with loads varying between 551 and 925 tons. Later, CC1 moved to the Portsmouth Direct line to haul freight trains. These trials were not without incident, for during one of these runs there was a failure of one of the boosters at Woking. Raworth was pleased with the locomotive's performance, offering his opinion that the 'formidable' problems of designing a third-rail electric locomotive had been overcome and that his design was 'entirely satisfactory'.[27]

From June 1943 a regular duty for the locomotive was the Norwood Junction to Chichester freight which was run at higher speeds and with heavier loads than had been possible with the replaced steam locomotives. A siding at Norwood was equipped with overhead conductors, and additional signalling. No sidings were electrified at Chichester at the other end of the route. The use of CC1 on heavy freight trains proved to be successful – so successful that the higher speeds resulted in the unexpected consequence of an outbreak of overheated axle boxes on some of the four-wheeled goods wagons. At a management meeting in August 1943, it was stated that early estimates of the savings accrued by using CC1 rather than steam power indicated that substantial economies were to be anticipated.[28] There were various teething troubles that had to be remedied, but CC1 began clocking up an increasing annual mileage: a total of 18,923 miles in 1942, 44,995 in 1943, 50,675 in 1944 and 74,415 in 1945.[29]

The squabbles between Raworth and Bulleid continued and on occasions Missenden had to intervene.[30] Raworth had built a virtual defensive wall around 'his' electrical department. For example, when on one occasion CC1 had a cracked tyre around one of its BFB wheels, Raworth insisted that the locomotive be removed from 'Bulleid's' Brighton works during one weekend so that it could be repaired at one of 'his' depots, despite the fact that Brighton works was capable of the work and moving a locomotive with a cracked tyre was undesirable.[31] When CC1 was regularly hauling the heavy goods trains between Norwood and Chichester, should it be unavailable for any reason, a steam locomotive had to be found and the timing of the train modified. Bulleid was not impressed by this arrangement and keen to show how good his locomotives were, so a trial using a Q1 to haul the train to the 'electric' schedule was arranged, but to his dismay it was proved that the electric locomotive easily outperformed the steam locomotive and no further attempts to use a single steam engine on that working were to be made.[32]

Despite some remaining concerns about the reliability of the boosters, in 1942 negotiations began between English Electric, Raworth and Bulleid concerning the possibility of constructing further electric locomotives to a smaller design. The new traffic manager, R.M.T. Richards, believed that he could make good use of a smaller electric locomotive with a single four-wheeled power bogie. These smaller locomotives would be fitted with a train heating boiler and capable of hauling 250-ton passenger trains or 500-ton freight trains and they could be used in multiple unit pairs when necessary. While Bulleid and Raworth rather uncharacteristically agreed that the six-wheel bogies as used on CC1 should be standard, they characteristically disagreed as to how this would be achieved: Bulleid wanted to fit an unpowered four-wheel bogie either side of the power bogies (numbering sequence commencing 2C2) while Raworth supported a single six-wheeled unpowered bogie after the power bogie (numbering sequence commencing C3).[33]

But where did Richards propose using these smaller locomotives? A clue might be found in that in April 1942 consideration was given to the possibility of using electric locomotives on the London to Gillingham route and this would require some minor additional electrification work. Despite CC1's ability to cope with gaps in the conductor rail, standing stationary off the conductor rail at the buffer stops of a terminus would not be possible. Therefore, expenditure was approved to extend the conductor rail closer to the buffer stops at two of the platforms at three Southern Railway London termini – Cannon Street, Charing Cross and Victoria. Also, the unelectrified platform 8 at Cannon Street was to receive the conductor rail and sidings at Chatham and Gillingham were to be electrified.[34] There appears to be no record however of the CC1 being used on these routes during wartime. Given that in 1942 the likely route for an invasion by Allied forces into Europe was to be from north-east Kent across the straits of Dover – not the actual, much longer crossing further south to Cherbourg – a possible military request might have been the original reason for both the additional electric locomotives and for the reinforcement of electrification infrastructure in north Kent.

Raworth eventually decided that while the concept of a smaller locomotive for use on shorter trains and a larger locomotive for heavier trains was superficially attractive, he considered that the practical advantages of having all locomotives of the same type far outweighed the relatively small first cost savings achieved by introducing a smaller, second type of electric locomotive.[35] In this respect he was in unison with Bulleid who too was only to build large locomotives which could haul any train, and there were to be many instances when his Pacifics were used to haul short one- or two-coach local trains in between duties on heavier trains. So Raworth and Bulleid on this occasion agreed that the CC design was all that was needed for main line work, but what was required for shunting was a small locomotive weighing only 64 tons with four coupled axles (numbering sequence commencing B). This proposal would have resulted in a small locomotive with a single cab and a pantograph to collect power from overhead wires in the marshalling and goods yards. All this activity however was to prove to be a series of wasted meetings and design effort, for the final decision was that only a third electric locomotive, CC3, should be built as a prototype for all future electric locomotives.[36]

In 1943 the Southern Railway directors examined the 10-year agreement with English Electric for the supply of motors and control gear that was soon to be due for renewal. One of the benefits of the rolling programme of electrification had been that the railway could obtain a significant procurement advantage by entering long-term contracts with essential suppliers. Unfortunately, this plan took no account of a war situation suddenly curtailing electrification projects and the provision of new trains. The agreement with English Electric had run for only four years before the war intervened and, apart from the electric locomotives and the very few early Bulleid era 4-SUBs, the demand for electrical equipment had dried up. This had left the unfortunate English Electric with development costs that could not be passed on to the railway as absorbed costs when production commenced. The company had worked on the development of equipment for the electric locomotives, including the abortive projects for the smaller electric engines for which, on Bulleid and Raworth's instructions, they had prepared drawings. None of these costs could be recouped due to the lack of any resulting orders. For the railway's part the directors believed, perhaps unreasonably, that because of the war the Southern too was losing out because it had not been possible to make the fullest advantage of the cooperation and planning during the earlier years. It was decided that, in the light of the good service that English Electric had given to the railway, the existing agreement should be extended for a further 10 years.[37]

Railway executive poster issued at the end of the war celebrating the work of the railway companies in constructing the locomotives that helped win the war. The poster highlights locomotives from the USA, Bulleid's 40 Q1 0-6-0s and Raworth's only electric locomotive used during the war period. (Author's Collection)

CHAPTER 24

LOOKING TO THE FUTURE

In the early years of the Second World War plans were made as to how the nation could be better administered when hostilities ceased, as had been the case during the First World War when the 'Reconstruction Committee' was at work. Despite the grave international situation there was again the aspiration to build a 'land fit for heroes', and hopefully this time it would be more successful. In 1942, the senior civil servant and economist William Beveridge published his report aimed at purging society of the 'giant evils' of squalor, ignorance, want, idleness and disease. His report was to be the template for the post-war welfare state which included the National Health Service (NHS).[1] Meanwhile the Conservative politician R.A. Butler was preparing the 1944 Education Act,[2] proposing free education for all children, raising the school leaving age to 15 (with a view to later raising it to 16) and introducing the 'eleven-plus' examination to stream children into either grammar, technical or secondary modern schools.

In 1942 in the same spirit, but of less national or social importance of course, Sir Herbert Walker, now a director of the Southern Railway, requested that his successor Eustace Missenden consider the completion of the electrification of the Eastern and Central Sections and present schemes to the board for approval.[3] Given the pressures that all his staff were enduring, this was a big ask of Missenden, but Sir Herbert, sensing waning enthusiasm for electrification, wanted to fan the dying embers and jolt the company into making decisions for the future. The omens were not good: Sir Herbert's dream of further electrification was not shared by all of his fellow directors. He would have been aware that at the time when the application to install overhead wires in goods yards for the electric locomotives had been presented to the Ministry of War Transport, Cyril Hurcomb authorised only a limited use of such 'low voltage' overhead lines, and then only in the area east of a line that he had drawn from Reading to Pirbright. To the west of this line Hurcomb insisted that the Southern Railway must not carry out third-rail electrification without consulting the government.[4]

Due to all the wartime pressures it was to be two years before the result of Sir Herbert's enquiries arrived in the form of Alfred Raworth's *Report on Proposed Extensions of Electrification* which he completed in May 1944.[5] The delay was not only due to Raworth being fully engaged in his wartime duties, but also to the difficulties he had getting data from overworked clerical staff. Furthermore, at the end of 1943, Raworth had spent Christmas in hospital[6] for an undisclosed illness. Following his return to work in February 1944, Missenden applied pressure on his chief electrical engineer in an attempt to get his report completed.

For all the difficulties in production, the report was neatly bound between hard covers and would appear to be the only instance when Raworth made use of the letters MICE after his name. It clearly laid out his concept of an electric railway in much the same way that his report for the SECR had done a quarter of a century earlier. To confront the 'steam' lobby led by Bulleid, he pointed out the economic advantages of having large generating stations using efficient machines to provide power for the trains. This avoided the need for each train to have the capacity to produce sufficient energy to meet its maximum power demand – which might only be required for a very short period of time. He argued that the electric trains were simpler, more robust in construction, easier to maintain, and could be used almost continually without having to be taken to the sheds every few hours for the removal of ash and clinker and every few days for boiler washouts.

Significantly, Raworth claimed that it was now necessary to blur the previous distinction between 'suburban' and 'main line' electrification and he proposed a new definition, 'general electrification'. He was now prepared to concede some ground to Bulleid and others that the Southern Railway would soon run out of routes that would fit the 'suburban' traffic model. This had been the classic Walker era business plan based on the anticipated increase in receipts from rising passenger numbers following the electrification of a line. Any new electrification proposal must now be considered on a wider basis, particularly the benefits from reduced operating costs. Raworth viewed the Southern Railway as a network of two halves. In the east there was heavy traffic across a high mileage within a small geographic area (in part due to the fortuitous over-provision of lines by the SER and the LCDR); whereas in the west there was relatively little traffic along straggling main lines with many branches. He too drew a line on the map between two areas, using the original London and Southampton Railway main line between Basingstoke and Southampton Central as his reference point.

He proposed electrification of the main line from Basingstoke to Southampton and all lines geographically to the east thereof. Since this would include Woking to Basingstoke, this would complete the electrification of the LSWR main line from Waterloo to Southampton. Steam-hauled trains from Exeter would run for three miles over the electrified track between Worting Junction and Basingstoke station, where the steam locomotive would be removed and substituted for an electric locomotive of the CC1 type. Similarly trains from the Bournemouth direction would have their locomotive changed at Southampton. Raworth controversially argued that the increase in speed which was possible with electric trains would counteract the time lost due to the locomotive changes.

The proposed all-lines electrification of the eastern section would complete the Hastings scheme and the conductor rails would reach the Channel ports. Within the eastern area there were 1,482 route miles of railway; 714 had been electrified already leaving a further 768 to be electrified. Within the 768 miles of non-electrified track were some lines which carried little traffic and where arguably other forms of traction such as diesel locomotives might be used – particularly in the goods yards. Nonetheless Raworth argued that because of the additional operational convenience that would accrue from the use of full electric traction, he had included in his estimate the electrification costs of all lines except dock sidings and private sidings. He was again describing his vision for a fully electrified railway which would eliminate the smoke, ash and general filth associated with steam locomotives and would facilitate commercial development of the space above the stations, thereby providing a valuable new source of income for the Southern Railway. Within the report Raworth illustrated the potential for such development with 'before' and 'after' photographs of Grand Central Station, New York. The first picture taken in 1906 before electrification showed an open area of railway track with a pallor of smoke and steam hanging above the trains; while the later photograph taken in 1921 showed a clean modern station building and an elegant boulevard – Park Avenue – which had been built above the railway. Complete electrification would simplify railway operations, removing the need for such steam-related clutter as turntables, coaling stages and watering equipment. With only electric trains on the line, the maximum return would accrue from the investment in cables, substations and conductor rails. His plan would require new connections to the National Grid at the CEB substations at Folkestone, Canterbury and Maidstone and a new Southern Electric control room would have to be built at Canterbury.[7]

This 'all-line' proposal to include many secondary lines was in accordance with Raworth's earlier report to the directors when he promoted the cost effectiveness of rural electrification. His approach chimed with the musings of the LSWR chief engineer J.W. Jacomb Hood who in 1902 had contemplated that it might prove economical to electrify minor branch lines.[8] It would also have been applauded by Raworth's late father who also believed in electric traction for rural areas. If the 'all-lines' policy was to be implemented this would include such byways as the Hayling Island branch. This branch was short enough not to need any new rectifier substations, only requiring the laying of the conductor rail and track bonding – although the issue of maintaining the direct current connectivity across the swing bridge at Langston Harbour would need to be addressed. Rather than the existing steam shuttle service, the timetables could be recast to have 2-BIL units continuing from the branch beyond Havant as far as Portsmouth and, on the busy summer Saturdays, through trains from Waterloo could run to the island terminus from the junction at Havant.

Raworth concluded that the monetary savings from the complete substitution of steam locomotives for new electric traction throughout the Eastern Section would produce a 14 per cent return on the capital.[9] This was a controversial statement given that such a high return on capital was not deemed possible by the directors in 1939 when they rejected the Hastings proposals. He wrote that the total costs of operating steam trains in the area in the 12 months before September 1939 had been £2,799,099 and claimed that

The concept of an 'all-line' electrification policy and not just electrification of the busy main lines was not so outlandish an idea. Here in a semi-rural setting 2-NOLs 1809 and 1812 are photographed on the single track line between Mitcham and Morden Halt. This view shows how many single track lines such as the Hayling Island branch might have looked if Raworth's all-line electrification proposals had been accepted. (Meredith/Transport Treasury)

the equivalent electric locomotive costs for working the same traffic at the same speeds would have been £1,333,259 – a saving of £1,465,840. He conceded that the electric traction would have to bear the cost of maintaining the 33,000V distribution network, the substations and the extra track maintenance (including 'danger money' to permanent way staff) which amounted to £221,926. The net saving on traction costs therefore would be £1,243,914.[10]

In 1950, two-thirds of the steam locomotives then operating across the Eastern Section would be over 40 years old and therefore due for renewal. Raworth claimed that at pre-war prices the cost of replacing these locomotives on a like-for-like basis would be £5,600,000. He then claimed – without giving a precise amount – that the provision of equivalent electric trains and locomotives would be 'rather less' than this.[11]

Had Raworth prepared his report for Sir Herbert Walker, it is probable that the then general manager who had complete trust in Raworth would have duly accepted it – perhaps after a few insightful questions – and presented it to the board. But Missenden, with the likes of Bulleid offering the polar-opposite view, did not have the same degree of trust in his chief electrical engineer. Missenden's deputy John Elliot however remained loyal and supportive of Raworth, describing him as 'the architect of electrification',[12] and was most impressed by the report and offered his congratulations describing it as a 'fine piece of work'.[13] Others were less than impressed, and unsurprisingly one of the critics was Oliver Bulleid who sent out a memo claiming that there were several inaccuracies in the report. For example, he took issue with Raworth's claim that the electrically hauled trains from Waterloo to Basingstoke or to Southampton would be fast enough to permit a locomotive change without an increase in the journey time. Other unattributed comments pointed out that the third-rail electrification would be extended into areas outside the zone agreed with the Ministry of War Transport. It was also noted that locomotive changeovers would be required at many other locations besides Basingstoke and Southampton, such as at Eastleigh and Clapham Junction. The use of electric locomotives in goods yards was criticised on safety and practical grounds – the live wires might be off the ground on overhead poles, but were they not a danger to cranes loading and unloading wagons? More constructively the point was made that in electrifying the old LSWR main line as far as Southampton, an additional 28.5 miles of electrification would link Bournemouth into the Southern Electric network. Bournemouth had become one of the most prosperous places in the country with an enormous potential for an increase in traffic.[14]

Given the climate in which he was now working, Alfred Raworth could not have expected his electrification proposals to go unchallenged. He should not have felt too aggrieved by the criticisms since his report was a solo effort made under difficult conditions and it was easy for others to make adverse comments. Valuable opinions and views were being flushed out that might otherwise have been held back during the manic war period.

Missenden, who received a knighthood for his wartime service in the king's birthday honours list in June 1944, would not present Raworth's findings to the board but elected to resolve the issue and allay his doubts by ordering another report. So, after almost three months of prevarication, another new committee was to look again at the whole issue of electrification. Alfred Raworth was to be chairman and others around the table were the chief accountant, R.G. Davidson, the new chief engineer, V.A.M. Robertson, the traffic manager, R.M.T. Richards and, significantly, the unconvinced O.V.S. Bulleid, chief mechanical engineer.[15]

The remit of Raworth's new committee was to investigate two scenarios: the completion of the electrification of the Eastern and Central Sections – but 'Central' was to be extended to include all of the former LSWR lines that had already been electrified; and to go further than Raworth had advised in 1944 and investigate the electrification of all lines as far west as Salisbury on the West of England main line and beyond Southampton to Bournemouth. The Eastern and Central Sections scheme would require an additional 623 route miles of electrification and to electrify as far out as Salisbury – the 'complete scheme' – would add an additional 943 miles to the Southern Electric.

This more formal committee required a secretary. This was to be Stanley Warder, recruited by Raworth from ASEA in 1936. The two had first met professionally when Warder was responsible for electrification contracts at ASEA, and he was first appointed as Raworth's technical assistant. In 1943 Warder's official designation had become 'new works assistant to the electrical engineer'. Warder was a likeable man, even though he was seen by many as egotistic and full of his own importance, but Raworth had confidence in his abilities and he was later to become the chief electrical engineer for British Railways.[16] When Raworth could not attend the REC M&EE subcommittee Warder would often deputise for him. After Bulleid was appointed as chair of the M&EE committee (following the death of Gresley and the secondment abroad of Stanier) Raworth never attended the meetings and only rarely did Warder attend.

Raworth announced that he wished to retire, and the report when finally completed was to be signed 'Alfred Raworth – chief electrical engineer (retired)'. A succession plan was required. In December 1944, the Southern Railway directors overlooked Raworth's assistant Stanley Warder and appointed the 49-year-old Australian-born Charles Cock as deputy chief electrical engineer with a view to succeeding Alfred Raworth as chief electrical engineer the following April.[17] Cock was to be paid a salary of £2,500 per annum, which was to be increased to £3,500 a year when he took over from Raworth and further increased to £4,000 on 1 January 1946.

It can only be speculated as to why the directors decided not to promote Warder and chose instead to install an outsider who had spent his whole career building and operating the rival 1,500V direct current overhead line system, first with Merz and McLellan in Australia (the Melbourne suburban system that Charles Merz had proposed in 1912) and later in India. Cock had left Merz and McLellan and at the time of his appointment to the Southern Railway he was the divisional transportation (operating) superintendent in charge of the Bombay Division of the Great Indian Peninsular Railway. Alfred Raworth, a survivor of the 'Walker' era, had become marginalised and was not fully trusted by Sir Eustace Missenden, so perhaps the board and the general manager wanted a 'new broom' at the electrical department. In truth, similar electrical engineering expertise and knowledge was essential to manage both direct current electrification systems, enabling the directors to be reassured that Cock's qualifications and experience made him suitable for the post.[18] Warder may have been tainted by his association with Raworth, but it has been alleged that he had a different approach to his work to that of his master and was judged by some as to be unable to keep his own projects within budget and was forever having to write reports to explain over-expenditure.[19] If this was a fair assessment of Warder, Alfred Raworth – who did everything on time and to budget – would not have been able to recommend his assistant for promotion.

Raworth's last official day in service with the Southern Railway was 31 April 1945 when he retired with his SECR pension fund annuity of £1,756 which the Southern Railway generously topped up to £2,000 per year. Additionally, it was agreed that he should be appointed for one year as a consultant with a fee of £1,500[20] to complete the electrification report. The report was to take longer to complete than had been expected so in May 1946 it was agreed that his services should be further retained until the end of the year.[21]

With Raworth continuing to work for the Southern as a consultant, his formal retirement presentation did not take place until 9 November 1945. The event was duly reported in the staff magazine.[22] Those wishing him goodbye included Sir Eustace Missenden, John Elliot, Charles Cock, Mr V.A.M. Robertson and Oliver Bulleid. Sir Eustace, as is usual on such occasions, was full of unctuous praise for the retiring employee and said that he was very sorry that health considerations were the reason for Alfred Raworth's early retirement and that he 'would go down in history as a monument to electrification'. He then commented at length on how the Southern Electric system had helped the Southern Railway cope with all the difficulties that had been imposed on the company during the recent war. Missenden wished Alfred, Mrs Raworth and Miss Raworth (Ruby and Marjorie) many future years of happiness and as a 'token of friendship from himself and fellow officers' he presented a set of golf clubs to Raworth.

One of these 'friends', Oliver Bulleid in an apparently humorous tone, stated that although he was one of those whom Raworth described as 'steam dogs' he nonetheless admired Raworth's great electrification work. John Elliot too was full of praise for Raworth, citing his outstanding ability to explain technical problems to the layman in his reports which were a model of clarity. The recently appointed chief engineer V.A.M. Robertson thanked Alfred Raworth for the valuable advice he had recently provided. Hopefully even with the presence of 'OVSB' and Sir Eustace, Alfred Raworth was truly among his friends, indeed he certainly had a friend in John Elliot, for despite having to be tough on expenditure and waste – a stance not guaranteed to endear himself to others in the organisation – Raworth was 'in a quieter way than his father' a very sociable person.[23]

Thanking the gathering for the gift, Raworth noted that he had joined the railway some 33 years ago – this he claimed was a significant figure, a third of a century – usually judged to be the normal lifespan of a piece of electrical equipment, however he hoped to prove that with the exercise facilitated by his new golf clubs there was still plenty of life left in him.

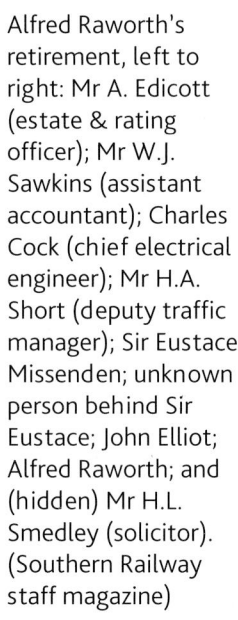

Alfred Raworth's retirement, left to right: Mr A. Edicott (estate & rating officer); Mr W.J. Sawkins (assistant accountant); Charles Cock (chief electrical engineer); Mr H.A. Short (deputy traffic manager); Sir Eustace Missenden; unknown person behind Sir Eustace; John Elliot; Alfred Raworth; and (hidden) Mr H.L. Smedley (solicitor). (Southern Railway staff magazine)

The official Southern Railway photograph of the occasion shows Sir Eustace flanked by Charles Cock and Alfred Raworth. Standing behind them, and obviously oblivious that the photograph is being taken, John Elliot shares a joke with H.L. Smedley – the solicitor who was standing directly behind Alfred Raworth. Oliver Bulleid is not to be seen in the picture. The photograph reveals that Raworth has lost some weight and this might well have been a consequence of whatever illness had caused him to be recently hospitalised. Publicly, no clue was ever given to exactly what the 'health considerations' were that resulted in Alfred's retirement. All that can be said at this distance in time is that whatever was ailing him it was not life threatening because he was to live on for over two decades after his retirement. There must be the suspicion – given that many illnesses can be caused by or be exacerbated by stress – that all was not well within the working environment that Raworth had latterly found himself. He had endured the stress of keeping the electric railway running in the face of air raids, wartime and the depressing effect of no longer being completely in step with the Southern Railway's management policy following the retirements of both Sir Herbert Walker and Edwin Cox, and the subsequent arrival on the scene of his nemesis Oliver Vaughan Snell Bulleid.

Another probable reason why the weary Raworth wished to retire was due to the changing political climate. He wished to avoid the inevitable policy changes and uncomfortable restructuring of the railway's management structures. The surprising result of the general election held on 5 July 1945 was a landslide victory for Clement Attlee's Labour Party. The Labour Party manifesto had promised massive social reform, notably the creation of the welfare state which would take good care of its citizens from the cradle to the grave, and diversion of 'industry to the service of the nation'.[24] All forms of transport – rail, road, and canal – were to be unified and brought into public ownership. Full nationalisation of the railways was clearly the aim of the new government but, in the meantime, the railways were to remain effectively under government control through the REC.[25]

The issue of railway nationalisation had been the subject of controversy for two decades. During the war, the coalition government had been divided on this issue. When there was discussion about how the railways might best be organised to efficiently serve the nation, some Cabinet members such as Deputy Prime Minister Attlee. and Minister of Labour Bevin proposed nationalisation, but Prime Minister Churchill cited the experiences of the railway administrations following the Great War and opined that it would be a 'misfortune' if the railways were nationalised.[26] This was a change of mind, for as a Liberal MP in 1918 he had supported nationalisation, but as a Conservative leader he now took the opposite view. Surprisingly perhaps, even some railway managers had a tacit support for nationalisation. For example, William Whitelaw, the first chairman of the LNER and grandfather to one of Margaret Thatcher's Cabinet ministers of the same name, declared when nearing his retirement in 1938 that if given fair terms in his opinion the LNER shareholders would not be averse to nationalisation. His view was based on the belief that the imposition of the Railway Rates Tribunal and the National Wages Council imposed by the 1921 Railways Act meant that the railways could not operate as truly independent private companies, as significant costs and income were out of their hands.[27] When railway nationalisation was being discussed during the war years, the Southern chairman Robert Holland Martin had let his views be known on this issue at the AGM meeting in February 1942. He led a profitable railway and was thus able to set himself firmly against nationalisation, unlike William Whitelaw at the unprofitable LNER for which nationalisation was a lifeline. He gave his shareholders a rousing defence of the system of private enterprise, claiming that his railway was an efficient organisation that had been ready for the wartime difficulties, despite what he declared to have been 'many years of unsympathetic treatment' by the government.[28]

Following the death of Hollland Martin in 1944, Sir Eric Gore-Browne, the previous deputy chairman, was elected chairman. Publicly, in line with its membership of the Railway Companies Association, the Southern Railway board led by Gore-Brown agreed to take part in a campaign to oppose railway nationalisation. This campaign however was arguably a sham, concealing the fact that its overarching motive was to ensure the best possible compensation arrangement for the shareholders following what was to be an inevitable state takeover.[29] The Railway Companies Association duly set up a committee to fight the cause, but the Southern Railway directors, showing some concern about the direction that this committee might take, established a small committee of four directors to monitor the national campaign.[30]

As expected the Railway Companies Association campaign did stir up a public debate. Generally, most of the press was anti-nationalisation and offered the public a negative view. Badges were being manufactured with the slogan: 'State transport means "take it or leave it"'. A St Helens newspaper did however report the Lancashire town's Labour MP, Sir Hartley Shawcross, defending nationalisation, but he conceded that 'only the Southern Railway has made any attempt to cater for the amenities of the travelling public by an up-to-date system of electrification'.[31] The recommendation of Raworth's committee would now have less of a commercial or engineering significance but would potentially form the basis of a credible electrification policy that could be held up to the world to underline the railway's suitability to remain a prosperous independent private company. The fact that the Southern Railway had been of late less enthusiastic about electrification was irrelevant. In the context of maximising potential earnings, the railway had to be *seen* to be prepared to modernise. Unfortunately, Raworth was again finding it difficult to obtain the required statistics from the Southern Railway offices and in 1946 the general manager began writing short letters to 'Ladywood' in Cobham – headed 'My dear Raworth' – chasing a completion date for the report that he now needed to present to the board for political reasons.

Raworth submitted his committee's report to Sir Eustace in February 1946.[32] The committee had a different term of reference to that which Sir Herbert had proposed in respect to the area of proposed electrification, but essentially the content and conclusions of his solo report were vindicated. As requested, two options were considered, the first being the complete electrification of the Eastern and Central sections and the other a very ambitious scenario that would electrify all lines as far west as Salisbury on the West of England main line and also beyond Southampton to Bournemouth. The map within the report again suggested electrification of even the most insignificant sections of railway – including for example the short Bisley branch which was only used once a year for the National Rifle Association's tournament, but a cheap electrification installation would eliminate all the infrastructure needed to support steam operation. For the electrification further west there was no certainty that the improved services on the minor lines would increase passenger numbers, but the traffic manager, R.M.T. Richards, believed that timetable improvements could possibly create an additional 4.5 million train miles a year in the Eastern and Central Sections. The report confirmed that beyond the traditional suburban areas, electrification of lines should be on the basis that all traffic – not just passenger trains – should be by electric trains. The success of CC1 had proved that there was no technical reason why third-rail electric locomotives should not be used for all types of traffic. It was confirmed that the operating costs for the electric railway were sufficiently less than for the steam railway justifying investment in electrification, and this time these important calculations were above the signature of the railway's assistant accountant. As for Raworth's earlier report, the conclusion reached was that 'should financial advantage be considered sufficient, extension of electrification should be adopted as the settled policy of the Company', but as in the days of Sir Herbert Walker, a cautious step by step approach must be taken examining each section on merit based on the financial situation at the time.

Again Raworth brought in the concept of 'general' electrification for all traffic, main line and local. The operating costs and the savings detailed in each report cannot be compared directly due to the different scope, but it appears that the full committee believed that previously Raworth had overestimated the saving from reduced operating costs with electric traction, but not sufficiently to change the overall argument. But there was disagreement over details. Whereas Raworth had based his calculations on post-war traffic being at 1939 levels, the committee as a group were more timid, fearing that peacetime traffic would not be fully resumed, but in the event immediate post-war traffic generally matched 1939 levels. Raworth had predicted a 14 per cent return on capital; the committee less optimistically expected a return of between 7 and 10 per cent only. Another area of uncertainty which Bulleid and Raworth disagreed over was the degree of savings to be had from not replacing steam locomotives The former chief electrical engineer envisaged that the expenditure on electrification would be spread over 10 years while the chief mechanical engineer – optimistically as subsequent events were to show – saw a 40-year renewal period for his steam locomotives, a difference which influenced the calculations. About 80 per cent of the cost of any new electric trains could be financed from the 'locomotive renewal' budget if over 1,000 steam locomotives were not replaced.

The report concluded that the cost of fully electrifying only the Eastern and Central Sections would be £7,144,249 and the electrification out to Salisbury would cost £10,708,465 (this compares with Raworth's earlier total of £8,768,461 for his lesser proposal). It was conceded in the report that the actual likely cost of the electrification infrastructure was to an extent unknown since all the prices were at 1939 rates, and it was not possible to judge what the availability of electrical plant and resources would be in the post-war world. Raworth showed an early outline of the document to the new chief electrical engineer who would eventually be responsible for the expenditure. Cock immediately wrote a memo to Sir Eustace stating that while he accepted the basic principles he had seen, he was insistent that he did not agree with the cost calculations in the report and so concerned was he that he requested that while the carbon copy of his memorandum outlining his views should be held in the open file, the signed top copy should be locked away in the office safe.[33]

Sir Eustace too had some reservations of his own. He queried the amount of non-passenger traffic in the electric area being sufficient to justify the building of new electric locomotives; the excessive distance from steam locomotive depots of the steam to electric changeover points; and that insufficient thought had been given to the costs of new locomotive-hauled rolling stock for joint steam and electric services.[34] Rumours were circulating concerning the Southern Railway's future electrification policy, and after a speculative article in the *Daily Express* Sir Eustace issued instructions to all his staff insisting on absolute secrecy in this matter. Those at the centre of Southern Railway management at the time held the view that Sir Eustace did not trust Raworth and he also received a 'My dear Raworth' communication demanding secrecy. It is possible that Sir Eustace suspected a disloyal Raworth of doing some clandestine 'leaking', emulating his former master Sir Herbert who had issued a press statement just before his retirement in 1937 boosting public anticipation of further electrification.

Following the hurried completion of the report and the formal 'goodbye' gathering, Raworth left the service of the Southern Railway to become only an interested observer in all these plans. On 29 March 1946 he had met Sir Eustace to give him the report and discuss its contents. Sir Eustace, more emollient when Raworth was about to depart forever, complimented him on the way that it had been prepared and expressed his sincere thanks for his efforts, appreciating that while the report was a joint effort of all those on the panel, he appreciated that it always fell to the chairman to do the lion's share of the work. Mischievously Raworth unofficially attached to the report he handed over to Sir Eustace details that he had received from Bulleid concerning the relative projected timings of steam and electric trains on the main lines from Waterloo. This clearly showed that the projected electrically hauled express trains were faster than the steam equivalent, thereby indicating that Bulleid had been wrong in his earlier statements. Raworth explained to Sir Eustace that other members of the committee did not want this information to be included in the report, but he was to give it to the general manager anyway.[35] This was the last spat between Raworth and Bulleid taking place on the day he left, bookending the shot across the bows fired at Bulleid by the speeding of the 4-COR train on the day they first met.

On leaving the Southern, Raworth did carry out some consultancy work for the railway's electrical equipment supplier English Electric. Alfred worked from their head office in Kingsway[36] in the West End of London. He would have contributed some value to English Electric in the field of electric traction but at the time the company was diversifying into the new electronic age by acquiring some of Marconi's business interests.[37]

Retired, Alfred Raworth could now give more time to family matters. In 1949 his daughter Marjorie, now aged 40, married Mr Graham E. Lovell at Westminster. Leaving 'Ladywood' after living there with her parents for 20 years, Marjorie was to start a new life living at West Harptree, Somerset.[38] Alfred's new son-in-law was the managing director of the Lovell family-owned Bristol Steam Navigation Co. Ltd.[39] This company, originally founded in 1821 under another name, became the Bristol Steam Navigation Co. Ltd in 1877 and provided shipping services from Bristol to Ireland and to other parts of Great Britain. Originally trade was both in cargo and passengers, but due to competition from the GWR's Fishguard services, the line was to carry only cargo after 1914.[40] At the time of Marjorie's marriage to Graham Lovell the Bristol Steam Navigation Co.'s general cargo trade was mainly from the Bristol Channel to the Netherlands and Belgium.[41] The company continued to operate until 1980. Given that the Bristol Steam Navigation line was based in Bristol without any connection with Southampton, the company's flag displaying a red diagonal cross on a white background never adorned a nameplate on one of Bulleid's Merchant Navy class locomotives.

The electrification report was duly submitted to the directors at a meeting on 15 August 1946.[42] The directors were not unimpressed but being aware that they had only an 'electric' advocate, Raworth, and a 'steam' advocate, Bulleid, advising them, they were rather concerned that outside the local Southern Railway enclave the rest of the world was looking towards diesel-electric traction. In search of enlightenment the directors decided that a fact-finding delegation should visit North America. Just as in earlier decades pilgrimages across the Atlantic had been made in the quest for knowledge about electric railways, now engineers were making the journey to get the very latest information on diesel power. The four men sent on this mission were Sir Eustace's assistant, J.L. Harrington, S.A. Fitch, an assistant superintendent of operation, Bulleid's assistant engineer M.S. Hatchwell and Alfred Raworth's former assistant, Stanley Warder, who now held the same position under Charles Cock. This delegation therefore comprised a manager, an operator, a mechanical engineer and an electrical engineer. Eustace Missenden gave instructions that they report on the latest diesel traction technology to determine if this could be either a full or partial alternative to further electrification.[43]

On arrival in the USA the delegation received a culture shock, for the American railroads were unlike the British railways in terms of distances and traffic density. They observed that the US railroads which had been for decades exemplars of good railway management appeared to be overmanned. They also noticed the rising appeal of private motoring, with the American public happily travelling 100 miles or more in their own cars rather than travel by train. They witnessed the striking advances in diesel locomotive technology that had recently taken place in the USA. The railroads, they were told, believed that compared to steam locomotives, diesel operating costs were dramatically less, and despite the high first cost of the new units a significant financial saving was always achieved. This still held true, even though it was expected that the new locomotives would have a shorter working life, as they were likely to be early victims of obsolescence due to the continued rapid improvements in design.

On their return home the delegation produced its report which was dated 20 September 1946. The Southern Railway directors were offered three high-level options. The first was to do nothing – the technological advances were so rapid that holding back in anticipation of a much better solution was a viable and possibly sensible course of action. The second option was to follow the American lead and immediately renew all life-expired plant with expensive diesel-electric locomotives, despite the expectation of a relatively short working life for these new machines. The third option was to simply proceed with electrification extensions in line with Alfred Raworth's committee's recommendations.

These three options were, the delegation conceded, a bit blunt and a more nuanced course of action should be considered. An alternative, mixed solution was therefore offered. This was that further electrification should be carried out on lines with a high traffic density using multiple unit passenger trains and electric locomotives for freight trains running between the larger marshalling yards. Within this geographically expanded electrification area diesel traction should be employed on lines on which the level of traffic did not fully justify electrification. Within the electrified area, even on electrified routes, diesel locomotives should be used for the local pick-up goods trains, thereby removing the need for overhead lines in the station goods yards. For main line use beyond the electrified area – such as the former LSWR line to the West Country – 2,500 horsepower (1,865 kW) diesel-electric locomotives should be provided, capable of hauling passenger trains weighing 500 tons running at speeds of up to 100mph. For shunting in the marshalling yards, the recommendation was a type 'A' diesel shunting locomotive and for use on the pick-up goods and for all trains on the secondary non-electrified lines a larger type 'B' diesel locomotive should be provided. This type 'B' could be a hybrid running as an electric locomotive with third-rail pick up but would also have a diesel engine on board to provide power away from the conductor rail. This would be an 'electro-diesel' as first proposed by Raworth 15 years earlier, but the report did not tally with his vision of an 'electric railway'.

In October 1946, the directors approved a modernisation plan which was presented to the press by Sir Eustace Missenden. He announced a £15, million, five-year programme that would eliminate all steam traction east of Portsmouth by further electrification and the introduction of new diesel traction. This was a rushed proposal based in part on Raworth's committee report but owed more to the report by the delegation to the USA. This sudden showcase proposal might cynically be viewed as part of the dubious campaign by the Railway Companies Association and the late inclusion of diesel traction enabled the railway to be presented as forward-looking and modern – even though the railway's most senior engineers such as Bulleid and Raworth had not been protagonists for this form of motive power. The news was well received and seems to have had the desired impact. A potential beneficiary of the plan, the English Electric Company, wrote a glowing letter of congratulation to the Southern Railway directors, enthusing about their farsightedness and claiming that such an excellent plan could not have been prepared so promptly had the railway been under government control.[44]

More details of the plan were to emerge later. The former LCDR main line to Dover and Margate was to be electrified. Electric locomotives would have been used on the boat trains to the Channel ports and Charles Cock believed that ultimately electric locomotives larger than the CC1 type would be required to haul boat trains weighing 500 tons at speeds of up to 100mph.[45] Later the third-rail electrification would be extended beyond Sevenoaks through Ashford to Folkestone and Dover; from Maidstone (West) to Paddock Wood (the junction with the main line) and from Ashford through Canterbury (West) to Ramsgate. The electrification of the SER route to Hastings and Bexhill and from South Croydon to Horsted Keynes would follow. At the same time the now closed former LBSCR route from Christ's Hospital – just south of Horsham – to Shoreham would be electrified. Concurrently the former LSWR main lines from Waterloo to Exeter and Southampton would receive new diesel-electric locomotives of a type then being designed (perhaps reluctantly) by Bulleid in association with the English Electric Company. Only three of these main line diesel electrics were ever to be built and all appeared after nationalisation.

So, did the Railway Companies Association campaign result in a satisfactory settlement with the shareholders? It is difficult to tease out the true value of the compensation given the financial intricacies – not least the generally accepted belief that the assets of the railways were grossly undervalued. At nationalisation the shareholders were compensated with the issue of British Transport Stock valued at £1,132,000,000 held in 1,250,000 separate holdings.[46] Significantly these bonds promised to pay a guaranteed 3 per cent annual interest for 40 years. In the first year of nationalisation the railways made a commendable operating profit of almost £24 million to be eliminated to an £8 million *loss* after the interest had been paid to the erstwhile shareholders.[47] The former shareholders therefore were to pocket between them £32 million – much better than they might have expected and unaffordable by the railway. Had the Big Four not been nationalised a chunk of the £24 million profit would have fallen to the relatively prosperous Southern Railway which would have been able to raise any further necessary capital making the **£1** million required each year for the traction modernisation realisable. The actuality turned out to be a government policy of minimal investment.

CHAPTER 25
EPILOGUE

Following his retirement, the weary Alfred Raworth avoided being involved in the inevitable internal political manoeuvrings resulting from the transfer of power from the Southern Railway's board of directors to the nationalised British Transport Commission (BTC). The 1947 Transport Act had established a two-tier management structure, with the BTC above six executives entrusted to manage the railways, London Transport, the docks, inland waterways, the railway hotels and road transport. Of the six, the Railway Executive was the largest and operated 'British Railways' – a trading name only without any legal basis – taking control over the six railway regions.[1] The BTC chairman was the former senior civil servant at the Ministry of Transport, Cyril Hurcomb. The full-time commission member for mechanical and electrical engineering was Robert (Robin) Riddles, a steam locomotive engineer with an LMS background. To assist him his chief officers were mostly ex-LMS engineers with 'steam' experience. Riddles' only non-LMS assistants were Charles Cock who was appointed chief officer electrical engineering and, from the GWR, A.E.C. Dent, executive officer, road motor engineering. Sir Eustace Missenden was appointed as the Railway Executive chairman – generally regarded as a less than satisfactory choice because most of the other suitable senior railwaymen had decided to retire. The Southern Railway became the Southern Region of British Railways and its first general manager was to be John Elliot. The mechanical and electrical functions were combined, and it was Bulleid who was to be the Southern Region's first chief mechanical and electrical engineer – and we can only speculate what frictions would have occurred had Raworth not retired. Following Bulleid's retirement soon after nationalisation, Raworth's former right-hand man, Stanley Warder, was appointed as the Region's 'CM&E'.[2] The announcement that Warder was to replace Bulleid was greeted with derision by many in the railway industry for he was an electrical, not a mechanical, engineer, and this only reinforced the widespread prejudice that the Southern Region was nothing more than a 'tramway'.[3]

From his remote vantage point as a retired railway engineer, Alfred Raworth was to enjoy over the subsequent years the quiet satisfaction of witnessing many of his ideas on railway traction that had previously been dismissed or overruled being accepted and put into practice. The most obvious of these is the resolution of the disagreement between himself and the 'steam dog' Oliver Bulleid, even though for several more years than many would have wished, steam remained supreme on the newly nationalised British Railways and electrification was only a vague aspiration. While a steam traction policy was to be at the fore at the time of Raworth's retirement, within his lifetime he was to witness the elimination of almost all steam from Britain's railways.

In 1946 Bulleid was elected president of the Institution of Mechanical Engineers following in the footsteps of such notables as George and Robert Stephenson, as well as Sir Nigel Gresley and Sir William Stanier. This appointment was a worthy tribute to a man who had been devoted to the cause of steam locomotive development and was a great engineer despite being the subject of much adverse comment over the years. He continued production of his Merchant Navy locomotives, resulting in a total of

30 being completed, to which he added a further 110 'lightweight' versions of his creation that became the West Country and Battle of Britain class Pacifics which were used on the majority of lines operated by the Southern Region.

The new British Railways traction policy may have been to continue with steam but straightforward reliable steam, unencumbered by any of the unorthodoxy created by Bulleid. At the time of nationalisation Bulleid was developing a very unorthodox steam locomotive to be subsequently known as the 'Leader' class.[4] This locomotive had been inspired to some extent by the electric CC1 design in that it had a fully enclosed external superstructure with driving cabs at both ends and was mounted on two steam driven power bogies producing 'total adhesion' as all the locomotive's weight bore down on the 12 powered wheels. Within the superstructure, hidden away behind the panels was the boiler, built to a new and innovative design, the coal store, water tanks and a hot and uncomfortable compartment where the hero of a fireman was expected to feed the firebox with fuel.[5] Emotions ran high among the engineers at the new BTC, as the trials of the first and only locomotive proved disastrous and eventually the whole programme was axed.[6] This may have been a rash decision. When the cancellation was announced, the prototype had just received a complete overhaul of the power bogies and on the final day of testing in November 1950, the condemned locomotive hauled 15 coaches on an uphill gradient from Eastleigh towards Basingstoke at an average speed of 50mph and on the return journey the Leader, unencumbered by the carriages and running as smooth as any passenger coach, achieved 90mph near Winchester. But it was too late – the decision had already been made and the prototype and the four part-completed locomotives were all duly scrapped.[7]

Bulleid's last design was a failure, but what about his magnificent Pacifics? Following testing at the new British Railways testing plant at Rugby, the locomotives were found to have a low overall thermal efficiency despite the excellent steaming qualities of the Bulleid boiler. This was attributed to inefficient steam distribution through the valves operated by his unconventional valve gear. This inefficiency had led to relatively high coal consumption figures when the Merchant Navies were compared to other locomotives. The ingenious valve operating gear, placed out of sight within a fully enclosed oil-filled bath between the locomotive's main frames, frequently failed in service and oil leaks were blamed for contaminating the lagging between the boiler and the 'air-smoothed' casing, resulting in 38 fires in 1953 alone. The outer casing made many serviceable parts difficult to access which substantially increased maintenance costs as depot fitters took longer to attend to routine work. The maintenance repair costs for the Bulleid Pacifics were found to be on average *five times* higher than for the Lord Nelson locomotives – £2,677 per repair compared to £542 for the older Maunsell engines.[8] Oliver Bulleid had bequeathed to British Railways his Pacific locomotives, which despite being virtually brand new were costing a considerable amount of money in running and maintenance costs. Therefore the drastic decision was taken to rebuild these locomotives in a more conventional form. This rebuilding included the complete renewal of the inside cylinder and the substitution of more conventional valve gear.[9] Most noticeably the air-smoothed casing was to disappear, thereby completely transforming the locomotive's appearance by revealing the cylindrical boiler. All 30 Merchant Navies were rebuilt between 1956 and 1959, and half of the 'light Pacifics' between 1957 and 1961.[10]

The very structure of Riddles' mechanical and electrical engineering department indicated that a steam not a diesel or electrification policy was to be followed by the new regime. The LMS had built two 1,600 horsepower (1,200kW) main line diesel-electric locomotives and the Southern Railway contributed the three Bulleid diesel-electrics. The LNER had wanted 25 diesel-electric locomotives for use on King's Cross to Edinburgh services but the order had been placed close enough to nationalisation to enable the new Railway Executive to promptly demand cancellation.[11] At first things looked very bleak for Raworth and others who wished to see further extensions to the Southern Electric, for despite the post-war travel boom which had generated for British Railways a healthy gross profit, the Labour – and the subsequent Conservative government – failed to adequately invest in new equipment for the railways. It was not that the BTC was against electrification, but it feared risking millions of pounds on schemes that were to prove to be not remunerative when the receipts did not meet expectations due to changes in traffic patterns, route rationalisation or the continuing drift of trade away to road transport. Cyril Hurcomb, now ennobled as Lord Hurcomb, publicly espoused this rather timid and uninspiring official policy in 1950, but his personal view was that the Southern Electric should be extended if 'the Commission's economic position were safeguarded'.[12]

The anti-electric, pro-steam arguments aired following the second Weir Report in 1931 reared their head again in 1950. Robin' Riddles proposed to construct a series of new locomotives designed to take full advantage of all the recent improvements to steam locomotive design which were intended to reduce running and maintenance costs. To do this he proposed to introduce a new range of standard steam locomotives, taking the best features from the pool of designs operating around the country. Simplicity of design – but not austerity – reliability, economy and ease of maintenance was what was required on the post-war railway. Unlike Bulleid, Riddles proposed a series of locomotives of different sizes for different duties.

Riddles has been criticised for maintaining steam traction, but it is easy with seven decades of hindsight to overlook the economic and political situation in Britain in the middle of the twentieth century. Diesel locomotives were seen as more expensive to buy from outside suppliers such as English Electric when compared to the much lower cost of steam locomotives built in-house at the railway's own workshops, and there was little confidence that the railway's steam-trained fitters would be able to maintain the complex electrical and mechanical equipment installed on the new form of motive power. Furthermore it would have been necessary to use up valuable foreign exchange obtaining fuel oil from the Middle East while steam locomotives could be supplied with coal from the newly nationalised British mines. This policy made sense at the time given the urgent need for reliable motive power and the general economic climate riven with rapid inflation, the effects of the unfortunate devaluation of the pound sterling and the unsettling international crisis due to the ongoing Korean war. Also the government had placed restrictions on capital investment because of the necessity to direct manufacturing effort towards exports and rearmament. Riddles was also to argue at meetings of the Railway Executive[13] that the new diesel locomotives were still being trialled and there was only a *hope* that this new form of transport might break even with fuel and other costs when compared to steam locomotives.[14]

Assuming that more new steam locomotives were required, it has often been asked why it was necessary to introduce new designs and not perpetuate the existing designs. Post-nationalisation 2,542 new steam locomotives were to be built. Of these 1,543 were modern designs constructed as part of the former railways' existing building programmes. After some careful deliberation, the executive officer for design, E.S. Cox, was adamant that there were to be a total of 12 *new* designs and that these would consist of various combinations of standard items such as boilers, cylinders and wheel sizes, etc.

and that only entirely new designs would be put forward if there was a definite step forward in availability or efficiency. He claimed that there were to be no new designs undertaken for their *own sake*, and therefore if an existing design was available it was to be perpetuated with only minor detail changes.[15] As a result of this controversial decision, 999 British Railways designed steam locomotives were built between 1951 and 1960.

Meanwhile little progress was being made extending electrification. The only electrification to be undertaken was the completion of the LNER schemes for its main line through the Pennines and the suburban electrification from Liverpool Street – both using the Ministry of Transport approved 1,500V direct current overhead lines. The Railway Executive's negativity continued when a committee rejected the 20-year-old Weir Report's conclusion that railway electrification might be used as a catalyst for the expansion of the nation's use of the available electricity supply following the creation of a National Grid. This was considered to be no longer valid due to the rapid rise in the consumption of electricity across the country for non-railway use, and because diesel traction would probably prove to be more economical on many lines.[16]

There was however some important investment in the Southern Electric which was largely unnoticed and hidden from the passengers' view. In 1950 the BTC approved a long-term plan to modernise the Southern Electric electricity distribution in the London area at a cost of £12 million.[17] This scheme, referred to as the 'frequency change' project, was a rolling programme to remove the original 11,000V 25-cycle network and replace it with a 33,000V network at the national frequency of 50 cycles. This new distribution network would follow the 'Raworth' principles with new Grid connections (at 66,000V, the National Grid voltage in central London) to be taken from a re-equipped Deptford power station. A total of 71 rectifier substations were to be built, 45 of these on completely new sites, 24 within existing buildings replacing the rotary converter installations and a further two to be installed within railway arches below the tracks. These new substations were so designed such that the direct current traction supply could be uprated from 660V to the 'approved' 750V when required.[18] The new 33,000V network was extended progressively throughout the London suburban area, eventually allowing the closure of Durnsford Road generating station following the provision of another new Grid connection from the nearby Wimbledon substation. Durnsford Road was demolished in 1965 after almost a half century of operation. This important investment ensured the survival of the Southern Electric network, as it avoided the fate that befell the former NER Tyneside electrification when British Railways' failure to replace obsolete substation equipment resulted in the withdrawal of the third-rail electric trains and the substitution of diesel trains.

In 1948, Charles Cock was appointed as the chairman of a new committee to review the 20-year-old Pringle Committee's recommendations and determine if a single system of electrification should be employed for all future schemes.[19] This should have been an opportunity for a dispassionate and objective review of the much maligned Southern Electric third-rail system but, apart from representatives from London Transport, the committee was composed of engineers in the '1,500V' camp. These included the chairman himself assisted by E.W. Rostern, the operating superintendent of the North Eastern and Eastern regions who was soon to become the operator of the new trans-Pennine 1,500V electrified railway, and, from Merz and McLellan, the 'outsider' Mr F. Lydall, who had carried out detailed design work for the NER's aborted York to Newcastle 1,500V project and had also worked on similar schemes in South Africa and India for Merz and McLellan.[20]

The committee proposed to compare the Pringle Committee's recommended 1,500V direct current preferred system with alternatives. To do this various scenarios were prepared and costed not in precise monetary values but by percentages, allotting the 1,500V system a 100 per cent value. The compared systems were 750V direct current third rail (the Southern Electric was expected to uprate to 750V), 3,000V direct current overhead and the rather odd choice of an alternating current system whereby the National Grid three-phase supply was used to power three-phase motors driving single-phase generators to produce electricity at 20,000V alternating current with a frequency of 16 cycles.[21]

None of the hypothetical railway routes chosen matched the Southern Railway routes in southern England in terms of route length, operational difficulty or traffic pattern except arguably the fourth. This hypothetical line had easy gradients and only multiple unit trains were to be used with a 'suburban' traffic pattern. Since the report deemed that, unlike the Portsmouth Direct, the gradients were easy, it was not a perfect comparison. The committee concluded that for this line the capital costs for electrification would be roughly the same – the third rail at 100.2 per cent of the 1,500V cost, but running costs would be much higher for the third rail at 104 per cent of the 1,500V alternative.

These 'higher' costs must be placed in the context of a completely new electric railway, not an extension to an existing electrified line. Part of Raworth's advocacy for the extension of the third rail was based on the reality that each extension was not a discrete stand-alone project but an extension to an existing railway already electrified using the third rail, and this alone – unless costs were wildly different – would have been justification enough to continue with the third-rail electrification. The only quite marginal increase in costs determined by Cock's committee would not offset the need to carry out either the conversion of the 'London' end of any line to 1,500V overhead, or alternatively installing additional overhead lines above the existing third-rail electrification to have both types of electrification on the same tracks. Furthermore, existing electric trains would not be able to operate on the new line without the additional expense of carrying out the necessary modifications.

Despite recommending that the national standard for railway electrification should remain as 1,500V direct current overhead and opining that the low voltage third-rail scheme was inferior, there was an appreciation that nearly 17 per cent of all passenger mileage on British Railways was by Southern Electric trains.[22] The committee had to grudgingly concede that in the case of the Southern Region, the dense passenger traffic and the limited overhead clearances at certain locations meant that the policy there should be to retain the third-rail electrification.[23] As to extending the Southern Electric, the Cock Committee recommended extensions to the system not only into Kent, as the Southern Railway had proposed, but also – provided that the economic case could be made for electrification rather than 'dieselisation' when the time came to replace steam completely – further west as far as Salisbury and even as far as Weymouth. Shortly after his report had been completed Charles Cock resigned from his post at the Railway Executive to become general manager of the traction department of English Electric.[24] His departure resulted in a further promotion for Raworth's former technical assistant Stanley Warder who became the BTC's chief electrical engineer.[25]

Meanwhile British Railways' financial situation was becoming increasingly more desperate. Following the early years when a modest gross profit was achieved, disastrously this became an unsustainable series of annual losses. Between 1952 and 1956 there was an increase in the number of passengers carried on trains but by only 3 per cent, compared to the 36 per cent increase in those travelling by motorised private road transport.[26] In the same period the freight traffic on the roads had increased by 18 per cent, while rail freight fell by 1 per cent, an abysmal performance during a period of increasing economic activity. In 1955 the loss was over £38 million[27] – and there remained crippling interest charges on the transport bonds. This woeful situation could not last long. The BTC in 1955 was faced with not only having to borrow money to fund the deficit, but also potential strike action by its workers who were seeing the real value of their wages depleting each year. Faced with politically damaging industrial action the Conservative government had to act. To justify the cost of the resulting 1955 wages settlement, the government performed what Charles Loft has described as a 'conjuring trick' when it was argued that the cost of the forced substantial pay settlement would be paid for by short-term productivity gains and by longer-term savings from proposed railway modernisation.[28]

To facilitate this political objective, the BTC was instructed to produce a report entitled *Modernisation and Re-equipment of British Railways*, which is universally referred to as the 'Modernisation Plan'. In the shameful opening words of the report, the blame for the railway network not working to its full efficiency was placed upon the railways themselves, due to 'their past inability to attract enough capital investment to keep their physical equipment fully up to date'.[29] In other words, it had not been due to any government neglect. By some warped logic it was the railways' own fault that they had been starved of investment. The report proposed the expenditure of over £1.2 billion. The outstanding feature of the report noted by the public was that it heralded the demise of the steam locomotive – but not immediately, only after completion of the 1956 locomotive building programme and no new locomotives designed for express passenger or suburban passenger duties would be built.[30] The report proposed that £345 million would be spent on the abolition of steam and its substitution for diesel and electric traction.[31]

Modernisation of the railways in 1955 implied investment in new modern equipment. What perhaps modernisation in 1955 should have meant was a reappraisal of what the nation *needed* from its railways, and the appropriate capital spending provided to achieve what was subsequently deemed to be necessary. It had to be left to Dr Beeching in 1963 to carry out such a review, but unfortunately his remit was to achieve financial equilibrium and not to look at either the 'social railway' or to place rail transport in the context of the drive for carbon reduction which was to be become one of the political considerations in the following decades. The sudden turning-on of the investment tap resulted in urgent requests to the regions for any suitable schemes that could be instigated as soon as possible. This tended to produce inappropriate proposals from ancient 'wish-lists'. From the North Eastern Region came the proposal to build a new motive power depot with facilities for steam locomotives at Thornaby, and the Scottish Region requested a new marshalling yard at Thornton in Fife. The Southern Region also had a ready plan approved by both the Southern Railway directors and the Cock Report which was to expand the Southern Electric into Kent to reach the Channel ports. Helpfully, the BTC Modernisation Plan made the claim that it had always been the intention to electrify all the main routes of the Southern Region east of a line drawn from Reading to Portsmouth.[32]

Time had moved on. In 1955 the Southern Region had an urgent need to improve the lines into the Medway area and beyond due to the rise in commuter traffic following the dramatic increase in housing in that part of Kent. Furthermore it was still having to cope with the legacy of the old cheaply built LCDR main line with its sharp curves and steep gradients. The revised electrification plan was therefore to include significant signalling and track improvements.

The Kent electrification scheme was to be carried out in two stages and was completed ahead of schedule by 1962. Both the former SER and the LCDR main lines were electrified.[33] The Kent electrification extension was the first to use a 750V nominal conductor rail voltage rather than the 660V in use since 1923. The electricity distribution system was once again to the Raworth principles of a 33,000V ring arrangement supplying a string of rectifier substations with track paralleling huts interconnecting the substations via the conductor rails. New Grid supplies were taken from substations at Canterbury, Maidstone and Folkestone. The design of the substations had advanced due to progress in the development of high voltage switchgear since the 1930s allowing 'indoor' 33,000V switchgear to be included within the substation buildings. The only significant plant visible outside was the main power transformer.

With the conductor rails reaching into Kent there was now a need for more electric locomotives to haul the boat trains such as the *Golden Arrow*, the *Night Ferry* and the heavy coal trains from the Kent coal field. The third 'Bulleid-Raworth' locomotive, CC3, had been destined to be the standard design for all future Southern Railway electric locomotives. It had at first been put on hold, which must have annoyed Charles Cock, particularly as at the time he was witnessing the progress of Bulleid's ambitious 'light Pacific' building programme producing new locomotives from both Ashford and Brighton works every month. Cock took his chance however by bizarrely getting agreement for the construction of CC3 as a replacement for a steam locomotive that had been destroyed during an air raid. The Southern Railway had a budget for replacing war damaged rolling stock, but nonetheless it is surprising that he was able to get agreement for CC3 to be the replacement of an ancient former LSWR locomotive, a T14 4-6-0 number 458, which had been destroyed at Nine Elms depot during the Blitz in 1940.[34] A more suitable replacement for such a venerable locomotive would have been one extra Bulleid Pacific so this appears to be an instance when an engineer pulled the wool over the eyes of the accountants to get 'justification' for a pet project.

20003 at Three Bridges. Behind the Southern Railway built signal box is the National Grid substation with the first 132,000V transmission tower on the overhead line to Brighton just visible. In the distance on the far left of the picture is the Southern Railway 33,000V substation. (Mike Morant Collection)

With more electric locomotives now required, 23 were built for the Southern Region at the former GNR Doncaster works where Oliver Bulleid had begun his career. Some development work had taken place on the control systems but nonetheless they were Raworth 'booster' locomotives. They were arguably the best looking of the electric locomotives, benefiting from the recently created BTC Design Panel, the input of which had resulted in a simple and uncluttered body shape. The locomotives had, as Raworth had originally envisaged before the arrival of Bulleid, two four-wheel power bogies containing four traction motors each rated at 638 horsepower giving a total of 2,552 horsepower (1,900kW). In common with the earlier electric locomotives a pantograph mounted centrally on the roof was provided to collect current from the overhead wires in the sidings. Locations where the overhead lines were installed included the sidings at Faversham Quay and Snowdown colliery in the Kent coal field. The original very smart livery was the standard Southern Region green, plain and unadorned apart from the numbers (E5001 to E5023) displayed below the cab side windows and the latest version of the British Railways heraldic crest halfway along the body side.[35] In later years they were to be classified as Class 71.

In addition to the continued development of the 'booster' locomotives, another concept first proposed by Raworth in the early 1930s was to be at last brought to fruition. In 1958 the BTC ordered six experimental dual-powered locomotives. These locomotives, delivered in 1962, could either be run as a 1,600 horsepower (1,200kW) straight electric locomotive or be switched to diesel-electric power using the on-board 600 horsepower (450kW) diesel generator set. The main electrical equipment was manufactured by English Electric, and could be used in multiple-unit mode with other electric trains on

A representative of the British Railways variant of the 'booster' locomotive, E5005, stands below Percy Tempest's magnificent overall roof at Dover Marine. (Mike Morant Collection)

The use of electric locomotives in Kent necessitated the electrification of several sidings using overhead conductors. An example is shown here at Faversham Quay. (James Boudreau/Southern Notebook (Southern Railways Group))

the Southern Region and with certain conventional diesel-electric locomotives. It was claimed that these 'electro-diesels' were the world's first dual powered locomotives for main line use, but it was conceded that this concept had been used elsewhere for shunting locomotives. The driver of an electro-diesel has the option of using either electric power from the conductor rail or electricity from the on-board diesel generator. This must be selected manually – there is no automatic change-over,[36] but trains hauled by these locomotives could pass through junctions where they might otherwise be gapped using diesel generation. Also the original issue of sudden jerks to the forward movement of freight trains was becoming less applicable as the archaic slack three-link couplings were disappearing from the railway system. The first six electro-diesels – E6001 to E6006 – were built by the Southern Region's own workshops at Eastleigh and allocated to Stewarts Lane depot in south-east London for use on the former SECR main lines.[37] They were built with straight-sided narrow bodies to allow them to be used – diesel powered of course – on the former SER line to Hastings. These electro-diesels were successful, and the design was perpetuated with some improvements with eventually 49 locomotives being built. Despite the fundamental design being 50 years old in 2011, 14 electro-diesels – now all classified as Class 73 – remained in service. The company leasing most class 73s, GB Railfreight, uses them principally for engineering work trains in the third-rail area.[38] Following the success of the electro-diesels, when it later became apparent that the dual power concept provided much better operational flexibility compared to the Raworth 'booster' electric locomotives, 10 of the Doncaster built locomotives were rebuilt during 1967 and 1968 at Crewe Works as electro-diesels receiving Paxman 650 horsepower (485 kW) diesel engines.

Despite approving the Kent coast electrification, the report maintained the

Electro-diesel number 73 201 with a Network Rail track inspection train. The locomotive was named *Broadlands* in 2009. (Peter Winchester)

BTC's lukewarm enthusiasm for main line electrification, concluding that the high cost of the associated civil and signalling works required to install the overhead electrification put wide scale main line electrification beyond the scope of the plan. It was proposed that as an alternative diesel traction would be introduced on the main lines as soon as possible.[39] Nonetheless two trunk routes were proposed for electrification using the 1,500V direct current system: the former LNER main line from King's Cross to 'Doncaster, Leeds (and possibly) York' and the former LMS lines northwards from Euston to Birmingham, Crewe, Manchester and Liverpool. Also proposed for 1,500V electrification was an extension of the LNER's Great Eastern suburban electrification further east to reach Ipswich, Clacton, Harwich and Felixstowe.[40] New 1,500V overhead line suburban electrification schemes included: the 85-mile-long former London, Tilbury and Southend routes from Fenchurch Street station; an additional 36 miles of lines of the former Great Eastern routes from Liverpool Street station to Enfield, Chingford, Hertford and Bishops Stortford; from King's Cross and Moorgate to Hitchin and Letchworth; and finally an ambitious scheme to electrify 190 miles of suburban lines around Glasgow.[41]

Raworth surprisingly found an unexpected ally in respect of his long-standing objection to Charles Merz's 1,500V electrification originally proposed by the BTC for its much-vaunted modernisation of the railways. This was Robin Riddles, the steam locomotive engineer, who was ultimately responsible for the overturning of the '1,500V direct current' policy and effectively instigated the use of the 25,000V, single-phase alternating current overhead system that was to become the standard on Britain's railways today. Raworth had advocated the use of single-phase alternating current at a 'much higher voltage' when quizzed by the colonels at the Ministry of Transport in 1935.[42] John Elliot attributed the change of heart to Riddles,[43] but the eminent mechanical engineer would have had to take advice from his electrical engineer who was, of course, Stanley Warder, Raworth's former assistant. Nonetheless, some credit should be given to the 1951 Cock Report which did at least mention and give consideration to the possible use of the French system.[44] The objections cited by Cock were that while there would be reduced installation costs of the electrification infrastructure, the cost of the locomotives and multiple unit rolling stock would be excessive. Other concerns included the effect of the large, unbalanced loads which might cause instability on the National Grid circuits – but no definite conclusions had been elicited from the electricity authorities. Furthermore, the report was concerned that the alternating current system would generate interference with the railway's communication systems. Nonetheless a trial installation was suggested.[45]

Riddles (probably accompanied by Warder) visited France to see the SNCF system himself, and consequently initiated a trial on the former Midland Railway branches to Morecambe and Heysham. These had been electrified in 1908 using a system similar to that provided by Sir Philip Dawson for the LBSCR. The electric trains continued in service until 1951 when the life-expired stock was removed, and steam traction reinstated. With much of the overhead supports remaining in situ Riddles decided that these two short lines would be the ideal testing ground for a modern alternating current installation and for his trial a 6,000V supply was to be used. This 'new' electric railway was opened in 1953 with Riddles characteristically taking charge in the driver's cab.[46] To trial the various possible control systems and determine the design of pantograph, four converted three-car sets that had previously been used on the former LNWR's Willesden Junction to Earls Court service were rescued from storage where they had stood unwanted since 1940.[47] While many of the original overhead line structures remained in place, evaluation of new designs for supports were also carried out, including a short section constructed by British Insulated Callender Cables (BICC) which provided some new masts at its own expense. These new structures were fully rated for 25,000V operation when required. The tests were successful and early in 1956 the BTC formally announced that for the Modernisation Plan it was proposed to use single-phase alternating current at the 'industrial frequency' of 50 cycles for future railway electrification. Meanwhile the electrification of the trans-Pennine route between Manchester, Sheffield and Wath using the 1,500V direct current system was finally completed. It was a challenging task for the British Railways press office to announce the completion of Britain's first main line electric railway using locomotives for both passenger and freight working without mentioning the now obsolete electrification system being employed. The irony of this would not have been wasted on the retired Alfred Raworth, perhaps reading the newspaper reports describing the 'new' electric railway while relaxing and taking refreshment in the clubhouse after a pleasant round of golf.

Despite the rather badly conceived modernisation plans, the railways' finances continued to deteriorate. In 1962 the then Conservative government abolished the BTC and the responsibility for operating the railways was vested in the British Railways Board (BRB). The BTC had already woken up to the need to close some minor lines and branch lines, but it took an outsider to the railway industry, Dr Richard Beeching, to advocate the most draconian of economy measures in his notorious – or wise, depending on your point of view – report *The Reshaping of British Railways* which was published in 1963. Recommending the slimming down and modernising of the railway network it was made clear that the retention of steam power until as late as 1980 was not going to happen, for the total elimination of steam traction was now planned to be within the decade. Bulleid may have envisaged an average 40 years for his steam locomotive but he did not foresee this purge. The longest surviving unmodified Bulleid locomotives – the ugly duckling Q1s – were in service for only about 22 years – but given that these were 'Austerity' locomotives, that might be deemed praiseworthy. The number of years in service for some of the Bulleid Pacifics after rebuilding can be counted on the fingers of one hand. Steam traction ended on the Southern Region in 1967, and on the whole of British Railways in the summer of 1968.

The former LSWR main line to the West Country was 'rationalised' and a reduced service operated with Western Region diesel locomotives. The LSWR main line from Waterloo to Bournemouth and Weymouth became the last in the country to be worked predominantly by steam locomotives. Latterly the locomotive fleet consisted of Bulleid Pacifics and British Railways standard locomotives. Many observers speculated as to what would follow the demise of steam and given that so much capital expenditure was taking place electrifying the former LMS lines in the north-west of England, the fear was that there would only be sufficient 'modernisation' funds to permit the introduction of diesel traction on this route. To the surprise and delight of many, the decision was made to extend the third rail from Brookwood to Bournemouth.

The execution of this new electrification scheme showed that the former Walker-Raworth-Cox ethos was alive and well. The scheme was to cost only **£15** million,[48] and the use of the third rail allowed the work to be carried out in a timely fashion with a minimal amount of civil works. This contrasted with the accelerating cost and slow pace of the West Coast electrification which was the subject of many rumours that the scheme would be abandoned in favour of dieselisation. The Bournemouth scheme used the 'Raworth' basic design of electricity distribution, that is, to have single rectifier substations interconnected through the conductor rails via the track paralleling huts. New Grid connections were obtained from Basingstoke, Southampton and Bourne Valley (Bournemouth) substations. There was innovation such as the introduction of modern solid-state rectifiers (thyristors), and the introduction of the first of the modular substations, which are factory assembled units containing all the electrical equipment in a metal prefabricated building. These are craned onto a previously prepared concrete base on site. From Brookwood onwards, the conductor rail voltage was to be 750V. There was one very 'Walker' feature reintroduced in a modern and more practical form. This was the economic rebuilding of existing carriages into unpowered trailer sets. Unlike the dangerous two-car sets of old, these three- and four-car sets (3-TC and 4-TC) had driving cabs at either end. Furthermore they were not ancient vehicles but were former steam stock to a recent specification (BR Mk 1s). The steam express service that was operated up to 1967 usually had two portions which divided at Bournemouth Central. One portion would continue to Weymouth, the other continued down the short branch to the terminus at Bournemouth West. The recent total closure of the Somerset and Dorset line meant that there were no longer any of the train services in the Bath direction which had previously terminated at Bournemouth West. Since Bournemouth West was geographically close to Bournemouth Central, it was argued that Bournemouth West could be closed, creating little inconvenience to passengers. With this rationalisation, the Southern Region operators worked out a neat solution for not only operating an electric service to Bournemouth but also to integrate this with the through services beyond Bournemouth to Weymouth. Seven unpowered coaches – a 3-TC set and a 4-TC set – were propelled from Waterloo by a high powered four-car 4-REP set as far as Bournemouth Central. The 4-REP would be detached, and the seven unpowered coaches were then hauled on to Weymouth by one of the Southern Region's Type 3 diesel-electric locomotives which had been suitably modified for 'pull-push' operation. The return journey would be achieved by the diesel-electric locomotive, now at the rear of the train, propelling the unpowered coaches as far as Bournemouth where they would be attached to the rear of a waiting 4-REP four-car unit. The train would leave the diesel-electric locomotive at Bournemouth and with the powered set now leading, journey back to Waterloo. Herbert Walker and Edwin Cox would have regarded this arrangement as a very elegant and economic use of assets.

Raworth would have approved too, and there was another electrification project at the time which finally put into practice his views on the electrification of minor routes which he had recommended in both his reports in the 1940s. The once extensive steam operated network on the Isle of Wight, which had once received thousands of holidaymakers every summer, most of whom had first journeyed south via the Portsmouth Direct, had suffered successive closures since the Second World War. When the Beeching Report was published in 1963 all that remained were the passenger services between Ryde Pier and Ventnor and between Ryde Pier and Cowes – lines Beeching had proposed for total closure.[49] Inevitably this did take place in part with the closure of the Cowes line and the Ventnor line being cut back to Shanklin. The 8 surviving miles of the island network, which had resembled a wonderful working steam railway museum, was to be electrified using the third-rail system. Costs were kept low; the electricity supply did not come from a Grid substation but from three intake points where the Southern Electricity Board had convenient 33,000V distribution lines. The three single rectifier substations were interconnected using the conductor rails without any intermediate track paralleling huts. 'Second hand' trains were used in the shape of former London Underground 'standard stock' that had been built between 1923 and 1934. Six four-car and three six-car sets were reformed and refurbished at the Southern Region's Stewarts Lane depot. These strange visitors to the Isle of Wight entered service in 1967 and would remain on the island for 25 years, before being replaced by more 'modern' former London Underground trains. To minimise the reconditioning work on the former tube trains the Isle of Wight line still operates at only 600V direct current. Despite some movement towards Raworth's vision of an all-electric railway there was much more to be achieved, as the long-distance freight trains (such as oil trains from the Fawley refinery) and the semi-fast trains to Exeter continued to be hauled by diesel locomotives and diesel-electric multiple unit trains were in use on all the secondary routes.

After the Bournemouth and Isle of Wight electrification there was to be a 20-year wait before there was any more electrification in southern England. With the demise of steam, the surviving minor routes that were not electrified were operated with diesel power. This was not what Raworth had envisaged in his reports but was exactly in line with the proposals heralded by Missenden at the end of the Southern Railway's existence. The passenger trains on these secondary routes were composed of Southern Region diesel-electric two- or three-car sets which outwardly looked like the contemporary electric units, but part of the interior of one carriage was given over to a diesel generating set. On the Hastings line, six-car express units were built to the narrow restriction 0 specification. When these products of the original modernisation proposals showed signs of becoming obsolete and life-expired, like-for-like replacement costs had to be weighed against possible 'cheap' electrification which would, as Raworth had advised in his reports, allow integration of services by avoiding inefficient enclaves of diesel operated zones bounded by an otherwise all-electric area.

Perhaps the most obvious 'infill' was the Hastings line. The issue of the narrow tunnels was overcome by reducing the number of tracks through the offending tunnels from two to one. Modern signalling systems made this a practical solution. Electric trains from London Bridge to Hastings began running in 1986. The long

The 'basic' electric railway on the Isle of Wight. A former London Underground Class 483 leaves Ryde for Shanklin on 6 April 2017. These trains (now withdrawn) were the second generation of former tube stock to be used on the Isle of Wight. They were originally built in 1938 and converted for use on the island in 1988. (Peter Winchester)

awaited 'Oxted' electrification came to partial fruition a year later when the conductor rails were extended from East Croydon but only as far as East Grinstead. Beyond East Grinstead the line had been closed in the 1950s. In 1988 the line from Bournemouth to Weymouth via Dorchester was electrified. The Dorchester to Weymouth line became the first and only part of the old GWR to receive the 'Southern' conductor rail. Also to be electrified by 1990 was the trio of former LSWR secondary lines which meet at Fareham, namely: Fareham to St Denys (on the main line just outside Southampton); Fareham to Portsmouth via Portchester; and Fareham to Eastleigh via Botley (the surviving remnant of the original London and Southampton route to Gosport). With the Eastleigh to Portsmouth section electrified there is now a good alternative electric service operating from Waterloo to Portsmouth Harbour station via Basingstoke, Eastleigh and Fareham. These schemes carried out after the Bournemouth electrification bear the hallmarks of being minimum cost electrification – another instance of inexpensive third-rail electrification or no electrification at all in a climate when most engineers were of the view that electrification should be at the standard 25,000V alternating current. The relatively inexpensive third-rail electrification was often achieved by using an 11,000V supply from the local electricity distribution network.

Sadly Alfred was not to see any of this. Work on the Bournemouth and Isle of Wight schemes was in progress when he died late in January 1967 aged 85 and, as is often the case with survivors of a long marriage, his wife Ruby died just a few months later in December 1967.

Imagine for one moment that Sir Herbert Walker, Edwin Cox and Alfred Raworth were to be transported back from the oblivion of the afterlife and temporarily deposited at some location on the Southern Electric network – say at the London end of Guildford station. What would these ghosts from the past make of the privatised and franchised British railway scene? The initial thoughts from Alfred Raworth might be that here at last was his all-electric railway, his evidence being that the locality has been shorn of all locomotive facilities – the former steam motive power depot is now a car park. This would not of course be correct, there is now no regular freight traffic to Guildford thereby removing the need for any locomotives – the all-electric railway is an illusion as far as Guildford is concerned, for if there was any regular freight traffic this would be diesel powered. If the three gentlemen from the past were unfortunate enough to linger long enough to witness the diesel service from Reading to Redhill arriving at Guildford, I hope that they would not notice the lettering 'GWR' on the train. This former SECR line has not been electrified as an 'infill' project, despite much of the route already being electrified, the railway's 33,000V distribution system being easily accessible. Furthermore, half of the trains reverse at Redhill and continue down the electrified Brighton line to Gatwick airport.

The tetchy Alfred Raworth would not be impressed by the fact that all the jumper cables appear now to be laid on top of the ballast in a very untidy fashion and all three men might initially recoil from the rather garish colour schemes applied to some of the carriages perversely branded 'South Western Railway'. But otherwise they would be impressed. Alfred might guess – correctly – that the electric trains are purpose built, not conversions from older stock. Significantly too, multiple unit trains are still being used and continue to use the third rail. Sir Herbert and Edwin Cox would quickly appreciate that the basic 'clock face' timetable is still in force with regular fast trains running down the Portsmouth Direct interspersed with regular stopping trains, and that commuter trains are still operating to Waterloo on the New Line – all very close to the original Southern Electric pattern. The half hourly service to Aldershot and Ascot is still running, but unfortunately not continuing to Waterloo. In short, a very recognisable scene would be on view to these ghosts from the past. Raworth would have expected – and hoped – that technology would have advanced sufficiently to allow all the electrical plant to have been replaced and updated (at least once!) as by now all that he had installed would have far exceeded the 33.3 years of his useful life dictum. The very sound of traction motors would tell him that technology has moved on from the traditional direct current series motors. The Guildford substations are out of sight beyond a bridge, the oldest – the rotary convertor building – out of use, but the later 1937 building is still in service stripped of all its original equipment and now surrounded by metal buildings containing the high voltage switchgear and the power electronics to rectify the alternating current supply. He would, I am sure, suspect that the conductor rail voltage was now 750V.

These three gentlemen from another age would be well pleased that their basic concept for the Southern Electric is still being maintained in the twenty-first century. Perhaps then, despite the updating of all the equipment, what we see around us in southern England today remains Alfred Raworth's electric railway.

ENDNOTES

1 Textiles, Electric Lighting and Trams

1. *Southern Railway Magazine* January 1933.
2. Ibid.
3. Information provided by Chris Raworth.
4. John Smith Raworth's application to join the Institution of Civil Engineers.
5. The National Archives (TNA) 1851 census THO 107 2147.
6. Ferneyhough, Frank, *Liverpool and Manchester Railway* (Book Club Associates 1980), pp. 150–151.
7. Riden, Philip, *George Stephenson's Last Home at Risk*, Backtrack, February 2020.
8. TNA 1861 Census RG 9/2530.
9. Internet: mkheritage.co.uk/mkm/stonystratford/docs/hayes, accessed Jan. 2013.
10. TNA BT 31/2346/11477.
11. JSR application to join Telegraph & Electricians Society (forerunner of IEE).
12. TNA BT 31/2346/11477.
13. TNA 1861 Census RG 11/3571.
14. Riden, Philip, *George Stephenson's Last Home at Risk*, Backtrack, February 2020.
15. TNA Baptism Records.
16. J.S. Raworth obituary *Engineering*, March 301917.
17. Wilson, John F., *British Business History 1720–1994* (Manchester University Press 1995), p. 121.
18. Robinson, S.J.T. and Markham, J.D., *The Regenerative Braking Story* (Scottish Tramway & Transport Society/Venture Publications 2006), p. 14, citing *Tramway and Railway World*, 8 October 1903.
19. J.S. Raworth's application to join Institution of Mechanical Engineers.
20. Internet: Birminghamhistory.co.uk, accessed March 2013.
21. Internet: gracesguide.co.uk/British Electric Traction, accessed July 2013.
22. Alfred Raworth's application to join Institution of Mechanical Engineers.
23. Klapper, C.F., *Herbert Walker's Southern Railway* (Ian Allan, London,1973), p. 95.
24. Dulwich College Archive, courtesy of archivist Soraya Cerio.
25. Robinson, S.J.T. and Markham, J.D., *The Regenerative Braking Story*, p. 21.
26. Internet: tramroma, accessed Jan 2013.

2 Raworth's Traction Patents

1. US Department of Energy, *Determination of the Effectiveness and Feasibility of Regenerative Braking*, 1978, p. 15, online, digitised by Google Books.
2. Internet: Cornell University Archive, *Dictionary of Electricity*, p. 435.
3. Robinson, S.J.T. and Markham, J.D., *The Regenerative Braking Story* (Scottish Tramway & Transport Society/Venture Publications 2006), p. 16.
4. The National Archives (TNA) BT 31/10863/82432.
5. TNA BT 31/10863/82432. (Raworth's Traction Patents prospectus.)
6. Raworth, Alfred, regenerative control of electric tramcars and locomotives, *Journal of the Institution of Electrical Engineers*, vol. 38, issue 182, 1907, p. 379.
7. TNA BT 31/10863/82432.
8. Robinson, S.J.T. and Markham, J.D., *The Regenerative Braking Story*, Appendix 1, p. 190 (resumé of Board of Trade report).
9. John Edward Raworth, application to join Institution of Mechanical Engineers.
10. TNA BT 31/10863/82432.
11. Ibid.
12. News item, *Sheffield Daily Telegraph*, 4 November 1904.
13. Raworth, A., Regenerative control of electric tramcars and locomotives, *Journal of the Institution of Electrical Engineers*, vol. 38, issue 182, 1907, pp. 37–379.

14. Ibid.
15. Ibid.
16. Internet; Wikipedia (Central London Railway), accessed May 2014.
17. R.A.S. Hennessey, The British steeple cab, *Backtrack*, March 2013.
18. A.R. application to join institution of Mechanical Engineers.
19. Benest, K.R., *Metropolitan Electric Locomotives*, Lens of Sutton, p. 10.
20. John Smith Raworth, discussion after A.R. lecture: Regenerative control of electric tramcars and locomotives, *Journal of the Institution of Electrical Engineers*, vol. 38, issue 182, 1907, p. 395.
21. Robinson, S.J.T. and Markham, J.D., *The Regenerative Braking Story*, p. 41.
22. Raworth Alfred, Regenerative control of electric tramcars and locomotives, *Journal of the Institution of Electrical Engineers*, vol. 38, issue 182, 1907, p. 381.
23. Robinson, S.J.T. and Markham, J.D., *The Regenerative Braking Story*, p. 42.
24. Internet: Patents Espacenet, GB190524810 and GB190712443.
25. *The Manchester Courier*, 1 May 1908.
26. TNA 1901 Census ref RG 13/4058.
27. TNA Birth deaths and marriages records.
28. Dulwich College Register entries.
29. TNA 1911 Census ref RG 14/21556.
30. TNA BT 31/10863/82432.
31. Ibid.
32. Robinson, S.J.T. and Markham, J.D., *The Regenerative Braking Story*, includes local newspaper report, pp. 47–51.
33. Ibid, *Report* from Board of Trade, pp. 47–51.
34. TNA BT 31/10863/82432.
35. *London Gazette*, 10 December 1912.
36. Internet: Patents Espacenet, GB191119541 (A).

3 Alfred Raworth Becomes a Railwayman

1. Klapper, C.F., *Herbert Walker's Southern Railway* (Ian Allan, London 1973), p. 38.
2. Simmonds, Jack, South Western vs Great Western, *Journal of Transport History*, No. 4. 1959.
3. Mitchell, Vic and Smith, Keith, *Reading to Guildford*, (Middleton Press 1988).
4. Harvey, Michael G. and Rooke, Eddie, *Railway Heritage Portsmouth* (Silver Link Publishing 1997), p. 8.
5. Mitchell, Vic and Smith, Keith, *Branch Lines Around Ascot* (Middleton Press 1989).
6. South, Raymond, *Crown College and Railways* (Barracuda Books 1978), p. 118.
7. Faulkner, J.N., To Guildford via Cobham, *Railway Magazine*, September 1959.
8. Strutt, Ron, Scooter of the South Western, *Backtrack*, April 2014.
9. Weddell, G.H.R., *LSWR Carriages in the 20th Century* (Oxford Publishing Co. 2001) pp. 16, 34.
10. Chivers, Colin, *The Riverside Electric*, (South Western Circle 2010), p7,8.
11. Ibid., full text of paper Appendix 1, pp. 171–180.
12. Klapper, C.F., *Herbert Walker's Southern Railway* (Ian Allan, London 1973), pp. 10–11; *Modern Transport* 5 July 1937.
13. *Railway Gazette*, 4 December 1936.
14. Internet: Wikipedia accessed March 2019.
15. Internet, dnw.co accessed March 2019.
16. Internet, Ancestry.Co, 1901 Census accessed March 2019.
17. The National Archives (TNA) RAIL 411/366.
18. Lassalle, T.S., Diamond Jubilee of the Waterloo and City Railway, *Railway Magazine* August 1958.
19. Internet: Wikipedia, Sir Alexander Kennedy accessed Jan. 2014.
20. Lassalle, T.S., Diamond Jubilee of the Waterloo and City Railway, *Railway Magazine* August 1958.
21. TNA LSWR clerical staff character book No. 4 1859–1920.
22. TNA RAIL 411/196 LSWR subcommittee meeting 22 February 1912.
23. TNA RAIL 411/196.
24. Census records via Ancestry.com.
25. National Railway Museum (NRM) Hawes Collection ref 2012-7274 A, Alfred Raworth's personal notebook.

26. Ibid.
27. Internet: ellisisland accessed January 2014.

4 A Design for an Electric Railway

1. Harding, C. Francis, *Electric Railway Engineering*, 2nd edition (McGraw Hill 1916 from Forgotten Books), p. 374.
2. Travers, Ian, The Lancashire and Yorkshire Railway North Liverpool Electrification, *Backtrack*, December 2013.
3. Internet: Wikipedia, Mersey Railway, accessed January 2014.
4. Internet: Suburban Electric Railway Website accessed January 2014.
5. Electrification of the LBSC Railway, *Railway Magazine* June 1911.
6. Lee, Charles E., London's 'Elevated Electric', *Railway Magazine* December 1959.
7. Internet: Wikipedia, Philip Marsh accessed June 2014.
8. Dawson, Phillip, *Electric Traction on Railways* (The Electrician Printing and Publishing Co. 1909), Preface.
9. *Railway Magazine* July 1909.
10. Dawson, Phillip, *Electric Traction on Railways*, p. 729.
11. Ibid.
12. Ibid., p. 538.
13. TNA MT 6/3128.
14. Electrification of the LBSC Railway, *Railway Magazine* June 1911.
15. Internet: John Spellers web pages accessed January 2014.
16. Fell, M.G. and Hennesey, R.A.S., Buxton: What might have Been, *Backtrack* November 2013.
17. NRM Hawes collection ref 2012-7274 ALS3; 36B Electrification Report.
18. Ibid., p. 7.
19. Ibid., p. 8.
20. Ibid.
21. NRM Hawes collection ref 2012-7274 A, Hawes Collection Alfred Raworth's personal notebook.
22. TNA RAIL 411/39.
23. Ibid.
24. *Tramway and Railway World*, 18 November 1915.
25. *Railway Gazette*, 13 December 1913.

5 The LSWR Electrification

1. The National Archives (TNA) MT 2166/2.
2. Ibid.
3. Ibid.
4. Ibid.
5. TNA MT 2166/2 R2075.
6. TNA MT 6132.
7. TNA MT 6/3025, minute note 12 Dec 1919.
8. TNA MT 6/3124.
9. *Tramway and Railway World*, 18 November 1915.
10. TNA RAIL 411/931.
11. *Tramway and Railway World*, 18 November 1915.
12. *Railway Gazette*, 13 December 1913.
13. *Tramway and Railway World*, 18 November 1915.
14. National Railway Museum (NRM) ref 2012-7274 ALS3; 36; B.
15. Ibid.
16. Ibid.
17. TNA RAIL 411/196, LSWR subcommittee meeting 24 July 1913.
18. Weddell, G.H.R., *LSWR Carriages in the 20th Century* (Oxford Publishing Co. 2001), pp. 61–76.
19. Ibid.
20. NRM ref 2012-7274 ALS3; 36; A.
21. Klapper, C.F., *Herbert Walker's Southern Railway* (Ian Allan, London 1973), p. 113.
22. *Tramway and Railway World*, 18 November 1915.
23. Ibid.
24. TNA RAIL 411/64, LSWR subcommittee meetings.

6 Raworth RN

1. The National Archives (TNA) RAIL 1017/1/66.
2. Entry 4826 Dulwich College Register.
3. TNA ADM-337-120-0-99 AR service record.
4. Internet: theaerodrome.com/forum, accessed June 2014.
5. Internet: history.co.uk accessed June 2014.

6. Internet forces-war-records.co.uk accessed, June 2014.
7. *Modern Transport*, 5 November 1938.
8. *The Engineer*, 13 March 1917.
9. TNA ADM-337-120-0-99 AR service record.
10. Wikipedia: Sir Philip Dawson, accessed June 2014.
11. Wikipedia: *Ministry of Munitions*, accessed June 2014.
12. Wikipedia: *David Lloyd George*, accessed April 2013.
13. Klapper, C. F., *Herbert Walker's Southern Railway* (Ian Allan, London 1973), p. 59.
14. Chivers, Colin, *The Riverside Electric* (South Western Circle 2010), pp. 88–104.
15. TNA MT 6/2453/6.
16. *Railway Gazette*, 18 March 1918.
17. Findmypast.
18. TNA ADM-337-120-0-99 AR service record.

7 A Railway Under Stress

1. Clark, Jeremy, James Staats Forbes and the London, Chatham and Dover Railway, *Backtrack*, May 2011.
2. Kidner, R.W. *The Reading to Tonbridge Line* (Oakwood Press 1974), pp. 6–10.
3. Jackson, Alan, *London's Termini* (Pan 1972), p. 254.
4. Klapper, C.F., *London's Lost Railways*, (Routledge & Keegan Paul Ltd 1976), pp. 66–67.
5. Jackson, Alan, *London's Termini*, p. 254.
6. Ibid., p. 199.
7. Greaves, John Neville, *The Last of the Railway Kings: The Life and Work of Edward Watkin 1819–1901* (Durham University e-thesis), pp. 109–132.
8. *Modern Transport*, 4 December 1922.
9. The National Archives (TNA) 1911 Census.
10. Internet: Wikipedia, H. Cosmo O. Bonsor, accessed Feb. 2013.
11. Klapper, C.F., *Herbert Walker's Southern Railway* (Ian Allan, London 1973), p. 117.
12. Treby, Edward, From SECR to GNR: a Cross-London Journey of 1906, *Trains Annual* (Ian Allan 1965), p. 69.
13. Dendy, C.F., *A History of the Southern Railway* (Southern Railway 1936), p. 498.
14. TNA RAIL 633/12 SECR minutes 2 May 1902.
15. Ibid., 12 December 1902.
16. *London Gazette*, 21 November 1902, p. 7754.
17. London Metropolitan Archive, transcript of Electricity Commissioners hearing May/June 1922.
18. TNA MT 6 3124 extract from Act of Parliament.
19. TNA RAIL 633/12 SECR minutes 25 March 1903.
20. Jackson, Alan, *London's Termini*, pp.194–203.
21. Parliamentary Papers: Returns of number of Private Bills.
22. TNA MT 6 3124 extract from Act of Parliament.
23. TNA RAIL 633/30 SECR minutes 12 October 1910.
24. Internet, *Science Museum Group Collections Online – People*, accessed April 2013 and TNA census 1911.
25. *Railway Gazette*, 18 January 1919.
26. *Railway Gazette*, 21 February 1919.
27. Internet:1centenary Oxford University, accessed January 2016.
28. Internet: metadyne/subesitrrsa/rsapresidents, accessed April 2013.
29. *Southern Notebook* (Southern Railways Group), Summer 2012 p. 261.
30. TNA RAIL 633/52 SECR minutes 24 January 1912.
31. Internet: Wikipedia accessed January 2015.
32. TNA RAIL 633/54 SECR minutes 27 June 1913.
33. Glover, John, *Southern Electric*, (Ian Allan 2001), p. 21.
34. Ibid.
35. Moody, G.T., *Southern Electric* (Ian Allan 1958), p. 19.
36. *Railway Gazette*, 2 June 1922.
37. Taylor, A.J.P., *The First World War. An Illustrated History* (Penguin Books 1966), pp. 26–29.
38. "*A History of the Southern Railway*" C.F. Dendy Southern Railway 1936 p494 – p497.
39. *Railway Gazette*, 24 January 1919.

40. Dendy, C.F., *A History of the Southern Railway* (Southern Railway 1936), pp. 494–497.
41. *Railway Gazette*, 24 January 1919.
42. Ibid., 17 January 1919.
43. Ibid.
44. Taylor, A.J.P., *The First World War: An Illustrated History*, p. 150.
45. Ibid., p. 159.
46. *Railway Gazette*, 29 January 1918.
47. Ibid, 29 January 1918.
48. TNA RAIL 633/44 SECR minutes 12 November 1917.
49. Ibid, 29 November 1917.
50. *Railway Gazette*, 21 June 1918.
51. Klapper, C.F., *Herbert Walker's Southern Railway*, p. 85.
52. Ibid., p. 162.
53. TNA MT 6/2960, letter Percy Tempest to MoT.
54. Emblin, Robert and Longbone, Bryn, When push comes to shove – some effects of the Great War on the railways of Britain, *Backtrack*, May 2013.
55. Bonavia, Michael R., *The Channel Tunnel Story* (David & Charles 1987), p. 40.
56. *Railway Gazette*, 25 January 1918, p. 87.
57. TNA RAIL 633/42 SECR minutes 8 November 1916.
58. TNA RAIL 633/44 SECR minutes 12 December 1917.
59. Ibid, 1 January 1918.
60. Internet:firstworldwar.com/bio/geddes-eric, accessed April 2013.
61. Grieves, Keith, *Sir Eric Geddes. Business and Government in War and Peace* (Manchester University Press 1989), p. 13.
62. Ibid., 53.
63. Irving, R.J., *The North Eastern Railway – An Economic History* (Leicester University Press, 1976), p. 216.
64. Ibid., p. 264.
65. Grieves, Keith, *Sir Eric Geddes. Business and Government in War and Peace*, p. 8.
66. TNA RAIL 633/44 SECR minutes 7 February 1918.
67. LMA – transcript of Electricity Commissioners hearing May/June 1922.
68. Hennessey, R.A.S., The British steeple cabs, *Backtrack*, March 2013.
69. *Railway Gazette*, 14 June 1918.
70. Moody, G. T., *Southern Electric*, p. 14.
71. Taylor, A.J.P. *The First World War: An Illustrated History* (Penguin Books 1966), pp. 220–238.
72. Dendy C.F., *A History of the Southern Railway*, p. 497.

8 Alfred Raworth's Electric Railway

1. Internet: findmypast, accessed January 2016.
2. Internet: ellisisland, accessed October 2013.
3. Ibid., accessed October 2013.
4. *Electrical Age*, July 1916, p. 27 (via internet Michael Sol Collection).
5. Trewman, H.F., *Railway Electrification* (Pittman 1924), p. 6.
6. Internet: trains.com, accessed Nov. 2013.
7. Plague in Grand Central Station, viewed in November 2018.
8. Ancestry.co.uk, incoming passenger lists.
9. The National Archives (TNA) RAIL 633/214.
10. TNA MT 6/2960.
11. *Railway Magazine*, December 1925.
12. Whitaker, F.P., *IEE Journal*, vol. 60, No. 309, May 1922, p. 502.
13. Internet: Wikipedia Lancashire and Yorkshire Railway, accessed Feb. 2013.
14. Steamindex.com accessed Jan 2013.
15. TNA RAIL 633/214 – map in report.
16. Ibid., p. 13, Appendix 3.
17. *Railway Gazette*, 17 January 1919, 7 February 1919.
18. Internet: economicshelp.org accessed July 2013.
19. TNA RAIL 633/48 SECR minutes 12 November 1919.
20. *Railway Gazette*, 25 May 1922.
21. Bonavia, Michael R., *The Channel Tunnel Story* (David & Charles 1987), p. 43.
22. Patent Office GB141488 (A), Improvements in the control of electrically propelled vehicles or trains.

9 Meeting the Men From the Ministry

1. Klapper, C.F., *Herbert Walker's Southern Railway* (Ian Allan, London 1973), p. 96.
2. Wikipedia.org/wiki/United_Kingdom_general-election,1918, accessed May 2013.
3. Grieves, Keith, *Sir Eric Geddes: Business and Government in War and Peace* (Manchester University Press 1989), p. 72.
4. Ibid., p. 53.
5. Ibid., p. 73.
6. Ibid., p. 53; Wringley, Chris, *Lloyd George and the Challenge of Labour* (Harvester/Wheatsheaf, London 1991), p. 252.
7. Watts D., On the Causes of British Railway Nationalisation: A re-examination of the Causes 1866–1921, *Contemporary British History*, vol. 16, no. 2, pp. 5–8.
8. *Railway Gazette*, 3 January 1919.
9. Dyos, H.J. and Aldcroft, D.H., *British Transport: An Economic Survey from the 17th Century to the Twentieth* (Penguin Books 1974), p. 312.
10. Grieves, Keith, *Sir Eric Geddes. Business and Government in War and Peace*, p. 76.
11. Ibid., p. 78.
12. Internet: legislation.gov.uk/ukpga/Geo5/9-10/50/contents, accessed April 2013.
13. Bagwell, Phillip and Lyth, Peter, *Transport in Britain From Canal Lock to Gridlock* (Hambledon 2002), pp. 74–76.
14. Wikipedia, Ministry of Transport, accessed April 2013.
15. Emblin, Robert and Longbone, Bryn, When push comes to shove – some effects of the Great War on the Railways of Britain, *Backtrack*, May 2013.
16. *Railway Gazette*, 21 June 1918.
17. The National Archives (TNA) RAIL 633/44 SECR minutes 26 November 1919.
18. Klapper, C.F., *Herbert Walker's Southern Railway*, p. 105.
19. TNA RAIL 633/48 SECR minutes 21 March 1920.
20. TNA MT 6/2960.
21. Moody, G.T., *Southern Electric* (Ian Allan 1958), p. 26.
22. *Southern Railway Magazine*, October 1938.
23. *Railway Gazette*, 19 May 1922.
24. King, Mike, *Southern Vans and Coaches in Colour* (Noodle Books 2015), p. 47.
25. TNA T161 1149.
26. TNA MT 6/2960 (extract from Advisory Committee meeting minutes).
27. Ibid.
28. Ibid.
29. Clark, Peter, Draft article for SECR Society magazine *Invicta*.
30. TNA RAIL 1124/204.
31. Wikipedia, accessed February 2016.
32. TNA MT 6/2960.
33. Ibid.
34. TNA T161 1149.
35. *Modern Transport*, 29 July 1922.
36. Internet: Denman, J. and McDonald, P., Unemployment trends from 1886 to the present day, *Labour Market Trends*, January 1996.
37. TNA RAIL 633/44 SECR minutes 26 October 1919.
38. TNA MT 6/2960.
39. TNA RAIL 633/52 SECR minutes 14 December 1921.
40. TNA RAIL 633/52 SECR minutes 11 January 1922.
41. TNA MT 6/2960.
42. London Metropolitan Archive (LMA) – transcript of Electricity Commissioners hearing May/June 1922.
43. TNA MT 2166/2.
44. TNA MT 6/2960.

10 Making Three Fit into One

1. Faulkner J.N., The Triumph of the Third Rail *Railways South East*, Summer 1988.
2. The National Archives (TNA) MT 6/3124 – Electrification Committee minutes.
3. Ibid.
4. TNA MT 6/2960 – meeting note, Trades Advisory Committee.
5. TNA RAIL 633/52 SECR minutes 11 January 1922.

6. TNA MT 6/2960 – letter K. & D. to Raworth.
7. Ibid. – copy of Kennedy and Donkin report.
8. Ibid. – letter Percy Tempest to Ministry of Transport.
9. Ibid.
10. Ibid – Ministry of Transport note.
11. TNA MT 6/3124 – Electrification Committee minutes.
12. Ibid – letter Percy Tempest to Ministry of Transport.
13. TNA MT 6/2960 – Ministry of Transport note.
14. TNA MT 6/3137 – Ministry of Transport notes of meeting.
15. *Electric Railway Journal*, 2 September 1922.
16. TNA MT 6/2960.
17. TNA MT 6/3124 – letter from Theodore Stevens.
18. TNA MT 6/3137 – letter Sir William Forbes to Phillip Dawson.
19. Ibid. – notes from first meeting of committee.
20. Ibid. – letter from Sir Herbert Walker.
21. Ibid. – committee minutes.
22. TNA MT 6/2960.
23. Ibid.
24. TNA T161 1149, TNA RAIL 633/54 SECR minutes 10 January 1923.
25. TNA MT 6/3124 – letter Kennedy to Nash.
26. TNA MT 6/2960.
27. Ibid.
28. *Railway Review*, 9 October 1915 (via internet Michael Sol Collection).
29. TNA MT 6/3137.
30. Ibid.
31. Internet: www.arrts-arrchives.com/electrification.html May 2019.
32. Hobbs, George, *Power, Poles and Platelaying, Keeping the MER on Track* (Loaghtan Books 2019), p. 63.
33. Internet: www.arrts-arrchives.com/electrification.html May 2019.
34. TNA MT 6/2960.
35. Internet accessed September 2013.
36. *Electric Railway Journal*, 2 September 1922.
37. Trewman H.F., *Railway Electrification* (Pittman 1924).
38. Klapper, C.F., *Sir Herbert Walker's Southern Railway* (Ian Allan 1973), p. 113.
39. TNA RAIL 633/52 SECR minutes 12 April 1922.
40. Patent specification 3701/22;192,922 dated 8 February 1922.
41. Smith, Robert Henry, *Railway Electric Traction* (Harper Brothers from Forgotten Books 1905), p. 215.
42. TNA MT 6/2960.
43. Ibid. MoT internal memo.
44. Denman, J. and McDonald, P., Unemployment trends from 1886 to the present day, *Labour Market Trends*, January 1996.
45. TNA RAIL 633/52 SECR minutes 22 March 1922.
46. TNA RAIL 633/52 SECR minutes 12 April 1922.
47. Dendy, C.F., *A History of the Southern Railway* (Southern Railway 1936), p. 492.
48. TNA T161 1149.
49. *Railway Gazette*, 26 May 1922.
50. TNA RAIL 633/50 SECR minutes 7 June 1922.
51. TNA RAIL 633/54 SECR minutes 20 December 1922.
52. *Southern Railway Magazine*, January 1939.
53. TNA RAIL 633/54 SECR minutes 20 December 1922.

11 The Battle of Angerstein's Wharf

1. The National Archives (TNA) MT161 1149.
2. Brown, Joe, *London Railway Atlas* (Ian Allan 2012), p. 41.
3. TNA RAIL 633/214 Appendix 11.
4. London Metropolitan Archives (LMA) – transcript of Electricity Commissioners hearing May/June 1922.
5. *Railway Gazette*, 21 March 1919.
6. Rowland, John, *Progress in Power* (Merz and Mclellan/Newman Neame 1959), p. 66.
7. Hennessey, R.A.S., 'Sparks': The Electrical Consultants Part 1, *Backtrack*, July 2008.
8. Rowland, John, *Progress in Power*, pp. 72–73.

9. TNA MT 2166/2.
10. Ibid.
11. TNA MT 2166/2, 6/2960.
12. TNA MT 6/2960.
13. TNA MT 2166/2.
14. *London Gazette*, 12 May 1922.
15. TNA MT 2166/2.
16. TNA RAIL 633/214 Appendix 11.
17. LMA transcript of Electricity Commissioners hearing May/June 1922.
18. TNA MT 6/3128.
19. Raworth J.S., during discussion after *Regenerative Control of Electric Tramcars and Locomotives, Journal of the Institution of Electrical Engineers*, vol. 38, issue 182, 1907, p. 387.
20. TNA RAIL 633/52 SECR minutes 14 June 1922.
21. TNA RAIL 633/214.
22. TNA MT 6/3128.
23. TNA RAIL 633/52 SECR minutes 26 July 1922.
24. TNA MT 2166/2.
25. TNA RAIL 633/54 SECR minutes 6 September 1922.
26. Ibid., 28 September 1922.
27. TNA MT 2166/2.
28. TNA MT 6/3025.
29. TNA MT 2166/2.
30. Ibid.

12 The Birth Pangs of the Southern Railway

1. Faulkner, J.N., The Triumph of the Third Rail, *Railways South East*, Summer 1988.
2. The National Archives (TNA) RAIL 633/54 SECR minutes 9 August 1922.
3. Ibid.
4. Ibid., 11 October 1922.
5. Ibid., 25 October 1922.
6. Ibid., 9 August 1922.
7. Letter Eric Geddes to Henry Babington Smith, Box 42 private papers of Sir Henry Babington Smith – Trinity College Library, Cambridge.
8. Note in Box 42, Trinity College Library, Cambridge.
9. Letter Eric Geddes to Henry Babington Smith, Box 42 Private papers of Sir Henry Babington Smith – Trinity College Library, Cambridge.
10. Ibid.
11. Ibid.
12. Dendy, C.F., *A History of the Southern Railway* (Southern Railway 1936), p. 511.
13. *Modern Transport*, 29 July 1922.
14. Ibid., 12 August 1922.
15. TNA RAIL 633/54 SECR minutes 9 August 1922.
16. Box 43 Trinity College Library, Cambridge (Henry Babington Smith's own copy annotated 'draft').
17. Klapper, C.F., *Sir Herbert Walker's Southern Railway* (Ian Allan 1973), p. 101.
18. Box 43 Trinity College Library, Cambridge (Henry Babington Smith's own copy annotated 'draft').
19. Faulkner, J.N., The Triumph of the Third Rail, *Railways South East*, Summer 1988.
20. TNA RAIL 633/54 SECR minutes 25 October 1922.
21. Churella, Albert J., *The Pennsylvania Railroad, Volume 1: Building the Empire 1846–1917* (Pennsylvania University Press), p. 784.
22. *Electric Railway Journal*, 2 September 1922.
23. TNA MT 6/2960 Alfred Raworth letter to Ministry of Transport.
24. Moody G.T., *Southern Electric* (Ian Allan 1958) p. 22.
25. TNA RAIL 633/54 SECR minutes 8 November 1922.
26. TNA MT 6/2960 Alfred Raworth letter to Ministry of Transport.
27. TNA RAIL 633/54 SECR minutes 22 November 1922.
28. Dendy, C.F., *A History of the Southern Railway* (Southern Railway 1936), p. 512.
29. TNA MT 6/2960 extract from proceedings of Court of Directors.
30. TNA MT 6/2960 letter Alfred Raworth to Ministry of Transport.

13 Completing the Legacy Schemes

1. Klapper, C.F., *Sir Herbert Walker's Southern Railway* (Ian Allan 1973), p. 111.
2. Elliot Sir John, *On and Off the Rails* (George Allen & Unwin 1982), p. 43.

3. Alfred Raworth's application to join the Institution of Civil Engineers.
4. The National Archives (TNA) RAIL 645/3 SR minutes – various dates.
5. Klapper, C.F., *Sir Herbert Walker's Southern Railway*, p. 169.
6. TNA T161 1149, TNA RAIL 633/54 SECR minutes 30 December 1922/10 January 1923.
7. Parris, Henry, *Government and the Railways* (Routledge & Keegan Paul 1965), p. 32.
8. TNA MT 6/3406.
9. Wikipedia accessed August 2014.
10. TNA MT 6/3406 – extract from minutes of 1923 conference.
11. TNA MT 6/3406.
12. TNA MT 6/2960 extract from proceedings of Court of Directors.
13. TNA RAIL 645/3 SR minutes 30 April 1925.
14. *Southern Railway Magazine*, January 1924.
15. *Railway Gazette*, 11 October 1925.
16. *Southern Railway Magazine*, June 1923.
17. TNA RAIL 633/54 SECR Committee minutes 30 April 1923.
18. TNA RAIL 645/16 SR Engineering Subcommittee minutes, 30 April 1924.
19. Klapper, C.F., *Sir Herbert Walker's Southern Railway*, p. 170.
20. TNA MT 2166/2.
21. Nock, O.S., *Fifty Years of Railway Signalling* (The Institution of Railway Signalling Engineers 1962), p. 67.
22. Kitchenside, G.M. and Williams, Alan, *British Railway Signalling* (Ian Allan 1975), p. 27.
23. Nock, O.S., *Fifty Years of Railway Signalling*, p. 68.
24. Ibid., p. 74.
25. TNA RAIL 645/16 SR Engineering Subcommittee minutes 18 December 1924.
26. Moody G.T., *Southern Electric* (Ian Allan 1958), p. 22.
27. *Railway Gazette*, 20 May 1927.
28. Pryer, G., *A Pictorial Record of Southern Signals* (Oxford Publishing Co. 1977).
29. *Railway Gazette*, 5 March 1922.
30. *Southern Railway Magazine*, April 1926.
31. Moody G.T., *Southern Electric*, p. 35.
32. *Railway Gazette*, 26 Feb. 1926.
33. Moody G.T., *Southern Electric*, p. 36.
34. TNA RAIL 645/3 SR minutes 1 November 1923.
35. TNA RAIL 645/3 SR minutes 6 November 1923.
36. *Southern Railway Magazine*, August 1925.
37. TNA RAIL 645/33 SR Engineering Subcommittee minutes 18 December 1924.
38. *Southern Railway Magazine*, April 1927.
39. Brown, Joe, *London Railway Atlas* (Ian Allan 2012), p. 70.
40. *Southern Railway Magazine*, August 1925.
41. TNA RAIL 645/28 SR Locomotive and Electrical Committee minutes 28 Nov. 1925.
42. *Railway Gazette*, 31 December 1926.
43. TNA RAIL 645/28 SR Locomotive and Electrical Committee minutes 25 Nov. 1925.
44. Moody G.T., *Southern Electric*, p. 93.
45. *Southern Railway Magazine*, June 1923.
46. Robertson, Kevin, *First Generation Southern Region EMUs* (Noodle Books 2006), p. 6.
47. *Railway Gazette*, 11 Oct. 1926.
48. TNA RAIL 645/33 SR Engineering Subcommittee minutes 29 January 1924.
49. TNA RAIL 645/33 SR Engineering Subcommittee minutes 29 January 1925.
50. Pryer, G., *Historical Survey of Southern Stations* (Oxford Publishing Co. 1980), p. 54.
51. Lee, Charles, *London's Elevated Electric, Railway Magazine*, December 1959.
52. TNA MT 6/3128.
53. *Railway Gazette*, 26 May 1922.
54. TNA MT 6/3128.
55. TNA RAIL 118/11.
56. TNA RAIL 118/11.
57. TNA RAIL 645/33 SR Engineering Subcommittee minutes 1 November 1923.
58. TNA RAIL 645/3 SR minutes 6 December 1923.
59. Klapper, C.F., *Sir Herbert Walker's Southern Railway*, p. 177.
60. TNA RAIL 645/71 Traffic officers conference 23 March 1925.

61. Bennet Alan, The Southern Railways Electric Imagery, *Backtrack*, April 2003.
62. Ibid.
63. David Brown talk at Southern Railway Group AGM 18 October 2014.
64. TNA RAIL 645/3 SR minutes 9 October 1924.
65. Moody G.T., *Southern Electric*, p. 35.
66. TNA RAIL 645/3 SR minutes 29 March 1925.
67. Ibid.
68. Ibid., 26 April 1928.
69. TNA RAIL 645/74 Traffic officers conference 20 February 1928.
70. *Southern Railway Magazine*, June 1928.
71. TNA RAIL 645/74 Traffic officers conference 26 March 1928.
72. Ibid., 1 March 1928.
73. Robertson, Kevin, *First Generation Southern Region EMUs*, p. 6.
74. Moody G.T., *Southern Electric*, p. 25.
75. Diagram in *Railway Gazette*, 4 September 1925.
76. Trewman H.F., *Railway Electrification* (H.F. Pittman & Sons 1924), p. 209.
77. Raworth, Alfred – article in *Modern Transport* 17 July 1939.
78. Bonavia, Michael R., *Railway Policy Between the Wars* (Manchester University Press 1981), p. 129.
79. Report of Ministry of Transport Railway Electrification Committee (1927).
80. *Railway Gazette*, 30 September 1938.

14 'The World's Greatest Suburban Electrification'

1. Elliot, Sir John, *On and Off the Rails*, (George Allen & Unwin 1982), pp. 15–19.
2. Ibid.
3. *The Times*, 22 Jan. 1925.
4. Elliot, Sir John, *On and Off the Rails*, p. 41.
5. Klapper, C.F., *Sir Herbert Walker's Southern Railway* (Ian Allan 1973), pp. 176–177.
6. Elliot, Sir John, *On and Off the Rails*, p. 22.
7. Ibid., pp. 32–35.
8. Wikipedia: London Transport/Southern Railway Art Deco, accessed Sept. 2014.
9. Internet: Wentworth studio.com, accessed Oct. 2014.
10. Internet: Scottisharchitects.org.uk, accessed Sept. 2014.
11. Bonavia, Michael R., *The History of the Southern Railway* (Unwin Hyman 1987), p. 52.
12. Elliot, Sir John, Early Days of the Southern Railway, *The Journal of Transport History*, November 1960.
13. The National Archives (TNA) RAIL 645/3 SR minutes 24 April 1928.
14. TNA RAIL 645/74 Traffic officers conference 1929.
15. Ibid., 1 October 1929.
16. Ibid., 28 April 1930.
17. *Railway Gazette*, 22 April 1927.
18. Elliot, Sir John, *On and Off the Rails*, p. 37.
19. TNA RAIL 645/29 SR Locomotive and Electrical subcommittee minutes 27 Apr. 1927.
20. *Railway Gazette*, 22 April 1927.
21. Klapper C.F., *Sir Herbert Walker's Southern Railway*, p. 18.
22. TNA RAIL 645/29 SR Locomotive and Electrical subcommittee minutes 20 May 1927.
23. *Hansard*, 26 April 1927.
24. TNA RAIL 645/29 SR Locomotive and Electrical subcommittee minutes 26 Jan 1928.
25. Ibid., 20 Dec 1933.
26. Findmypast '1939' lists and electoral roles.
27. *Evening Telegraph and Post*, 5 June 1943.
28. Rutherford, Michael, Provocations Bulleid versus Raworth, *Backtrack*, October 1994.
29. TNA RAIL 645/3 SR minutes 8 Oct. 1936.
30. Ibid., 27 October 1927.
31. Brown, Joe, *London Railway Atlas*, pp. 52, 65.
32. *Railway Gazette*, 11 June 1926.
33. Klapper, C.F., *Sir Herbert Walker's Southern Railway*, p. 197.

15 A New Design for an Electric railway

1. Parsons, R.H., *The Early Days of the Power Station Industry* (Babcock & Wilcox 1939), p. 197.
2. Ibid., p. 198.

3. Ministry of Transport, *National Problem of the Supply of Electrical Energy*, 1925, table 2, p. 5.
4. The National Archives (TNA) CAB/24/173.
5. Internet: gracesguide.co.uk, accessed October 2014.
6. TNA RAIL 645/29 SR Locomotive and Electrical subcommittee minutes 20 May 1931.
7. TNA RAIL 645/3 SR minutes 23 January 1931.
8. *Southern Railway Magazine*, August 1932.
9. Ministry of Transport, *Report of the Committee on Main Line Electrification*, 1931, pp. 5–6.
10. Ibid., Appendix 4.
11. Ibid., Appendix 5.
12. Crompton, Gerald and June, Robert, An awkward fence to cross: railway capitalization in Britain in the inter-war years, *Accounting, Business & Financial History*, 12, 3.
13. *Railway Gazette*, 15 December 1933.
14. *Railway Gazette*, 9 March 1934.
15. Institute of Civil Engineers, obituary for Alfred Raworth.

16 The Brighton Electrification

1. Bradshaw, George, *1861 Bradshaw's Handbook* (Collins Harper 2014), p. 56.
2. Wikipedia: London, Brighton and South Coast Railway, accessed Oct. 2014.
3. Lee, Charles, By Pullman to Brighton, *Railway Magazine*, November 1958.
4. National Railway Museum online, Pullman Car Company 1996–7911.
5. Lee, Charles, By Pullman to Brighton, *Railway Magazine*, November 1958.
6. Kidner, R.W., *Pullman Trains in Britain* (Oakwood Press 1998) p. 69.
7. The National Archives (TNA) RAIL 645/3 SR minutes 15 Oct. 1931.
8. Internet: southernelectric.org.uk, accessed November 2014.
9. King, Mike, An illustrated History of Southern Coaches (Oxford Publishing Co. 2003), various drawings, *Railway Gazette*, 30 Dec. 1932.
10. Internet: southernelectric.org.uk, accessed October 2014.
11. Brown, David, Fifty Years of the Portsmouth Direct Electrics, *Railways South East*, winter 1987/88.
12. *Railway Gazette*, 1927.
13. *Railway Gazette*, 22 July 1932.
14. Ibid., 30 Dec. 1932.
15. Internet: brightonbelle.com, accessed November 2014.
16. Internet: southernelectric.org.uk, accessed November 2014.
17. *Railway Gazette*, 25 June 1937.
18. Ibid., 22 July 1932
19. Ibid., 30 December 1932.
20. TNA RAIL 645/29 SR Locomotive and Electrical subcommittee minutes 28 May 1931.
21. Ibid., 26 Nov. 1930.
22. Trewman, H.F., *Railway Electrification* (Pitman & Sons 1924), p. 55.
23. TNA RAIL 645/29 SR Locomotive and Electrical subcommittee minutes 26 Nov. 1930.
24. *Railway Gazette*, 22 July 1932.
25. Ibid., 25 June 1937.
26. TNA RAIL 645/29 SR Locomotive and Electrical subcommittee minutes 26 Nov. 1930.
27. Ibid., 18 Dec. 1935.
28. Ibid., 20 May 1931.
29. Ibid.
30. *Railway Gazette*, 30 Dec. 1932.
31. TNA RAIL 645/29 SR Locomotive and Electrical subcommittee minutes 20 May 1931.
32. *Railway Gazette*, 30 December 1932.
33. Ibid., 22 July 1932.
34. TNA RAIL 645/29 SR Locomotive and Electrical subcommittee minutes 26 March 1930.
35. *Over the Points*, December 1932.
36. King, John, *Gilbert Szlumper and Leo Amery of the Southern* (Pen & Sword 2018), p. 11.
37. Wikipedia, accessed Jan. 2015.
38. *Southern Railway Magazine*, December 1932.
39. *Over the Points*, December 1932.
40. *Southern Railway Magazine*, November 1932.

41. Internet wbsframe.mste.co.uk/public/Brighton.html accessed November 2014.
42. *Over the Points*, September 1932.
43. *Southern Railway Magazine,* August 1932.
44. TNA RAIL 645/77 Traffic officers conference, 15 February 1932.
45. *Over the Points*, December 1932.
46. *Railway Gazette*, 23 December 1932.
47. Wikipedia, Great Depression UK, accessed April 2016.
48. Wikipedia, accessed March 2015.
49. King, John, *Gilbert Szlumper and Leo Amery of the Southern*, p. 27.
50. *Southern Railway Magazine*, January 1933.
51. Ibid.
52. *Southern Railway Magazine*, February 1933.
53. Ibid., April 1933.
54. TNA RAIL 645/77 Traffic Officers Conference 10 October 1932
55. *Railway Gazette*, 28 June 1935.
56. *Southern Railway Magazine*, July 1934.

17 Maintaining Momentum

1. Elliot, Sir John, Early Days of the Southern Railway, *The Journal of Transport History*, vol. iv, No. 4, November 1960.
2. Klapper, C.F., *Sir Herbert Walker's Southern Railway* (Ian Allan 1973), foreword by Sir John Elliot, p. 7.
3. Ibid., p. 139.
4. Elliot, Sir John, *On and Off the Rails* (George Allen & Unwin 1982), p. 46.
5. Ibid., p. 34.
6. Ibid., p. 27–29.
7. Elliot, Sir John, Early Days of the Southern Railway, *The Journal of Transport History*, November 1960.
8. Elliot, Sir John, *On and Off the Rails*, p. 43.
9. Elliot, Sir John, Early Days of the Southern Railway, *The Journal of Transport History*, November 1960.
10. The National Archives (TNA) RAIL 645/3 SR minutes 17 December 1936.
11. George Ellson obituary, *Journal of the Institution of Civil Engineers*, vol. 33, issue 3, January 1950, p. 258.
12. Elliot, Sir John, *On and Off the Rails*, p. 61.
13. King, John, *Gilbert Szlumper and Leo Amery of the Southern* (Pen & Sword 2018), p. 37.
14. Ibid., p. 47.
15. *Over the Points*, March 1934.
16. TNA RAIL 633/54 SECR minutes 8 Nov. 1922.
17. *Sussex Courier*, 30 September 1932.
18. Ibid.
19. TNA RAIL 645/29 SR Locomotive and Electrical subcommittee minutes 9 October 1936.
20. TNA RAIL 645/3 SR minutes 30 November 1933.
21. *Railway Gazette*, 28 June 1935.
22. *Southern Railway Magazine*, March 1935.
23. Moody, G.T., *Southern Electric* (Ian Allan 1958), p. 57.
24. *Railway Gazette*, 28 June 1935.
25. Moody G.T., *Southern Electric* (Ian Allan 1958) p. 57.
26. Ibid., p. 58.
27. TNA RAIL 645/80 SR Traffic managers conference 19 November 1934.
28. White, H.P., *A Regional History of the Railways of Great Britain, Vol 2: Southern England* (David & Charles 1969), p. 187.
29. Wikipedia, accessed March 2015.
30. King, John, *Gilbert Szlumper and Leo Amery of the Southern*, p. 13.
31. Poster on display at Bluebell Railway depicting Southern Railway Instruction 16a, 1935.
32. Moody, G.T., *Southern Electric*, p. 59.
33. TNA RAIL 645/75 SR Traffic managers conference.
34. *Southern Railway Magazine*, April 1933.
35. *Railway Gazette*, 28, 1935.
36. TNA RAIL 645/75 SR Traffic managers conference, 19 February 1934.
37. *Southern Railway Magazine*, May 1934.
38. TNA RAIL 645/75 SR Traffic managers conference, 19 February 1934.
39. *Southern Railway Magazine*, December 1933.
40. Moody, G.T., *Southern Electric*, p. 56.
41. National Railway Museum (NRM) – Ministry of Transport Accidents Reports 1937.

42. Ibid.
43. TNA RAIL 645/29 SR committee minutes 22 July 1936.
44. *Railway Gazette*, 26 October 1934.

18 Two Electric Routes to Portsmouth

1. Moody G.T., *Southern Electric* (Ian Allan 1958), p. 63.
2. TNA RAIL 1017/1/92.
3. White, H.P., *A Regional History of the Railways of Great Britain Vol 2: Southern England* (David & Charles 1969), p. 191.
4. The National Archives 28 (TNA) RAIL 645/4 SR minutes 30 June 1936.
5. TNA MT 6/3406.
6. TNA RAIL 645/4 SR minutes 27 June 1935.
7. Raworth, Alfred, *Report on Proposed Extensions of Electrification* (Southern Railway May 1944), p. 22.
8. *Railway Gazette*, August 1935, quoted in TNA MT 6/3406.
9. TNA MT 6/3406.
10. TNA MT 6/3406.
11. www.dailymail.co.uk, How the thirties saw Britain fall in love with the car.
12. TNA MT 6/3406.
13. National Railway Museum, *Prevention of Accidents to Staff Engaged in Railway Operation*.
14. TNA MT 6/3406.
15. TNA MT 6/3406.
16. TNA RAIL 645/4 SR minutes 30 June 1936.
17. *Railway Gazette*, 5 March 1937.
18. Ibid., 25 June 1937.
19. Ibid. and Google Earth shows a substation building with two 'towers'.
20. TNA RAIL 1017/1/92.
21. TNA RAIL 645/29 SR Locomotive and Electrical subcommittee minutes 22 July 1936.
22. Ibid., 9 October 1936.
23. TNA MT 6/3422.
24. *Southern Railway Magazine*, October 1935.
25. *Railway Gazette*, 25 June 1937.
26. Ibid.
27. TNA RAIL 1017/1/92.
28. Ibid.
29. *Railway Gazette*, 25 June 1937.
30. Ibid.
31. Klapper, C.F., *Sir Herbert Walker's Southern Railway* (Ian Allan 1973), p. 218.
32. Ibid., p. 95.
33. Ibid., p. 218.
34. Ibid.
35. TNA RAIL 645/83 Traffic officers conference 26 April 1937.
36. *Modern Transport*, 26 June 1937.
37. TNA RAIL 645/83 Traffic officers conference 26 April 1937.
38. TNA RAIL 1017/1/92.
39. Nash, Stephen, Paddle boxes and Double Arrows, *Backtrack*, November 2004.
40. Gray, Adrian, The Mid Sussex Railway, *Backtrack*, February 2000.
41. TNA RAIL 645/3 SR minutes 23 July 1936.
42. *Modern Transport*, 9 July 1938.
43. Ibid., 24 April 1938.
44. *Southern Railway Magazine*, July 1938.
45. *The Times*, 6 July 1938.
46. The National Archives (TNA) RAIL 645/84 Traffic officers conference 17 October 1938.

19 The Southern Electric in the Community

1. Elliot, Sir John, *On and Off the Rails* (George Allen & Unwin 1982), pp. 22–23.
2. Bennet, Alan, The Southern Railway's Electric Imagery, *Backtrack*, April 2003.
3. Quoted in *Railway Gazette*, 30 April 1937.
4. The National Archives (TNA) MT 6/3326.
5. TNA MT 6/3387 internal memo G.W. Rowntree to Ministry of Transport.
6. TNA MT 6/3387 internal minute 7 November 1934.
7. Ibid., internal memo G.W. Rowntree to MoT secretary.
8. *Railway Gazette*, 5 March 1937.
9. TNA MT 6/3387 internal minute 7 Nov.1934.
10. Ibid., internal memo G.W. Rowntree to MoT secretary.

11. Ibid., Sir Herbert Walker letter to MoT.
12. Ibid., internal memo G.W. Rowntree to MoT secretary.
13. *Sussex Courier*, 30 September 1932.
14. Ibid., February 1932.
15. *Railway Gazette*, 5 March 1937.
16. TNA MT 6/3406.
17. *Hansard*, 3 March 1937, vol. 321, pp. 436–479.
18. Ibid.
19. *Railway Gazette*, 5 March 1937.
20. *Hansard*, 3 March 1937, vol. 321, 436–479.
21. TNA MT 6/3387.
22. *Southern Railway Magazine*, January 1938.
23. *Hansard*, 3 March 1937, vol. 321, 436–479.

20 The End of an Era

1. *Railway Gazette*, 13 Nov. 1936.
2. *Hastings and St Leonards Observer*, 7, 22 and 29 August 1903.
3. Gould, David, *Maunsell's Steam Passenger Stock 1923–1939* (The Oakwood Press 1978), p. 25.
4. Norris, P.J., The Bexhill West Branch, *Railway Magazine*, July 1958.
5. *Hasting and St Leonards Observer*, 1 June 1935.
6. The National Archives (TNA) RAIL 118/127.
7. TNA RAIL 1188/292.
8. TNA RAIL 645/4 SR minutes 29 April 1937.
9. Elliot Sir John, Early Days of the Southern Railway, *The Journal of Transport History*, November 1960.
10. *Southern Railway Magazine*, July 1938.
11. TNA RAIL 645/4 SR minutes 27 May 1937.
12. Day-Lewis, Sean, *Bulleid: Last Giant of Steam* (George Allen & Unwin, 1964), p. 39.
13. Ancestry.com UK, Mechanical Engineer Records 1847–1930.
14. Day-Lewis, Sean, *Bulleid: Last Giant of Steam*, p. 68.
15. Ibid., p. 70.
16. King, John, *Gilbert Szlumper and Leo Amery of the Southern* (Pen & Sword 2018) p. 13, Sean Day-Lewis, *Bulleid*, p. 105.
17. Wikipedia: O.V.S. Bulleid, accessed July 2015.
18. TNA RAIL 645/4 SR minutes 17 December 1936.
19. King, John, *Gilbert Szlumper and Leo Amery of the Southern*, p. 47.
20. TNA RAIL 645/4 SR minutes 27 May 1937.
21. Elliot, Sir John, *On and Off the Rails* (George Allen & Unwin 1982), p. 45.
22. TNA RAIL 645/4 SR minutes 14 October 1937.
23. *Southern Railway Magazine*, December 1934.
24. TNA RAIL 645/4 SR minutes 22 July 1937.
25. Elliot, Sir John, *On and Off the Rails*, p. 48.
26. Klapper, C.F., *Sir Herbert Walker's Southern Railway* (Ian Allan 1973), p. 220.
27. *Modern Transport*, 13 November 1937.
28. Rutherford, M., Provocations: Bulleid versus Raworth, *Backtrack*, October 1994.
29. *Modern Transport*, 10 July 1937.
30. Sean Day-Lewis, *Bulleid*, p. 129.
31. *Southern Railway Magazine*, December 1945.
32. Sean Day-Lewis, *Bulleid: Last Giant of Steam*, p. 130.
33. *Southern Railway Magazine*, December 1945.
34. Bonavia, Michael R., *The History of the Southern Railway* (Unwin Hyman 1987), p. 33.
35. *Loco Profile 22 – Merchant Navy Pacifics* (Profile Publications 1972).

21 Chief Electrical Engineer

1. The National Archives (TNA) RAIL 645/4 SR minutes 21 July 1938.
2. Ibid., 20 October 1938.
3. *Modern Transport*, 5 November 1938.
4. TNA RAIL 645/3 SR minutes 8 October 1936.
5. Marshall, V. and Smith, K., *Branch Lines Around Ascot* (Middleton Press 1989).
6. Moody, G.T., *Southern Electric* (Ian Allan 1958), p. 75.
7. Robertson, Kevin, *First Generation Southern EMUs* (Ian Allan 2006), p. 6.

8. TNA RAIL 645/4 SR minutes 14 Oct. 1937.
9. *Modern Transport*, 8 July 1939.
10. Moody, G.T., *Southern Electric*, p. 76.
11. *Modern Transport*, 1 July 1939.
12. Moody, G.T., *Southern Electric*, p. 76.
13. Gould, David, *Maunsell's Steam Passenger Stock 1923–1939* (The Oakwood Press 1978), p. 39.
14. King, Mike, *An Illustrated History of Southern Coaches* (OPC/Ian Allan 2003), p. 25.
15. Cox, E.S., *Locomotive Panorama Vol. 2* (Ian Allan 1966), p. 16.
16. Ibid.
17. King, Mike, *An Illustrated History of Southern Coaches*, p. 16.
18. *Southern Railway Magazine*, July 1938.
19. Robertson, Kevin, *First Generation Southern EMUs*, p. 6.
20. Internet: southern-electric.org.uk, accessed June 2015.
21. *Modern Transport*, 1 July 1939.
22. Elliot, Sir John, *On and Off the Rails* (George Allen & Unwin 1982), p. 42.
23. *Southern Railway Magazine*, July 1939.
24. Moody, G.T., *Southern Electric*, p. 78.
25. TNA RAIL 645/29 SR Locomotive and Electrical subcommittee minutes 19 December 1934.
26. TNA RAIL 1188/143.
27. Stevens-Stratten, S.W., *Bulleid Coaches in 4mm Scale* (Ian Allan 1983), p. 42.
28. TNA RAIL 645/4 SR minutes 16 December 1937.
29. Elliot, Sir John, Early Days of the Southern Railway, *The Journal of Transport History*, November 1960.
30. TNA RAIL 645/4 SR minutes 30 March 1938.
31. *Hastings & St Leonards Observer*, 5 March 1938.
32. Ibid., 19 March 1938.
33. Ibid.
34. TNA RAIL 1188/127.
35. TNA RAIL 645/4 SR minutes 29 June 1939.
36. TNA RAIL 1188/292.
37. Raworth, Alfred, article in *Modern Transport*, 17 July 1939.

22 The Southern Electric at War

1. Bonavia, Michael R., *The History of the Southern Railway* (Unwin Hyman 1987), p. 101.
2. Wikipedia, 'Maundy Gregory' quoting Graham Stewart article in the *Sunday Times*, accessed July 2015.
3. Bonavia, Michael R., *The History of the Southern Railway*, 1987, p. 101.
4. Dendy Marshall, C.F., revised by Kinder, R.W., *A History of the Southern Railway Volumes 1 & 2 Combined* (Ian Allan 1968), p. 449.
5. Elliot, Sir John, *On and Off the Rails* (George Allen & Unwin 1982), p. 50.
6. Elliot, Sir John, Early Days of the Southern Railway, *The Journal of Transport History*, November 1960.
7. Bonavia, Michael R., *The History of the Southern Railway*, p. 103.
8. The National Archives (TNA) CAB 24/279/8 Government 'War Book', 14 September 1938.
9. TNA AN 3/32 REC M& EE subcommittee minutes 6 March 1939.
10. Ibid., 12 March 1940.
11. TNA AN 3/1 REC minutes 3 September 1939.
12. *Railway Gazette* 20 September 1939.
13. TNA AN 3/1 REC minutes 5 September 1939.
14. *Railway Gazette*, 20 October 1939.
15. TNA AN 3/1 REC minutes 11 September 1939, *Railway Gazette*, 15 September 1939.
16. *Railway Gazette*, 29 September 1939.
17. *Railway Gazette*, 20 October 1939.
18. *The Times*, 8 September 1939, 15 September 1939.
19. TNA AN 3/1 REC minutes 29 September 1939. *The Times*, 15 September 1939.
20. Moody, G.T., *Southern Electric* (Ian Allan 1958), p. 117.
21. *The Times*, 20 June 1940; *Belfast Telegraph*, 20 June 1940; *Railway Gazette*, 28 June 1940.
22. *Railway Gazette*, 12 July 1940.
23. TNA CAB 68-7-80.

24. TNA AN 3/5 REC minutes 23 May 1941 – paper attached as appendix.
25. *Railway Gazette,* 28 June 1940.
26. Tatlow, Peter, *Return from Dunkirk* (Oakwood Press 2010), table in Appendix 2, p. 168 – p. 175.
27. Darwin, Bernard, *War on the Line* (Southern Railway 1946), p. 42.
28. Internet: Ukweatherworld.uk, accessed February 2016.
29. *Railway Magazine,* February 1940; March 1940.
30. TNA RAIL 645/120 SR Company officers conference 17 January 1941.
31. TNA RAIL 645/5 SR minutes 24 September 1942.
32. TNA RAIL 645/120 SR Company officers conference, 6 June 1940.
33. Darwin, Bernard, *War on the Line,*1946, p. 56.
34. Ibid., p. 57.
35. TNA RAIL 1188/295.
36. Brooksbank, B.W.L., *London Main Line War Damage* (Capital Transport 2007), p. 77.
37. TNA RAIL 1188/295.
38. TNA RAIL 645/119 SR Company officers conference 17 October 1940.
39. Darwin, Bernard, *War on the Line,*1946, p. 64.
40. Brooksbank, B.W.L., *London Main Line War Damage*, p. 33.
41. Internet: www.dnw.co.uk/auction-archive/past-catalogues.
42. Brooksbank, B.W.L., *London Main Line War Damage*, p. 33, citing Southern Railway report.
43. TNA RAIL 645/119 SR Company officers conference 17 October 1940.
44. Ibid.
45. Darwin, Bernard, *War on the Line,*1946, p. 64.
46. Ibid.
47. TNA RAIL 645/120 SR Company officers conference, 10 January 1941.
48. Ibid., 17 Jan. 1941.
49. TNA RAIL 645/119 SR Company officers conference, 17 January 1941.
50. TNA RAIL 1188/326.
51. TNA RAIL 645/120 SR Company officers conference, 19 March 1941.
52. Ibid., 17 January 1941.
53. Moody, G.T., *Southern Electric*, p. 115.
54. Internet: dalyhistory.wordpress.com, accessed July 2015.
55. Elliot, Sir John, *On and Off the Rails* (George Allen & Unwin 1982), p. 52.
56. Hennessey, R.A.S., 'Sparks', The Electrical Consultants Part 1, *Backtrack*, July 2008.
57. *Evening Telegraph and Post,* 5 June 1943.
58. Internet: ww2today.com.
59. Ibid.
60. *Evening Telegraph and Post,* 5 June 1943.
61. TNA AN 3/32 REC Electrical subcommittee minutes 16 July 1941.
62. Ibid., 17 July 1941.
63. Ibid., 20 April 1942.
64. Ibid., 8 August 1940.
65. Alfred Raworth's Institution of Civil Engineers obituary.
66. Alfred Raworth's application for membership, Institution of Civil Engineers Archive.
67. Wikipedia, accessed September 2015.
68. TNA RAIL 645/121 SR Company officers conference 15 September 1944.
69. Moody, G.T., *Southern Electric*, p. 119.
70. *Southern Railway Magazine,* December 1945.

23 The Electric Locomotives

1. *Railway Gazette,* 13 November 1936.
2. Tyler, Arthur, T.H. CEng FIMechE, 600/750V DC Electric and Electro-Diesel Locomotives of the Southern Railway and its Successors, *Journal for History of Engineering and Technology*, vol. 68, issue 1 (1996).
3. Ibid.
4. Winkworth, D.W., Southern Railway Electric Locomotives, *Backtrack*, Jan 2005.
5. Tyler, Arthur, T.H. CEng FIMechE, 600/750V DC Electric and Electro-Diesel Locomotives of the Southern Railway and its Successors, *Journal for History of Engineering and Technology*, vol. 68, issue 1 (1996).
6. Internet: worldwide.espacenet.com.

7. The National Archives (TNA) RAIL 645/29 SR Locomotive and Electrical subcommittee minutes 7 October 1936.
8. Winkworth, D.W., Southern Railway Electric Locomotives, *Backtrack*, Jan. 2005.
9. Tyler, Arthur, T.H. CEng FIMechE, 600/750V DC Electric and Electro-Diesel Locomotives of the Southern Railway and its Successors, *Journal for History of Engineering and Technology*, vol. 68, issue 1 (1996). (1996) quoting H.A.V. Bulleid's book *Bulleid of the Southern*.
10. Internet Bulleid.org accessed July 2015.
11. TNA AN 3/4 REC minutes 19 January 1940.
12. 'Borderer', 10 Years of British Railways, *Railway Magazine*, January 1958.
13. Day-Lewis, Sean, *Bulleid: Last Giant of Steam* (Allen and Unwin 1964).
14. Fletcher, Barry, *Bulleid Pacific Names and Nameplates*, *Southern Notebook*, Summer 2015.
15. Allen, Cecil J. and Townroe, S.C., *The Bulleid Pacifics of the Southern* (Ian Allan 1976, reprint from 1951), p. 26.
16. Winkworth, D.W., Southern Railway Electric Locomotives, *Backtrack*, Jan. 2005.
17. Ibid.
18. Ibid.
19. Tyler, Arthur, T.H. CEng FIMechE, 600/750V DC Electric and Electro-Diesel Locomotives of the Southern Railway and its Successors, *Journal for History of Engineering and Technology*, vol. 68, issue 1 (1996).
20. TNA RAIL 645/120 SR Company Officers Conference 7 November 1941.
21. Raworth, Alfred, *Report on Proposed Extensions of Electrification*, Southern Railway, May 1944, p. 15.
22. TNA RAIL 645/121 SR Company officers conference 12 June 1942.
23. Ibid., 30 October 1942.
24. Tyler, Arthur, T.H. CEng FIMechE, 600/750V DC Electric and Electro-Diesel Locomotives of the Southern Railway and its Successors, *Journal for History of Engineering and Technology*, vol. 68, issue 1 (1996).
25. TNA RAIL 645/5 SR minutes 29 Jan. 1942.
26. TNA RAIL 645/121 SR Company officers conference 2 April 1943.
27. Raworth, Alfred, *Report on Proposed Extensions of Electrification*, Southern Railway, May 1944, p. 15.
28. TNA RAIL 645/121 SR Company officers conference 20 August 1943.
29. Internet Bulleidlocos.org accessed July 2015.
30. Day-Lewis, Sean, *Bulleid*, p. 129.
31. Ibid., p. 130.
32. Tyler, Arthur, T.H. CEng FIMechE, 600/750V DC Electric and Electro-Diesel Locomotives of the Southern Railway and its Successors, *Journal for History of Engineering and Technology*, vol. 68, issue 1 (1996).
33. Winkworth, D.W., Southern Railway Electric Locomotives, *Backtrack*, Jan. 2005.
34. TNA RAIL 645/5 SR minutes 30 April 1942.
35. Raworth, Alfred, *Report on Proposed Extensions of Electrification*, Southern Railway, May 1944 p. 15.
36. Winkworth, D.W., Southern Railway Electric Locomotives, *Backtrack*, Jan. 2005.
37. TNA RAIL 645/29 SR Locomotive and Electrical subcommittee minutes 19 December 1934.

24 Looking to the Future

1. Wikipedia: Beveridge Report, accessed July 2015.
2. Wikipedia: Education Act 1944, accessed July 2015.
3. TNA RAIL 645/5 SR minutes 26 March 1942.
4. TNA RAIL 1188/290.
5. Raworth, Alfred, *Report on Proposed Extensions of Electrification*, Southern Railway, May 1944, p. 1.
6. TNA RAIL 645/121 SR Company officers conference 31 December 1943.
7. TNA RAIL 1188/292.

8. Chivers, Colin, *The Riverside Electric*, (South Western Circle 2010), full text of paper Appendix 1, pp. 171–180.
9. Raworth, Alfred, *Report on Proposed Extensions of Electrification*, Southern Railway, May 1944, p. 6.
10. Ibid., p. 5.
11. Ibid., p. 4.
12. Bonavia, Michael R., *The History of the Southern Railway* (Unwin Hyman 1987), p179.
13. The National Archives (TNA) RAIL 1188/290.
14. Ibid.
15. TNA RAIL 648/124.
16. Internet: steamindex, accessed Oct. 2015.
17. TNA RAIL 645/5 SR minutes 21 December 1944.
18. Ibid.
19. Internet steamindex, accessed October 2015.
20. TNA RAIL 645/5 SR minutes 26 April 1945.
21. TNA RAIL 645/5 SR minutes 30 May 1946.
22. *Southern Railway Magazine*, December 1945.
23. Alfred Raworth's Institution of Civil Engineers obituary.
24. Internet: Labour Party manifesto, accessed September 2015.
25. 'Borderer', 10 Years of British Railways, *Railway Magazine*, January 1958.
26. TNA CAB 65/9/2 War Cabinet minutes 3 September 1940.
27. *Railway Gazette*, 1 January 1938.
28. Ibid., 17 September 1942.
29. Elliot, Sir John, *On and Off the Rails* (Georg Allen & Unwin 1982), p. 66.
30. TNA RAIL 645/5 SR minutes.
31. TNA RAIL 1188/303.
32. National Railway Museum (NRM) Hawes Collection ref 2012-7274 ALS5.
33. TNA RAIL 1188/291.
34. TNA RAIL 1188/292 – Eustace Missenden's annotated copy of Raworth report.
35. TNA RAIL 1188/291.
36. Alfred Raworth's Institution of Civil Engineers obituary.
37. Internet: Gracie's Guide accessed October 2015.
38. Findmypast.
39. *Hull Daily Mail*, 6 February 1946.
40. Wikipedia.
41. *Hull Daily Mail*, 6 February 1946.
42. TNA RAIL 1188/291, NRM Hawes Collection ref 2012-7274 ALS5.
43. TNA RAIL 645/90.
44. TNA RAIL 1188/291 letter from English Electric Co.
45. TNA RAIL 1188/291.
46. Hansard: Commons debate 1 December 1949.
47. Parker, David, *The Official History of Privatisation, Vol. II: Popular Capitalism, 1987–97* (Routledge), p. 434.

25 Epilogue

1. Bonavia, Michael R., *British Rail – the First 25 Years* (David & Charles 1981), p. 14.
2. Day-Lewis, Sean, *Bulleid: Last Giant of Steam* (George Allen & Unwin, 1964), p. 136.
3. Bonavia, Michael R., *The History of the Southern Railway*, (Unwin Hyman 1987), p. 160.
4. Cox, E.S., *Locomotive Panorama Volume 2* (Ian Allan 1974), p. 5.
5. Ibid., p. 17.
6. Ibid., p. 18.
7. Internet: bulleidlocos.org, accessed December 2015.
8. Derry, Richard, *The Book of the Merchant Navies* (Irwell Press 2001), pp. 33–35.
9. Ibid., p. 36.
10. Winkworth, *Bulleid's Pacifics* (George Allen & Unwin 1974), pp. 246–258.
11. Helm, John W.E., The Long Road to 1948, *Backtrack*, July 1999.
12. Internet: report from *Locomotive Magazine and Railway Carriage Review*, vol. 56 (1950), accessed Oct. 2015.
13. Elliot, Sir John, *On and Off the Rails* (George Allen & Unwin 1982), p. 84.
14. Internet: report from *Locomotive Magazine and Railway Carriage Review*, vol. 56 (1950), accessed Oct. 2015.

15. Cox, E.S., *Locomotive Panorama Volume 2* (Ian Allan 1974), p. 10.
16. British Railways, *Report of the Committee on Motive Power*, 1951, p. 47.
17. Moody, G.T., *Southern Electric* (Ian Allan 1958), p. 143.
18. Ibid., p. 14.
19. British Transport Commission, *Electrification of Railways*, p. 195.
20. Hennessey, R.A.S., 'Sparks' the Electrical Consultants, Part 1, *Backtrack*, July 2008.
21. British Transport Commission, *Electrification of Railways*, 1951, p. 42.
22. Ibid., p. 51.
23. Ibid., p. 47.
24. Internet: *Journal of Institution of locomotive Engineers*, 1952, 42, accessed Oct. 2015.
25. Internet: Steamindex, accessed November 2015.
26. Barker, Theodore and Gerhold, Dorian, *The Rise and Rise of Road Transport 1700–1990* (Cambridge University Press 1993), p. 72.
27. Aldcroft, Derek H., *British Railways in Transition* (Macmillan, 1968), p. 120.
28. Loft, Charles, *Government, the Railways and the Modernization of Britain: Beeching's Last Trains* (Taylor & Francis 2006), p. 39.
29. British Transport Commission, *Modernisation and Re-equipment of British Railways*, 1955, p. 5.
30. Ibid., p. 11.
31. Ibid., p. 6.
32. Ibid., p. 14.
33. Modernisation of the Kent Coast Main Line, *Railway Magazine*, May 1958.
34. Tyler, Arthur, T.H. CEng FIMechE, 600/750V DC Electric and Electro-Diesel Locomotives of the Southern Railway and its Successors, *Journal for History of Engineering and Technology*, vol. 68, issue 1 (1996).
35. New Electric Locomotives for Southern Region, *Railway Magazine*, April 1959.
36. 'Electro-diesel' Locomotives for the Southern Region, *Railway Magazine*, April 1962.
37. Casserley, H.C., *The Observer's Book of Railway Locomotives of Britain* (George Warne & Co 1964), p. 178.
38. Marsden, Colin, J., *Rail Guide 2011* (Ian Allan 2011).
39. British Transport Commission, *Modernisation and Re-equipment of British Railways*, 1955, p. 11.
40. Ibid., p. 15.
41. Ibid., p. 14.
42. TNA MT 6/3406.
43. Elliot, Sir John, *On and Off the Rails* (Georg Allen & Unwin 1982), p. 89.
44. British Transport Commission, *Electrification of Railways*, 1951, p. 36.
45. Ibid., p. 67/8.
46. Elliot, Sir John, *On and Off the Rails*, p. 89.
47. A Midland Electric Centenary, photo feature, *Backtrack*, July 2008.
48. *Surrey Herald & News*, 8 January 1967.
49. British Railways Board, *The Reshaping of British Railways* (the 'Beeching Report'), 1963, p. 107.

BIBLIOGRAPHY

Books

Aldcroft, Derek, H., *British Railways in Transition* (Macmillan 1968).

Allen, Cecil J. and Townroe, S.C., *The Bulleid Pacifics of the Southern* (Ian Allan 1976, reprint from 1951).

Barker, Theodore and Gerhold, Dorian, *The Rise and Rise of Road Transport 1700–1990* (Cambridge University Press 1993).

Benest, K.R., *Metropolitan Electric Locomotives* (Lens of Sutton 1963).

Bagwell, Phillip and Lyth, Peter, *Transport in Britain From Canal Lock to Gridlock* (Hambledon 2002).

Bonavia, Michael R., *Railway Policy Between the Wars* (Manchester University Press 1981).

Bonavia, Michael R., *British Rail – the First 25 Years* (David & Charles 1981).

Bonavia, Michael R., *The Channel Tunnel Story* (David & Charles 1987).

Bonavia, Michael R., *The History of the Southern Railway* (Unwin Hyman 1987).

Bradshaw, George, *1861 Bradshaw's Handbook* (Collins Harper 2014).

Brooksbank, B.W.L., *London Main Line War Damage* (Capital Transport 2007).

Brown, Joe, *London Railway Atlas* (Ian Allan 2012).

Chivers, Colin, *The Riverside Electric* (South Western Circle 2010).

Churella, Albert J., *The Pennsylvania Railroad Volume 1: Building the Empire 1846–1917* (Pennsylvania University Press 2013).

Cornell University Archive, *Dictionary of Electricity* (internet access).

Cox, E.S., *Locomotive Panorama Volume Two* (Ian Allan 1966).

Darwin, Bernard, *War on the Line* (Southern Railway 1946).

Dawson, Phillip, *Electric Traction on Railways* (The Electrician Printing and Publishing Co. 1909).

Day-Lewis, Sean, *Bulleid: Last Giant of Steam* (George Allen & Unwin, 1964).

Dendy, C.F., *A History of the Southern Railway* (Southern Railway 1936).

Dendy Marshall, C.F., revised by Kinder R.W., *A History of the Southern Railway Volumes 1 & 2 Combined* (Ian Allan 1968).

Derry, Richard, *The Book of the Merchant Navies* (Irwell Press 2001).

Dyos, H.J. and Aldcroft D.H., *British Transport. An Economic Survey from the Seventeenth Century to the Twentieth* (Penguin Books 1974).

Elliot, Sir John, *On and Off the Rails* (George Allen & Unwin 1982).

Ferneyhough, Frank, *Liverpool and Manchester Railway* (Book Club Associates 1980).

Glover, John, *Southern Electric* (Ian Allan 2001).

Gould, David, *Maunsell's Steam Passenger Stock 1923–1939* (The Oakwood Press 1978).

Grieves, Keith, *Sir Eric Geddes. Business and Government in War and Peace* (Manchester University Press 1989).

Harding, C. Francis, *Electric Railway Engineering*, 2nd edition (McGraw Hill 1916 from Forgotten Books).

Harvey, Michael G. and Rooke, Eddie, *Railway Heritage Portsmouth* (Silver Link Publishing 1997).

Hobbs, George, *Power, Poles and Plate laying, Keeping the MER on Track* (Loaghtan Books 2019).

Irving, R.J., *The North Eastern Railway: An Economic History* (Leicester University Press 1976).

Jackson, Alan, *London's Termini* (Pan 1972).

Kidner, R.W., *The Reading to Tonbridge Line* (Oakwood Press 1974).

Kidner, R.W., *Pullman Trains in Britain* (Oakwood Press 1998).

King, John, *Gilbert Szlumper and Leo Amery of the Southern* (Pen & Sword 2018).

King, Mike, *An Illustrated History of Southern Coaches* (Oxford Publishing Co. 2003).

King, Mike, *Southern Vans and Coaches in Colour* (Noodle Books, 2015).

Kitchenside, G.M. and Williams, Alan, *British Railway Signalling* (Ian Allan 1975).

Klapper, C.F., *Herbert Walker's Southern Railway* (Ian Allan, London,1973).

Klapper, C.F., *London's Lost Railways*, (Routledge & Keegan Paul 1976).

Loft, Charles, *Government, The Railways and the Modernization of Britain: Beeching's Last Trains* (Taylor & Francis 2006).

Marsden, Colin, J., *Rail Guide 2011* (Ian Allan 2011).

Mitchell, Vic and Smith, Keith, *Reading to Guildford* (Middleton Press 1988).

Mitchell, Vic and Smith, Keith, *Branch Lines Around Ascot* (Middleton Press 1989).

Moody, G.T., *Southern Electric* (Ian Allan 1958).

Nock, O.S., *Fifty Years of Railway Signalling* (Institution of Railway Signalling Engineers 1962).

Parker, David, *The Official History of Privatisation, Vol. II: Popular Capitalism, 1987–97* Routledge 2009).

Parris, Henry, *Government and the Railways* (Routledge & Keegan Paul 1965).

Parsons. R.H., *The Early Days of the Power Station Industry* (Babcock & Wilcox 1939).

Pryer, G.A., *A Pictorial Record of Southern Signals* (Oxford Publishing Co. 1977).

Pryer, G., *Historical Survey of Southern Stations* (Oxford Publishing Co. 1980).

Robertson, Kevin, *First Generation Southern Region EMUs* (Noodle Books 2006).

Robinson, S.J.T. and Markham, J.D., *The Regenerative Braking Story* (Scottish Tramway & Transport Society/Venture Publications 2006).

Rowland, John, *Progress in Power* (Merz and Mclellan/Newman Neame 1959).

Smith, Robert Henry, *Railway Electric Traction* (Harper Brothers from Forgotten Books 1905).

South, Raymond, *Crown College and Railways* (Barracuda Books 1978).

Stevens-Stratten, S.W., *Bulleid Coaches in 4mm Scale* (Ian Allan 1983).

Tatlow, Peter, *Return from Dunkirk* (Oakwood Press 2010).

Taylor A.J.P., *The First World War An Illustrated History* (Penguin Books, 1966)

Trewman, H.F., *Railway Electrification* (Pittman 1924).

Weddell, G.H.R., *LSWR Carriages in the 20th Century* (Oxford Publishing Co, 2001).

White, H.P., *A Regional History of the Railways of Great Britain, Vol 2: Southern England* (David & Charles 1969).

Wilson, F. and John F., *British Business History 1720–1994* (Manchester University Press 1995).

Winkworth, *Bulleid's Pacifics* (George Allen & Unwin 1974).

Wringley, Chris, *Lloyd George and the Challenge of Labour* (Harvester/Wheatsheaf, London, 1991).

Documents from the National Archives

AN series – Railway Executive Committee main and committee minutes.

BT series – details of John Smith Raworth's business, papers and correspondence relating to LSWR electrification.

CAB series – Cabinet papers.

MT series – Ministry of Transport papers and correspondence relating to SECR and Southern Railway electrification issues.

RAIL series – LSWR, SECR and Southern Railway main board and committee minutes.

Periodicals

Backtrack 1994–2020.
Electric Railway Journal 1922.
Modern Railways 1922–1939.
Over the Points (Southern Railway public magazine) 1932–1934.
Railway Gazette 1913–1940.
Railway Magazine 1909–1958.
Southern Railway Magazine (staff magazine) 1923–1945.
Tramway and Railway World 1915.

Internet

extra.southernelectric.org – (Southern Electric Group history pages)
www.steamindex.com.
www.spellerweb.net

INDEX

A
Accidents
  Battersea 213
  Halifax (tram) 26
  Haywards Heath 209
  Rosendale (tram) 32
  Swanley 213–215
Acts of Parliament
  Companies Consolidation Act 1908 32
  Consolidation Act 1862 16
  Electricity (Supply) Act 1919 127
  London and Southampton Railway Act 1834 57
  London and South Western Act 1913 59
  Ministry of Transport Act 1919 102
  Railways Act 1921 107, 165, 295
  Railway (clauses) Act 1845 36
  Railways (Electric Power) Act 1903 57/8, 130, 138
  Regulation of Forces Act 1871 67, 83
  South Eastern and London Chatham And Dover Companies Act 1903 79
  Tramways Act 1870 20
AEG (Allgermeine Elektricitats Gesellschaft) 53
alternator 46/7
Angerstein's Wharf 83, 96, 110/1, 125/6,
Anglo-American Brush Electric Light Corporation 18
Armstrong Whitworth 278
ASEA (Allmanna Svenska Elecktriska Aktie Bolaget) 175–177, 193/4, 221, 247
ASEA remote control system 194 -198
Ashfield, Lord 170
Aspinall, Sir John 40, 51, 86, 105, 107, 166
ASLEF (Association of Locomotive Engineers and Firemen). 284
Atlee, Sir Clement 295

B
Babington-Smith, Sir Henry 141, 142-144
Baring, the Hon Brigadier-General Everard 163/4, 175-177
Beamish, Rear Admiral MP 238
Belloc Hilaire 234/5
Bevin, Ernest 280/1,295
Bexley Heath and Dartford Council 235/6
Birmingham City Council 30–33
Birmingham Railway Carriage and Wagon Company 156, 200
Blitz 268, 270, 273
Board of Trade 33/4, 58, 67, 85
Bonsor, H Cosmo O 76, 80/1/9, 104, 109, 110/1, 124/5, 128, 131, 137, 140, 142, 241
booster for electric locomotives 278/9
Bowet, Lindley and Company 23
Boyce, Lesley MP 238
Brassey, Thomas 36, 226
Brighton Belle 203
Brighton Council delegation 160
British Electrical and Allied Manufacturers Association 175
British Electric Traction Company 20
British Insulated Callender Cables Limited 285
British Thomson Houston Limited 105, 176/7, 190, 225
British Transport Commission 301
Browett, Thomas 25, 34
Brown Boveri 193, 225
Brown, Brigadier-General Clifton MP 238, 240
Bruce Peebles 175, 225, 247
Brush Electrical Engineering Company 18–20
Brush Electric Traction Company 21
Bulleid, Oliver Vaughn Snell 148, 245/6, 247–249, 255–258, 260, 266, 280, 287, 292, 294, 298, 301

C
carriages
  'birdcage' stock (SECR). 80
  'block sets' (LSWR). 38
  4/6 wheeled (LCDR). 77
  'long tens' (SECR). 105
  Pullman cars (LBSCR). 188
  restriction 0 (SR Hastings line). 241/2, 244
Central Landowners Association 237
Central London Railway 29
Chadwick, Sir Burton 176
Challis, W 152
Chamberlain, Neville 220
Channel tunnel 87, 95/6, 100, Plate 5
Channel Tunnel Company 87
Chilston, Lord 115
Churchill, Sir Winston 69, 102, 182, 295
circuit breaker 61, 194/5
Clarke, Lieutenant-Colonel MP 237
Claughton, Sir Gilbert 85
Clode, W.C. KC 149, 130/1, 136, 151
Cock, Charles 293–295, 297, 299, 300, 304/5
Cole, F G 250
colour light signals 152–154, 199, 200, 226, Plate 7
Committee of Twenty-One 141/2
control centres
  Havant 221, 224, 267
  Ore 207, 267
  Swanley 211, 267
  Three Bridges 194 -196, 199, 267
  Woking 197, 221–224, 267
Cox, Edwin Charles 114, 146, 172, 204/5, 210
Cox, E, S. 13, 290/1, 337, 241, 303

D
Davidson F G 292
Dawson, Sir Phillip 52/3, 70, 105, 117/8, 123, 135, 161, 166, 187
Deepdene Hotel 265/6
demi-cars (trams). 23–29, Plate 1
Dent, Sir Francis 81/2, 84, 86, 88, 104, 131, 146, 246
Devonport Tramway Company 28/9
diesel-electric multiple units 313
Dorman Smith Switchgear 16
Dover Marine 81
Druitt, Lieutenant-Colonel 27, 33/4
Drummond, Sir Hugh 39, 40-45, 116, 124, 144,146, 148,159, 163
Dulwich College 22, 151
dynamo 4, 18
Dyson, Sir Frank 105

E
East Sussex County Council 236
East Sussex Federation of Ratepayers' Associations 236
Electricity Commissioners 105, 109, 126/7, 129/30, 137–140, 183
Electricity Commissioners' public inquiry 129–136, Plate 7
Electricity Companies
  Central Electricity Board 184, 192, 215, 232, 247
  County of London Electricity Supply Company 135, 136

INDEX • 337

Electrical Power Distribution
 Company 20
London Electricity Supply
 Corporation 52, 82, 126, 129/30,
 133–138, 149, 152, 160/1, 184/5
West Kent Electricity Supply
 Company 109/10, 129–135
Electrification Schemes
 BR 1500 Volt DC 310
 BR 25kV AC 310/1
 BR Southern Region Bournemouth
  312/3
 BR Southern Region Isle of Wight 313/4
 BR South Region Kent coast 307/8
 Hollentalbahn (Germany) 219
 Lancashire and Yorkshire
  Liverpool 51
 Lancashire and Yorkshire Bury 93,
  Plate 6
 LBSCR 6,700 Volt 52/3, Plate 6
 LSWR suburban 71/2, Plate 2, Plate 6
 Long Island Railroad (USA). 121/2
 Manx Electric Railway (IoM). 122
 Melbourne suburban (Australia).
  82, 106
 Mersey Railway 51
 Midland Railway Morecambe and
  Heysham 53
 New York Central Railroad Harlem
  Division (USA). 54
 North Eastern Railway Tyneside 51,
  127, 304
 North Eastern Railway Shildon to
  Newport 54, 106
 North Eastern Railway York to
  Newcastle (proposed). 106, 148
 SECR 1500/1500 Volt (proposed).
  92–97, Plate 3, Plate 4, Plate 6
 SNCF 25kV electrification
  (France). 311
 Southern Railway Brighton 189–202,
  Plate 10
 Southern Railway 6,700 Volt 160 -166
 Southern Railway Central Section
  (ex LBSCR). 180/1
 Southern Railway Chessington South
  259/60
 Southern Railway Eastbourne 206/7,
  Plate 14
 Southern Railway Eastern Section
  (ex SECR). 149–152, Plate 8
 Southern Railway Gravesend 174–179
 Southern Railway Hastings
  (proposed). 242/5, 260/1
 Southern Railway Horstead Keynes
  209–212
 Southern Railway Lewisham –
  Nunhead 215
 Southern Railway Maidstone
  253–257, 259
 Southern Railway Oxted (proposed).
  261–264
 Southern Railway Portsmouth Direct
  221–226, 247, Plate 15
 Southern Railway Portsmouth No. 2
  231/2
 Southern Railway Reading 251–255,
  Plate 16
 Southern Railway Sevenoaks 210-214
 Southern Railway Western Section
  (ex LSWR). 154–156, 158, Plate 9
 Southern Railway Windsor 174, 179
EMUs (Electric Multiple Units)
 2 – BIL 208/9, 228, 232, 252–255
 2 – HAL 258/9
 2 – NOL 203, 212, 229
 3 – SUB (Eastern Section). 156/7
 3 - SUB (ex LSWR 'block set'). 65/6
 3 – SUB (SR). 156, 213
 3 – SUB (Western Section). 157–159,
  Plate 9
 3 – TC 312
 4 – BUF 207/8, 232, 258
 4 - COR 226 -229, 232, 247/8, 273, 284
 4 – LAV 212, 223, 191, 200
 4 – REP 312
 4 – RES 326–328
 4 – SUB 240, 260, 287
 4 – TC 312
 5 – BEL 190–192
 6 – CIT 189/90
 6 – PUL 189–191, 208
 AC 2/3 car set (LBSCR). 53
 AC 5 car set (with 'milk van')
  (LBSCR). 161/3
 class 444 (Siemen Desiro). 222
 class 450 (Siemen Desiro). Plate 11
 class 455/7 155
 'Jones' 3 – set proposal (LSWR). 65,
  Plate 2
 London Transport sets for Isle of
  Wight 313/4
 'Raworth' 3 – set proposal (SECR). 97,
  Plate 3
 Unpowered 2 – sets (LSWR and SR).
  65, 213
Electrification of Railways Advisory
 Committee 105 -108
Elliot, Sir John 170/1, 204, 234, 246, 266,
 294/5, 295, 301, 310
Ellson, George 114, 123, 205, 275
English Electric 176, 247, 278/9, 287,
 298, 300
evacuations 268/9

F
Fairburn, C.E. 275
Falcon Works, Loughborough 18, 19
Farly, Frank 125
fatalities after trespassing 235/6, 239

fences 239/40
Fitch J L 298
Forbes, James Staart 74, 77/8
Forbes, Sir William 86, 116, 146, 202
Fry, Edwin Maxwell 172

G
Geddes, Sir Eric 87, 101–105, 111,
 126/7, 141
General Electric Company 26, 29, 90, 176
General Manager's Committee on
 Suburban Traffic Operations (SECR).
 104/5
Gibbs, George 143
Gore-Browne, Sir Eric 246, 296
Grand Central Station New York 91, 290
Greenwich Royal Observatory 105,
 129, 147
Gresley, Sir Nigel 166, 245, 266
Group Electrification Committee
 112–115
Gustaf IV, King of Sweden 199

H
Hall, Colonel G.L 225
Hampshire Federation of Women's
 Institutes 237
Harrington J L 298
Hatchwell, M S 299
Hastings 241/2, 261
Hayes, Edward 14/5
High Court of Justice, Chancery
 Division 32
HMS President II 69
Hodgson, John 28
Holland-Martin, Robert 237, 246, 295/6
Hood, J.W. Jacomb 39, 291
Hopkinson, John 15
Hore-Belisha, Leslie 220
Houghton, Robert 114
House of Commons debate 268, 269
Howard, Robert 134
Hudson, Captain Austin MP 238
Hurcomb, Lord Cyril 220, 301, 303

I
Institution of Civil Engineers 19, 34, 275
Institution of Electrical Engineers 19, 23,
 28, 34, 107
Institution of Mechanical Engineers 19,
 28, 34, 301
Institution of Railway Signal Engineers
 152
Institution of Transport 241
Isle of Wight 230/1

J
Johnson-Lundell 25, 31, 34
John Smith Raworth and Company
 16, 17

Jones, Herbert William 42/3, 71/2, 115, 132, 134, 146–148, 159, 161–167, 205, 250, 279
Jones, Tyldesley KC 131–136

K
Kennedy and Donkin 41, 107, 113–115, 118, 125, 143
Kennedy, Sir Alexander 41, 105–107, 115, 117/8, 120
Kindersley, Sir Robert 109, 111
Klapper, Charles F 7, 142, 144, 228/9, 247

L
Leathers, 1st Viscount Frederick 266
Leigh-Bennet, E.P. 206, 234
Lloyd George, David 70, 84–88, 101/2
locomotives – diesel
    BR type 3 (class 33/3) 312
    Bulleid 1-Co-Co-1 300,303
    LMS 10,000/10,001 Co-Co 303
    LNER proposed main line locomotives 303
locomotives – electric
    BR Southern Region E5000 class (class 71) 308
    BR Southern Region electro-diesel (class 72/73) 278, 309/10
    Bulleid-Raworth 'booster' locomotives (class 70)
        CC 1 282–288
        CC 2 284/5
        CC 3 (20003) 287, 307
    Central London Railway/ Metropolitan gearless locomotives 29/30
    Chicago, Milwaukee, and St Paul 3000V locomotives. 90/1, 121
    LBSCR 6,700V 'milk van' 161–163, 166
    LNER 2-B-2 148
    LSWR proposed GE locomotives 44
    LSWR shunter (BR DS74) 59
    SR proposed 'small' electric locomotive 286/7
locomotives – steam
    BR standard class locomotives 303/4
    Bulleid Leader 302
    Bulleid Merchant Navy 4-6-2 281/2, 302, 312
    Bulleid Q1 0-6-0 281, 285/6, 288, 312
    Bulleid West Country/Battle of Britain 4-6-2s 302, 312
    Drummond T14 4-6-0 (LSWR). 307
    Maunsell H15/N15(King Arthur)/ S15 4-6-0s 242, 166, 249
    Maunsell Lord Nelson 4-6-0 249
    Maunsell N and U 2-6-0s 249
    Maunsell Q 0-6-0 249
    Maunsell Schools 4-4-0 242, 244, 249, 260

Loder, Gerald (Lord Wakehurst). 12, 201/2, 210
Long Island Rail Road (LIRR) (USA). 121, 122
Lydall, F. 304
Lynes, Lionel 190

M
Macrae, Charles 116, 124, 140, 159
Mather & Platt 175
Maunsell, Richard E.L 81, 146, 190, 213, 226, 242, 245
mercury arc rectifiers 184/5, 225, 279
Merz and Mclellan 54, 81, 82, 106, 304
Merz, Charles 54, 81–83, 136, 106, 166, 127, 135, 274
Metropolitan Cammell 247
Metropolitan Carriage Wagon and Finance Company 130, 156/7, 200
Metropolitan Railway 29, 30
Metropolitan Vickers 157, 175, 190, 278
Midland railway Carriage and Wagon Company 157
Ministry of Munitions 70, 86
Ministry of Transport 103, 106–108, 11–114
Ministry (Office) of Works 57/8
Missenden, Sir Eustace 247, 260, 266, 272, 276, 289, 293/4, 295–299, 301
Moore, W.C 125, 206, 250
Moore-Brabazon, Colonel J.T.C 281
Mount, Lieutenant Colonel 213, 215, 217–219, 277
Mowlem, John 41, 59

N
Nash, Sir Phillip 105, 113, 115, 117, 120, 122, 125, 128/9, 143, 159
National Farmers' Union 237
National Grid 136, 181, 183/4, 186, 215, 264, 271, 304, 307, Plate 10
National Physics Laboratory Teddington 58/9, 52
National Wages Council 108, 295
Neal, Arthur MP 110 -112, 115, 117, 123, 128, 159,
New Line 37

O
Oerlikon 105
Over the Points 234

P
parallel running 104/5
Patents
    control system for an electric locomotive ('booster' system). 278
    design of a 'cycle tractor attachment' 34

    'improvements' for regenerative control 1906 25
    Improvements in the regenerative control of electrically propelled vehicles or trains 1907 31
    Improvements in the regenerative control of electrically propelled vehicles or trains 1919 100
    Improvements relating to the control of Electric motors 31
    new design of controller for regenerative control 25, 26, Plate 1
    rail protector (SECR/Ellson) 123, Plate 6
    'universal' high-speed steam engine 17
Pirelli General 192, 206, 274
Pocock, Mary 4, 18
power (generating) stations
    Barking 130, 135
    Belvedere 129
    Deptford (Stowage Wharf). 52, 82/3, 126, 133, 136, 138, 149, 152, 161, 185, 272/3,275, 304
    Durnsford Road 59–61, 68,
Prince of Wales 199
Pringle, Colonel Sir John Wallace 147/8,166, 237
Pringle Committee 166/7
Pullman Car Company 188/9
Pybus, John MP 166

Q
Queen Anne's Gate Westminster 19, 29

R
Railway Amalgamation Tribunal 141
Railway Companies Association 85, 296, 300
Railway Executive Committee 267/8,272, 280, 295
Railway Executive Committee Mechanical and Electrical Engineering Sub Committee 266, 274, 280
Railway Finance Corporation 216,220, 242
Railway Rates Tribunal 108, 295
Railway (Standardisation) Order 166
R and W Hawthorne 15
Raworth (relationship to Alfred)
    Arthur Basil see – Basil (brother)
    Basil (brother). 26, 32, 34, 69, 179, 274
    Benjamin (uncle). 14, 16, 17, 19, 34
    Benjamin Joseph (grandfather). 13, 14
    Epenetes, née Walker (grandmother). 13, 14, 18, 19
    Gladys (sister). 19, 28, 33
    Harrison (uncle). 14
    John Ernest (cousin). 27, 34

John Smith (father). 14–21, 23–34, 70, 251
John Smith (nephew). 179, 274
Margaret Cannington, née Kershaw (mother). 17, 18, 26, 250
Marjorie (daughter). 32, 177, 294, 314
Phyllis (sister). 19, 26
Ruby, née Robinson (wife). 31, 171, 274, 294, 314
Winifred, née Robinson (sister-in-law). 32, 308
Raworth R3 controller 28, 31
Reconstruction Committee 85
regenerative braking 25/5
regenerative control 25 -27, 28–31 Plate 1
reports
   Committee Appointed to Review the National Problem of the Supply of Electrical Energy, Lord Weir 1925 183
   Diesel traction, Southern Railway 1946 299
   Electrification of Railways Advisory Committee
   Ministry of Transport 1922 115
   Electrification of Railways
   British Transport Commission 1951 305
   LSWR suburban electrification, Herbert Jones 1912 53–55
   Main Line Electrification, Lord Weir 1931 186, 238
   Modernisation and Re-equipment of British Railways,
   (Modernisation Report), British Transport Commission 1955 306
   Proposed Extensions of Electrification (first report)
     Alfred Raworth, Southern Railway 1944 289–292
   Proposed Extensions of Electrification (second report)
     Alfred Raworth, Southern Railway 1944 29/3, 296–298
   Railway Electrification Committee (Pringle Report), Ministry of Transport 1928 166/7, 238
   SECR Management Committee, Electrical Engineer's Report on Electrification, Alfred Raworth 1919 92–29
   The Reshaping of British Railways ('Beeching Report'), British Railways Board 1963 311

Three-Position Signalling,
   Institution of Railway Signal Engineers 1924 152
Richards, R.M.T 320, 330
Riddles, Robert (Robin). 301/2, 311
RNAS 69
Robert Stephenson and Company 279
Robertson, V.A.M. 292, 294
Rostern, E.W. 304
rotary converters 49, 175–178
Rowntree, E W 236

S
St. Aubynes' Preparatory School Lowestoft 21
Scott, James Robb 172, 225/6
Shawcross, Sir Hartley MP 296
Siemens Brothers 15–17
Smith, Roger T 107
Smithers, Sir Alfred 109, 115/6, 132
Snell, Sir John 105/7, 127–129, 131, 134, 166
South Eastern and Chatham Power and Construction Company 125, 138, 143, 146/7
Southern Belle 198, 202/3
Southport Tramway Company 23
Sparks, Charles Pratt 130, 132, 134
Sprague, Frank 25
Stamp, Sir Josiah 187/8, 274
Standardisation of Electric Cables and Wires for Government Use Committee 275
Stevens, Theodore 117/8, 120–123, 143
Stevenson, George 14
Stevenson, Robert 18
stray currents 50
substations
   London Brighton and South Coast Railway 161/2
   LSWR electrification 61–64
   Southern Railway pre 1932 149/50, 155/6
   Southern Railway post 1932 184/5, 193/4, 208/9, 222, 252/3, 257, Plate 11, Plate 12, Plate 13
Szlumper, Alfred W. 115, 246
Szlumper, Gilbert 12, 202/5, 246, 260, 266, 275

T
Tapton House 14
Tempest, Percy 80/1, 84, 88,105, 114, 116, 125, 131–134, 146

Thornton, Sir Henry 104
Thorrowgood, W.J. 152
three phase electricity 46/7
Three-Position Signalling Committee 152/3
track paralleling huts 194, Plate 11
Trades Facilities Committee 108/9,111, 115, 125, 159
transformer 48
Trench, Lieutenant Colonel 217 -219, 225, 236, 274
Trewman. H. F 165
Trotter, A.P. 33

U
United Electric Car Company 126

V
V1 flying bomb damage 275/6
V2 rocket damage 276

W
Wainwright, Henry 80/1
Walker, Sir Herbert 12/3, 39–44, 66–68, 72, 84–88, 112, 116–118, 124, 143–148, 152, 159, 163, 166, 170, 173–177, 202–205, 210, 216, 226, 234, 236–239, 241, 246/7, 255, 260/1, 279, 289, 295
Wallace, Euan MP 259
War Department 240
Warder, Stanley 293, 299, 301, 305, 310
Ward-Leonard motor control 278
Warner, Surrey 65/6
Waterloo and City Railway 41/2
Watkin, Sir Edward 74, 77
Ways and Communications Bill 102, 106
Weir Lord of Eastwood 183
Wellington Mill 16, 17
Westinghouse 26, 51, 53, 65/6, 143, 245, 270
West Sussex County Council 238
Whitelaw, William 295
Wilkinson, George 14, 18
Wimbledon and Sutton Railway 179
Windsor Lines 36/7
winter weather 1939-40 269/70
Wren and Hopkinson 15
Wringly, C.R 125

Y
Young W.G. 229